The Scientific Imagination in South Africa

South Africa provides a unique vantage point from which to examine the scientific imagination over the last three centuries, when its position on the African continent made it a staging post for Portuguese, Dutch, and British colonialism. In the eighteenth century, South African plants and animals caught the imagination of visiting Europeans. In the nineteenth century, science became central to imperial conquest, devastating wars, agricultural intensification and the exploitation of rich mineral resources. Scientific work both facilitated, and offered alternatives to, the imposition of segregation and apartheid in the twentieth century. William Beinart and Saul Dubow offer an innovative exploration of science and technology in this complex, divided society. Bridging a range of disciplines from astronomy to zoology, they demonstrate how scientific knowledge shaped South Africa's peculiar path to modernity. In so doing, they examine the work of remarkable individual scientists and institutions, as well as the contributions of leading politicians from Jan Smuts to Thabo Mbeki.

WILLIAM BEINART is Emeritus Professor, St Antony's College, University of Oxford and former Director of the African Studies Centre. His publications include *The Rise of Conservation in South Africa* (2003), *Environment and Empire* (2007) with Lotte Hughes, *Prickly Pear: The Social History of a Plant in the Eastern Cape* (2011) with Luvuyo Wotshela, *African Local Knowledge & Livestock Health* (2013) with Karen Brown and *Rights to Land* (2017) with Peter Delius and Michelle Hay.

SAUL DUBOW is Smuts Professor of Commonwealth History at the University of Cambridge. He is the author of books, edited collections, and articles about nineteenth and twentieth-century South African history including *Scientific Racism in Modern South Africa* (1995), *A Commonwealth of Knowledge* (2006), *South Africa's Struggle for Human Rights* (2012), and *Apartheid 1948–1994* (2014). He is the editor of *Science and Society in Southern Africa* (2000).

The Scientific Imagination in South Africa

1700 to the Present

WILLIAM BEINART
University of Oxford

SAUL DUBOW
University of Cambridge

CAMBRIDGE
UNIVERSITY PRESS

CAMBRIDGE
UNIVERSITY PRESS

University Printing House, Cambridge CB2 8BS, United Kingdom

One Liberty Plaza, 20th Floor, New York, NY 10006, USA

477 Williamstown Road, Port Melbourne, VIC 3207, Australia

314–321, 3rd Floor, Plot 3, Splendor Forum, Jasola District Centre, New Delhi – 110025, India

79 Anson Road, #06–04/06, Singapore 079906

Cambridge University Press is part of the University of Cambridge.

It furthers the University's mission by disseminating knowledge in the pursuit of education, learning, and research at the highest international levels of excellence.

www.cambridge.org
Information on this title: www.cambridge.org/9781108837088
DOI: 10.1017/9781108938198

First published 2021

A catalogue record for this publication is available from the British Library.

Library of Congress Cataloging-in-Publication Data
Names: Beinart, William, author. | Dubow, Saul, author.
Title: The scientific imagination in South Africa : 1700 to the present / William Beinart, University of Oxford, Saul Dubow, University of Cambridge.
Description: Cambridge, UK ; New York, NY : Cambridge University Press, 2021. | Includes bibliographical references and index.
Identifiers: LCCN 2020058042 (print) | LCCN 2020058043 (ebook) | ISBN 9781108837088 (hardback) | ISBN 9781108940085 (paperback) | ISBN 9781108938198 (epub)
Subjects: LCSH: Science–South Africa–History. | Science and civilization.
Classification: LCC Q127.S6 B45 2021 (print) | LCC Q127.S6 (ebook) | DDC 306.4/509680903–dc23
LC record available at https://lccn.loc.gov/2020058042
LC ebook record available at https://lccn.loc.gov/2020058043

ISBN 978-1-108-83708-8 Hardback
ISBN 978-1-108-94008-5 Paperback

Contents

Map

Map of South Africa pre-1994 with place names mentioned in text

ZIMBABWE
MOZAMBIQUE
River Limpopo
SOUTPANSBERG
Valdezia
Tzaneen
Phalaborwa
KRUGER NATIONAL PARK
WATERBERG
Potgietersrus
Gaborone
TRANSVAAL
Lydenburg
Onderstepoort
Rustenburg
Pretoria
Botshabelo
Hartebeesthoek
Pelindaba
Maputo
Delagoa Bay
Johannesburg
Mbabane
Vanderbijlpark
SWAZILAND
Potchestroom
Sharpeville
Sasolburg
Vaal River
Phongola River
AFRICA
UMFOLOZI
ORANGE FREE STATE
Isandlwana
Nkandla
Modder River
NATAL
Bloemfontein
Caledon River
Maseru
Pietermaritzburg
LESOTHO
Durban
Verwoerd/Gariep Dam
DRAKENSBERG
INDIAN OCEAN
Fish River
Fort Hare
East London
Grahamstown
Port Elizabeth
0 100 200 300 400 km
0 50 100 150 200 250 miles

Figures

Acknowledgements

The idea for this book grew out of an extended conversation between William Beinart, Saul Dubow and Patrick Harries (then at the University of Basel). Each of us was researching and writing separately on the significance of scientific ideas and practices in South African history. In the course of our work we met frequently in Britain, South Africa, and in Basel where Patrick was always a most generous host, including us in various symposia and lectures involving his research students.

Over the years, our work has taken shape in a number of articles, books and collections such as: Saul Dubow, *Scientific Racism in Modern South Africa* (1995), *Science and Society in Southern Africa* (2000) and *A Commonwealth of Knowledge* (2006); William Beinart, *The Rise of Conservation in South Africa* (2003) and (with Luvuyo Wotshela), *Prickly Pear: The Social History of a Plant in the Eastern Cape* (2011); Patrick Harries, *Butterflies and Barbarians* (2007) and (with David Maxwell), *The Spiritual in the Secular: Missionaries and Knowledge about Africa* (2012). Many conversations led us to conclude that important themes of knowledge and science have been neglected in recent South African historiography and that a synthetic overview, incorporating new work and overlooked material, would be of value. To this end we collaborated on a major proposal for European Research Council funding in 2012. This helped us to conceive the project, but the application failed. We nevertheless decided to go ahead with a joint book and sketched out twelve to fourteen chapters. Tragically, Patrick Harries died suddenly in 2016 before much writing was done.

We are deeply grateful to Patrick – friend, colleague and teacher – over the long term, for discussions and ideas, and we have been able to include some of his contribution to our joint research proposal in the introduction. His own focus was initially on the earliest written literature on south-east Africa and he wrote two short drafts that covered the sixteenth to eighteenth centuries; a little of this material is also incorporated in our first chapter on the eighteenth and early nineteenth centuries. His published work on missionary science is explored in Chapter 4. We wrote tributes to Patrick in the *Journal of Southern African Studies* (2017) and *Africa* (2017).

Saul Dubow is especially grateful to the Stellenbosch Institute for Advanced Studies where he was based as a fellow in 2015 in order to work on this project. In the congenial environment of STIAS, discussions between the three of us continued, and some of the first chapters were drafted. We would like to thank Elleke Boehmer, Keith Breckenridge, Siamon Gordon, Rob Iliffe, Simon Schaffer, Jules Skotnes-Brown, Stephen Sparks and William Kelleher Storey for their valuable comments on earlier versions of the text. We are grateful, too, to Russell Martin for editorial advice, Elaine Williams for the index and Angela Roberts for her copyediting.

Introduction: The Scientific Imagination in South Africa

South Africa offers an inviting vantage point from which to write on the scientific imagination. It was a regional rather than a world power, and in global terms it was not a major centre for invention or the generation of new scientific ideas. Yet the country possesses distinctive features that make it worthy of attention for a wide audience. Its geographic position on the African continent made it a staging post for Portuguese, Dutch and British colonialism; from the mid-seventeenth century, the Cape and subsequently all of South Africa became part of an expanded global imaginary. Colonisation by Britain brought the region into tight connection with one of the most powerful, and technologically advanced, world empires in the nineteenth and early twentieth centuries.[1]

From the sixteenth century onwards, marvels and curiosities reported by European travellers and explorers fed the European imagination. By the mid-eighteenth century, and following the publication of Linnaeus's ordered system of classification, South Africa was becoming an important site for botanical and zoological exploration by visitors with a scientific bent.[2] The Cape's natural endowments, notably its thousands of plant species, as well as the region's diverse fauna, attracted sustained attention. The quest for scientific order also prompted speculation about the country's repository of ancient fossil and hominin remains. By the early twentieth century there was

[1] D. R. Headrick, *The Tools of Empire: Technology and European Imperialism in the Nineteenth-Century* (Oxford: Oxford University Press, 1981) and *The Tentacles of Progress: Technology Transfer in the Age of Imperialism, 1850–1940* (Oxford: Oxford University Press, 1988); H. Tilley, *Africa as a Living Laboratory: Empire, Development, and the Problem of Scientific Knowledge, 1870–1950* (Chicago: University of Chicago Press, 2011).

[2] S. Huigen, *Knowledge and Colonialism: Eighteenth-Century Travellers in South Africa* (Leiden: Brill, 2009).

accumulating evidence that the country may have constituted the 'cradle' of modern human evolution.

South Africa was at key moments a significant incubator and testing ground of innovation, which had profound social and economic effects on the country. In the nineteenth century, improved agricultural technology underpinned exports of wool and ostrich feathers. Devastating new rifles developed for European warfare helped to change the balance of power in favour of colonial regimes. The mineral revolution of the late nineteenth century created the impetus for the final conquest of African societies and launched a phase of rapid industrialisation. This necessitated developments in applied geology and the chemistry of gold extraction. Mining delineated scientific practice in related fields from electrification and explosives to medicine and this shaped and thickened the web of science in South Africa. Some older strands, such as botany and geology, thrived and economic growth as a whole in the first half of the twentieth century generated momentum in fields such as veterinary medicine and oil from coal technology. Key advances were often a response to urgent economic and technological requirements, but the scientific imagination was also more exploratory with respect to curiosity-driven fields such as astronomy, palaeontology, and wildlife conservation.

As a crucible of racial politics in the twentieth century, South Africa has long been seen as a social laboratory for the study of 'race relations'. Its complex peopling and diversity gave birth to two powerful nationalisms – Afrikaner and African – which have both impacted profoundly on its scientific as well as political trajectory. White South Africans were also carriers of a darker intellectual tradition – the attempts to justify racial segregation in scientific terms.[3] In the relatively brief era of African nationalist rule after 1994, the state has espoused more universalist and developmentalist goals and the segregationist drive to account for differences in *kinds* of humans has yielded to an emphasis on common *humankind*.

This book does not examine, systematically, the growth of particular disciplines in South Africa. It is largely a history of individuals, associations, and institutions that were at the fulcrum of important scientific developments. We have used the 'scientific imagination' as a title

[3] S. Dubow, *Scientific Racism in Modern South Africa* (Cambridge: Cambridge University Press, 1995).

because we wish to capture linked thematic strands. We aim to illustrate the scientific imagination as an expression of human curiosity, ingenuity, and the ability to make unlikely connections. In this sense, scientists were exploring the frontiers of knowledge in relation to core areas of scientific and technological innovation.

The idea of a scientific imagination was long discussed amongst European scientists. In the seventeenth and eighteenth century, they explicitly debated whether their activities were primarily the pursuit of imagination, considering this in relation to scientific hypotheses and understanding how to observe.[4] The term has also been used to describe innovation by individual scientists in the twentieth century.[5] We have not found much evidence of such discussions in South Africa and therefore explore scientific imaginations in a slightly different sense. Many of the scientific traditions in which we are interested involved thinking about how science could shape society. Science was therefore imbricated in the exercise of power: bound up in the imaginary of an expanding British Empire, later of an ambitious Afrikaner state and, in the present, the hopes of post-apartheid democracy. We are concerned with scientific innovations and interventions that had broader socio-economic and political impact – or the potential to do so. As a result, our discussion has a strong focus on the politics of science.

The rise of segregation and apartheid established an extreme system of racial division that for many years excluded the bulk of the population from scientific training and opportunity. Nevertheless, black people were always carriers of knowledge: indigenous (or local, vernacular) knowledge informed their medical, veterinary, botanical, environmental, and linguistic practices, and their cultural traditions too. Especially in the period before the twentieth century, these knowledge systems were absorbed and appropriated in published texts by amateur and professional scientists, often with inadequate attribution. We discuss the interactions between indigenous and scientific knowledge especially in Chapter 1 and 7.

4 R. Iliffe, 'Science Fictions: The Triumph of the Imagination and the Invention of Scientific Creativity', inaugural lecture, University of Oxford, 8 November 2018.

5 G. Holton, *The Scientific Imagination* (Cambridge, MA: Harvard University Press, 1998); A.Levy and P. Godfrey-Smith (eds.), *The Scientific Imagination* (New York: Oxford University Press, 2020).

All knowledge organises or systematises information and observation, and in this sense has been 'cooked' or processed.[6] There are no hard and fast boundaries between science and indigenous or local knowledge. Especially in the eighteenth century, when scientific training and discrete disciplines were in their infancy, texts by scientific travellers interspersed authorial observation with hearsay and vernacular ideas (Chapter 1). Patrick Harries has illustrated in detail the botanical and entomological work of Swiss missionary Henri Junod in Mozambique and the Transvaal in the late nineteenth and early twentieth centuries.[7] Junod became an authority on butterflies, many of which were collected by skilled African assistants such as diviner Elias Libombo (Chapter 4). Interview-based research still draws on local knowledge. Following the political transition in the 1990s, there has been renewed scientific and political interest in what was specifically called indigenous knowledge (Chapter 7).

Although we mention examples of fertile interconnection, our concern is not to find the origins of science in indigenous knowledge. We focus largely on formal, disciplinary-based and scientific work. By this we mean published or written ideas by those with specialist training who were familiar with disciplinary developments and international thought. We cannot aim at encyclopaedic coverage. Our choice of examples is shaped by our own previous research as well as a subjective sense of important and interesting developments bearing on society. Scientific knowledge and its application has been central to South Africa's development; understanding its history is one essential element in explaining the country's complex and uneven path to modernity. This claim is not clearly evident in the broader historiography, where discussions have tended to focus on relations of conflict between coloniser and colonised, capital and labour. We argue that it is impossible to fully understand patterns of change or to imagine contemporary society, without consideration of the impact of science and technology.

Scientists, especially those who were educated elsewhere, often saw their work through eyes trained in a global, comparative frame.

[6] P. Burke, *A Social History of Knowledge* (Cambridge: Cambridge University Press, 2000), 11; J. M. Mackenzie (ed.), *Imperialism and the Natural World* (Manchester: Manchester University Press, 1990), 5.

[7] P. Harries, *Butterflies and Barbarians: Swiss Missionaries and Systems of Knowledge in Southeast Africa* (Oxford: James Currey, 2007).

Although there were notable contributors to pure and abstract sciences, a higher proportion of original work in South Africa emerged out of field sciences, that is from observation born of the desire to comprehend the country's natural phenomena. We evaluate the contributions made by relatively well-known individuals and illuminate some of those whose work has been overlooked. This includes self-trained specialists whose knowledge and enthusiasm fed into fields such as botany. We are also alert to the scientific associations in which they participated. Such institutions were key building blocks of colonial identity and deserve to be seen as additional constitutive elements in national and ethnic imaginaries.[8]

At another level, the scientific imagination refers to what Sheila Jasanoff calls 'socio-technical imaginaries'. These move beyond the ideas or visions of brilliant individuals to encompass broader collective views capable of 'uniting members of a social community in shared perceptions of futures that should or should not be realized'.[9] We concur with Jasanoff that the role of science and technology in national imagining has been underestimated by comparison, say, with cultural and linguistic production, whose importance to 'imagined communities' has received sustained attention.[10] In this book we focus on the impact of science in conceptions of national and ethnic identities. This had important implications for shaping the boundaries of political and social belonging and, reciprocally, in determining and legitimating grounds for exclusion or marginalisation.

While there is little evidence of scientists in South Africa imagining social utopias or dystopias, they were certainly active in conceiving how society could be shaped or improved. The relatively close proximity of scientists to politicians often gave individual scientists special leverage. In a country as divided as South Africa, this implied controversial interventions in the social domain geared to the defence or modification of racial privilege. But there were also important periods

[8] S. Dubow, *A Commonwealth of Knowledge: Science, Sensibility, and White South Africa 1820–2000* (Oxford: Oxford University Press, 2006).

[9] S. Jasanoff, 'Future Imperfect: Science, Technology, and the Imaginations of Modernity', in S. Jasanoff and Sang-Hyun Kim (eds.), *Dreamscapes of Modernity: Sociotechnical Imaginaries and the Fabrication of Power* (Chicago: University of Chicago Press, 2015), 6.

[10] B. Anderson, *Imagined Communities* (London: Verso, 1983); E. Hobsbawm and T. Ranger (eds.), *The Invention of Tradition* (Cambridge: Cambridge University Press, 1983).

when the medical sciences, for example, were a repository for imagining radically alternative futures.[11] Immediately before Afrikaner nationalists came to power in 1948, a broadly integrative, developmentalist conception of liberal progress (with African nationalist and socialist variants) was elaborated. Elements of this developmentalism were repurposed in techno-nationalist ways to feed apartheid conceptions of white supremacy and autarchic self-reliance. In the post-apartheid era, non-racial visions combined the discourse of universal human rights with hopes that participation in global science might bring broader social benefits as well as international standing. Such claims were counterposed by appeals to indigenous knowledge and a suspicion of Western science as an agent in maintaining, rather than eroding, existing racial hierarchies.

Who were the scientists? Until the mid-twentieth century, they were mostly European men born or trained outside of the country. Although they came from many different backgrounds and countries, including France, Germany, the Netherlands, Switzerland, and the United States, British-born scientists predominated. Many became key figures in government departments, institutions, and universities. These patterns of migration are, in themselves, clues to the globality and connectedness of South African science. As professional and scientific education gradually expanded in South African universities from the interwar years, English- and Afrikaans-speaking white South African-born scientists filled rapidly expanding opportunities in the state and universities. In the apartheid era (1948–1994) Afrikaners were favoured in state employment and benefitted from the mushrooming of new universities.

White women had few opportunities to train or to work as scientists in the nineteenth century, although a few, such as Mary Barber, managed to educate themselves and contribute as non-employed specialists to scientific endeavour.[12] In the first half of the twentieth century, a limited number were able to forge scientific careers, especially in the field of botany. Opportunities for black scientists were

[11] S. Marks, 'South Africa's Early Experiment in Social Medicine: Its Pioneers and Politics', *American Journal of Public Health*, 87, 3 (1997), 452–9; S. Dubow and A. Jeeves (eds.) *South Africa's 1940s: Worlds of Possibility* (Cape Town: Double Storey, 2000).

[12] T. Hammel, *Shaping Natural History and Settler Society. Mary Elizabeth Barber and the Nineteenth-Century Cape* (Cham: Palgrave Macmillan, 2019).

highly constrained. Jotello Soga trained as a vet in Scotland and was able to find government employment in the late nineteenth-century Cape (Chapter 3).[13] Institutions such as Lovedale and Fort Hare provided science courses for a small number of black students, and some were recruited to teach in African schools or gained medical qualifications in the United States and Britain.[14] But the rise of segregation excluded the bulk of the population from any opportunity to develop as professional or research scientists until posts were introduced in segregated universities during the apartheid era.

To mark the centenary of South Africa's elite Royal Society of scientists, A. C. Brown edited a collection, written by fellow scientists, entitled *A History of Scientific Endeavour in South Africa* (1977). The twenty-one disciplinary-specific chapters aimed to be comprehensive and stretched to about 300,000 words.[15] Even so, Brown felt that they could not achieve 'complete coverage' and noted that his collection of essays amounted to 'a history' and not 'the history' of science. At less than half the length, and covering another four decades, our volume must necessarily be selective. Our approach is to think across disciplines over the long term and to highlight important themes. The chapters follow a chronological sequence, consonant with major political eras, whose periodisation will be recognisable to historians of the country.

Eighteenth- and Nineteenth-Century Southern Africa in Global Science: A Summary

When the Dutch plunged into the Indian Ocean at the beginning of the seventeenth century, they brought with them both a new form of capitalism and fresh ways of looking at nature. Their sprawling seaborne empire turned Amsterdam and Leiden into centres for the collection of global knowledge. Colonised regions were systematically

[13] W. Beinart, *The Rise of Conservation in South Africa: Settlers, Livestock and the Environment, 1770–1950* (Oxford: Oxford University Press, 2003).

[14] K. Snedegar, *Mission, Science, and Race in South Africa: A. W. Roberts of Lovedale, 1883–1938* (Lanham: Lexington Books, 2015).

[15] A. C. Brown (ed.), *A History of Scientific Endeavour in South Africa* (Cape Town: Royal Society, 1977).

scoured for valuable natural resources and curiosities.[16] Jan van Riebeeck, Simon van der Stel, and Rijk Tulbagh, among other Cape governors, were keen amateur naturalists. While African food species found at the Cape made only a limited global impact, decorative plants and flowers were disproportionately important. In ways similar to world exhibitions and imperial or national museums, botanical gardens and herbaria presented the world, in all its exotic novelty, to their metropolitan visitors. Such institutions served as portals to, and gatekeepers of, knowledge. The late eighteenth and early nineteenth centuries produced a rich literature on South Africa, as scientific interest drew it into the orbit of educated Europeans. They and their published accounts have become the focus of intense recent academic interest.

We begin by exploring the curiosity and legacy of these scientific sojourners, linking the ideas they generated both to local knowledge and to European Enlightenment understandings (Chapter 1). Some of the writers became scientists of consequence. Luminaries who visited and wrote about the Cape included Carl Peter Thunberg, who became the leading botanist at Uppsala; Hinrich Lichtenstein, later professor of zoology in Berlin; John Barrow, a key scientific figure in the British Admiralty and the Royal Society; and William Harvey, who, after being appointed to Trinity College Dublin, compiled the most thorough list of Cape plants. These travellers often reported local knowledge and collectively created a literary tradition about the Cape that helped to define the region's character and interest to scientists.

The final transfer of the Cape from Dutch to British control in 1806, during the Napoleonic Wars, gave the region new geopolitical prominence as the British imperial meridian shifted eastwards, thus linking the Indian and Atlantic oceans.[17] For the next seventy years all shipping between Europe and Asia passed, and usually stopped at, southern African ports. In the case of the post-abolition slave trade, East

[16] H. Cook, *Matters of Exchange: Commerce, Medicine and Science in the Dutch Golden Age* (New Haven, CT: Yale University Press, 2007).

[17] C. A. Bayly, *Imperial Meridian: The British Empire and the World 1780–1830* (London: Longman, 1989).

African ports became connected to Brazil via the growing hub of Cape Town.[18]

The Cape sea-route became strategically ever more important just as the new colonial authorities sought to consolidate their control of the southern African hinterland. British colonisers legitimated their presence at the Cape, and in Natal, by insisting on their commitment to civilisation, progress, and commercial development. Writer J. M. Coetzee has argued that from the beginnings of European settlement, terms like 'indolence, sloth, laziness, torpor' suffuse descriptive accounts of indigenous Khoesan peoples; the British were also highly critical of the 'idleness of the Boers.'[19] They were keen to demonstrate that they were better colonists than their European competitors, measured by their industriousness, capacity for governance, and moral virtue. Scientific accomplishment and receptiveness formed part of this suite of civilisational skills.

Charles Somerset, autocratic governor from 1814 to 1826, oversaw a damaging war with the Xhosa (1818–19) coupled with the British settlement programme in 1820; he also came into intense conflict with British immigrants who tried to launch a free press along with robustly independent literary and scientific associations. Less frequently noted is his role in the foundation of a library and reforming medical institutions (Chapter 2), especially with the appointment of James Barry, a military physician whose life has recently attracted intense interest.[20]. Simultaneously, the Zulu kingdom was consolidating its strength, establishing links with British traders at Port Natal, and expanding its military organisation and weapons production. Other kingdoms were also changing rapidly as they absorbed new military technologies such as horses and firearms.

In the 1820s, the Cape was selected by the British Admiralty as the site of a new Royal Observatory. This was intended not only to expand knowledge of positional astronomy in the underexplored southern hemisphere, but also to help with practical matters such as navigation,

[18] P. Harries, 'Slavery, Indenture and Migrant Labour: Maritime Immigration from Mozambique to the Cape, c. 1780–1880', *African Studies*, 73, 3 (2014), 323–40.

[19] J. M. Coetzee, *White Writing: On the Culture of Letters in South Africa* (New Haven, CT: Yale University Press, 1988), 2, 18, 28–9.

[20] M. du Preez and J. Dronfield, *Dr James Barry: A Woman Ahead of Her Time* (London: One World, 2016).

mapping, and measurement. The handsome building housing the observatory was designed to be seen and admired, serving as a visible symbol of British imperial power and rational order. The independent arrival at the Cape in 1834 of gilded British scientist John Herschel gave southern astronomy added prominence. During his brief stay at the Cape, Herschel participated energetically in establishing scientific networks and associations with secure footholds in local colonial society.

In the first half of the nineteenth century these networks supported the establishment of a natural history museum (1825), the revival of the government's botanical garden, as well as new publications and zoological expeditions into the interior – notably by Andrew Smith.[21] As British rule expanded eastwards, fledgling institutions were created, such as botanical gardens at Durban and the Albany Museum in Grahamstown. Dr William Atherstone, a physician and polymath based at Grahamstown, became the centre of a committed group of interdisciplinary scientists writing on everything from geology and botany to medicine and ostrich breeding. Some had a strong interest in African medical remedies. Amongst them, Andrew Geddes Bain turned to geology in order to understand the curious stratigraphies he encountered as a road-builder in the 1830s. Bain discovered exciting fossils in the Karoo which caught the attention of T. H. Huxley and Richard Owen in London as well as the Royal Geographical Society. Ancient geological formations prompted new thinking about the age of the earth and the peopling of the country. Flints and middens found in the environs of Cape Town, whose meaning and significance proved puzzling, helped to catalyse archaeology as a new field of interest and source of wonder.

In the second half of the nineteenth century, the Cape became a place of brokerage for ideas as well as commodities such as wool and wine, diamonds and ostrich feathers. No longer only looking outwards to Europe, it also became a permanent base for scientific institutions, expeditions, and collecting. The German comparative philologist Wilhelm Bleek, who arrived to catalogue governor Sir George Grey's private library in 1856, revealed the complexity of indigenous languages and grammar and this prompted deeper interest in Bushman mythology and rock art. Coming at a time when Darwinian ideas were

[21] Dubow, *A Commonwealth of Knowledge.*

reshaping theories of evolution, and when Bishop Colenso in Natal began to challenge the orthodox truths of Christian scripture, controversies about human origins and hierarchies reverberated through settler as well as metropolitan societies. The science of race, which slip-streamed in Darwin's wake, was given particular impetus by imperial conquest in South Africa.

The conquest of the semi-arid Karoo and the neighbouring grasslands was largely completed by the mid-nineteenth century. These areas became the site of a largely settler-controlled, commercial pastoral economy, specialising in wool from merino sheep, angora, ostrich feathers and meat.[22] The value of colonial agricultural production exceeded that of diamonds throughout the nineteenth century. Landowners were responsible for much of the technical innovation that underpinned these developments. When the Cape received responsible government in 1874, led by sheep-farming prime minister John Molteno, the colonial scientific bureaucracy was gradually expanded to include a veterinary surgeon (1876), a Department of Agriculture with specialist officers, and an *Agricultural Journal* (1888–9). A state botanist, geologist, entomologist and marine biologist were also appointed at a time when few other colonies of this size had comparable positions.

The discovery of diamonds in the 1860s and Witwatersrand gold in the 1880s reshaped the contours of race and science (Chapter 3). Mining was reliant on increasingly sophisticated technology and on cheap black labour in the newly concentrated conurbations of Kimberley and Johannesburg. This coincided with growing imperial acquisitiveness as European nationalism and competitive industrialisation intensified the scramble for Africa. The geostrategic importance of southern Africa was a point of growing competition between Britain, Germany, Portugal, and the Boer republics. The borders of what came to be constituted as a unitary South African state in 1910 emerged out of wars of conquest against African societies and intense conflict between English- and Dutch-speaking citizens. Imperial conquest and expansion were in turn associated with rapid technological change: steamships, transatlantic cables, and devastating new rifles which, in combination, helped to change the balance of

22 Beinart, *The Rise of Conservation.*

power decisively.[23] Although many African societies were able to purchase firearms in the nineteenth century, imperial and settler armies stayed a step ahead with breech-loading rifles. Together with British financial resources and organisation, such technological advantages proved decisive in subduing the remaining independent African states by the end of the century.

Rapid growth in the mining sector has been analysed by historians in terms of the relationship between capital and labour. Scientific and technological innovations are less integrated into these accounts: applied geology, water pumps, explosives, and stamping gear. Expansion of underground gold mining would have been far more difficult and costly without the recently discovered MacArthur-Forrest cyanide extraction process, invented in Scotland and developed on the Witwatersrand, or the local manufacture of dynamite. William Storey, writing of Cecil Rhodes's interests in mining, agriculture, railways, telegraphs, and university education, characterises these progressivist impulses as elements of a 'full-blown socio-technical imaginary'.[24]

Private sector companies provided much of the new research momentum around the mining industry. But as in other settler societies, such as the United States and Australia, the state had already began to co-ordinate scientific research and policy in agriculture and invested in educational institutions that provided broader training. The South African School of Mines and Technology, established at Kimberley in 1896, developed into the Transvaal Technical Institute (1904) and then the University of the Witwatersrand, which opened in 1922 as South Africa's fourth university.[25] Disciplines such as geology, engineering and chemistry expanded from these bases – and in unpredictable directions: just as Bain found fossils that would inform understandings of evolution while engaged in road building, so Alex du Toit theorised continental drift while working as a consultant geologist to the mining industry.

23 W. K. Storey, *Guns, Race and Power in Colonial South Africa* (Cambridge: Cambridge University Press, 2008); Headrick, *Tools of Empire.*

24 W. K. Storey, 'Cecil Rhodes and the Making of a Sociotechnical Imaginary for South Africa', in Jasanoff and Kim (eds.), *Dreamscapes of Modernity*, 53.

25 B. K. Murray, *Wits: The Early Years* (Johannesburg: Wits University Press, 1982), 4.

Science, Race and Power in the Twentieth Century: A Summary

The South African War of 1899–1902 constituted a major political rupture, followed by the imposition of regional British hegemony with strong anglophone political and cultural ambitions. This was both resisted and accommodated by a revitalised Afrikaner nationalism. The new South Africa that emerged in 1910 was formally independent of outside control and increasingly geared to exclusive nation-building around racial segregation. In this period, science and technology were utilised in order to consolidate the embryonic state, which legitimated white authority by valorising rationality, order, and progress. Science was promoted as a modernising force embodying values that transcended identity and politics. (Chapter 4). New scientific associations founded after the war, such as the South African Association for the Advancement of Science, and the Royal Society, consciously portrayed themselves as agents of progress. Their leaders sought to locate South Africa as a self-sufficient country within the ambit of the British Commonwealth. Over the next two decades, members helped to build the research capacity of government departments and the post-Union university system.

Jan Smuts, the leading Afrikaner politician and statesman, was committed to active participation in the emergent British Commonwealth. Comfortable in the milieu of anglophone society, he saw science as a means to avoid narrow nationalisms dividing English- and Afrikaans-speakers. In the 1920s and 1930s Smuts and his political protégé, Jan Hofmeyr, became strong advocates of the 'South Africanisation' of science, deepening local intellectual capacity while maintaining strong connections with Britain. Smuts, who had broad scientific interests and a gift for cross-disciplinary synthesis, was attentive to forms of scientific thinking that illuminated a changing balance of power between southern and northern hemispheres. He skilfully deployed the expansion of locally based scientific knowledge to demonstrate South Africa's importance to universal science.

Investment in scientific agriculture was one way in which the new Anglo-Afrikaner compact was sealed. Onderstepoort, established in 1908 under Swiss-born Arnold Theiler, became a leading centre for African veterinary research and training. Charles Lounsbury, an American entomologist in the Cape, unravelled tick-borne diseases and headed an expanding entomology section in the Union

Department of Agriculture.[26] Vets, largely British-trained, initially drew on local knowledge of livestock diseases, poisonous plants, and treatments.[27] They named and researched many of the diseases that confronted them within this older but dynamic framework. For example, the Zulu word *nagana* was retained for the animal variety of trypanosomiasis, carried by tsetse flies. Vets increasingly applied advanced scientific thinking to control destructive diseases such as rinderpest, tick-borne East Coast fever, *lamziekte*, anthrax and African horse sickness. Their methods resulted in far-reaching interventions into rural life.

In the emerging field of social science, new disciplines such as comparative education, social anthropology, sociology and psychology were gaining a foothold in the university system by the 1920s. These provided a basis for understanding the growing subject of race relations; some ambitious proponents imagined South Africa as a living social laboratory worthy of international attention.[28] The Natal educationist and liberal segregationist Charles Loram became Sterling professor of education at Yale in 1931 on account of his South African expertise. Reciprocally, African-American expert Ralph Bunche, later a Nobel peace prize-winner, toured South Africa in the 1930s. American foundations like the Carnegie Corporation and the Phelps-Stokes Fund were actively engaged in these transnational research exchanges.

Social anthropology emphasised the role of culture in human development; physical anthropology and comparative anatomy focused on biology and evolution. Both approaches counterposed the modernity of Western civilisation to the supposed backwardness of African society while debating whether seemingly fundamental differences were inherent or shaped by environmental factors. New discoveries from the late 1930s, including important palaeontological finds at Sterkfontein, and the discovery of a coelacanth, a living 'fossil fish', drew further attention to South Africa as a formative place in the global evolutionary imaginary.

[26] W. Beinart and L. Wotshela, *Prickly Pear: The Social History of a Plant in the Eastern Cape* (Johannesburg: Wits University Press, 2011).

[27] Beinart, *Rise of Conservation*; K. Brown, 'Poisonous Plants, Pastoral Knowledge and Perceptions of Environmental Change in South Africa, c. 1880–1940', *Environment and History*, 13,3 (2007), 307–32.

[28] Tilley, *Africa as a Living Laboratory*.

By this time science was firmly tied to promoting white South African identity and, in the case of racial science, white supremacy (Chapter 5). Yet the ethnic configuration of white unity and identity remained open to contestation. Smuts saw the South Africanisation of science as a means to develop a unified sense of white identity. Conversely, Afrikaner nationalists, led by J. B. M. Hertzog, prime minister from 1924 to 1939, were hostile towards the British Empire and often suspicious of science that questioned scriptural authority.[29] Yet accounts of Afrikaner nationalist mobilisation have overlooked the significance of science and technology as a resource available to be tapped. Especially from the 1930s, an assertive and ambitious Afrikaner intelligentsia came to view science as a tool to advance the interests of the *volk*. The creation of scientific sections in the Suid-Afrikaanse Akademie vir Wetenskap en Kuns (South African Academy of Science and the Arts) offered alternatives to membership of anglophone institutions. Afrikaans-speaking universities and the state also provided new scope for employment and funding.

South Africa's entry into the Second World War on the side of the Allies intensified underlying tensions in white politics. The war years (1939–45) encouraged competing 'worlds of possibilities' in the political and scientific imagination.[30] Anglophone and liberal constituencies looked forward to reforms and post-war reconstruction. The African National Congress, established in 1912, found renewed impetus and confidence as it articulated demands for full citizenship and human rights, along with social benefits for black South Africans. Under the leadership of the medical doctor A. B. Xuma, and with the support of leading intellectuals such as Z. K. Matthews at Fort Hare, the ANC pursued a social-democratic political vision which was allied to new ideas of scientific progress and development in health, nutrition and agriculture.[31] Conversely, Afrikaner nationalists began to envisage an alternative political vision oriented towards republican

[29] T. Dunbar Moodie, *The Rise of Afrikanerdom: Power, Apartheid and Afrikaner Civil Religion* (Berkeley: University of California Press, 1974).

[30] Dubow and Jeeves (eds.) *South Africa's 1940s*.

[31] S. Gish, *Alfred B. Xuma: African, American, South African* (Basingstoke: Palgrave Macmillan, 2000); V. Padayachee and R. van Niekerk, *Shadow of Liberation: Contestation and Compromise in the Economic and Social Policy of the African National Congress, 1943–1996* (Johannesburg: Wits University Press, 2019).

independence and 'apartheid'. The promise of science in helping to deliver these competing outcomes was expressed by all three political traditions.

Smuts's government linked warfare to broader welfare, though without committing itself to expanded rights of citizenship. National mobilisation encouraged rational planning, ambitious schemes of development, and social welfare systems with a view to post-war 'reconstruction'. During and immediately after the war, an influential triumvirate greatly expanded the scientific and technological capacity of the state and its enterprises. Hendrik van der Bijl and H. J. van Eck developed national power supply through the state electricity commission (ESCOM); they also controlled the Industrial Development Council (IDC), a semi-autonomous government funding agency with considerable resources and ambition. Basil Schonland, a Cambridge-trained physicist who served in 1944 as General Montgomery's scientific adviser preparing for the Allied invasion of Europe, became the inaugural founding president of the Council for Scientific and Industrial Research (CSIR), charged with stimulating fundamental and applied science.[32] Schonland promoted research into nuclear physics, utilising South Africa's uranium deposits and Commonwealth connections. Science became the instrument of an increasingly interventionist state as it laid down detailed plans for post-war renewal in industry and agriculture as well as development in African areas.

The shock victory of Malan's National Party in 1948 represented a repudiation of the wartime government's outward-looking policies. Many English-speaking critics pilloried apartheid as backward-looking, insular, and resistant to modernity. Yet the Malan government proved adept at using science for its own instrumental purposes (Chapter 6). Afrikaner scientists achieved their own momentum across a range of disciplines. Malan's 1949 'African Charter' promoted science and technology as a means to secure regional hegemony and to cement South Africa's position as a bulwark of anti-communism. Under successive apartheid administrations, South Africa pursued a form of scientific diplomacy that, while less intellectually capacious and imaginative than Smuts's, included participation in institutions like the Scientific Council for Africa South of the Sahara and initiatives

[32] B. Austin, *Schonland: Scientist and Soldier* (Bristol: Institute of Physics, 2001).

like the International Geophysical Year. One result was the growth of a scientific and strategic presence in Antarctica. Some of these connections survived even after the 1960 Sharpeville massacre, a shocking event that gave momentum to the global anti-apartheid movement but did little to disturb existing international trade and security arrangements.

In the 1960s, under the purposive leadership of Hendrik Verwoerd, master-builder of apartheid, a major state-led technological programme emerged to counter sanctions and preserve white supremacy. IDC-sponsored state corporations, such as SASOL, extracting oil from coal, and FOSKOR, producing phosphate fertilisers, expanded rapidly.[33] In the 1970s, phosphate from the huge Phalaborwa complex, mapped by geologist Hans Merensky, satisfied domestic needs as well as being exported in large quantities. Uranium was enriched at a secret plant at Valindaba from the mid-1970s. Each of these activities, as well as dam construction, hydroelectric power, and irrigation, nurtured scientific networks and technological innovation. Resources were lavished on the CSIR and Afrikaner tertiary institutions. Apartheid also led to seven new segregated universities catering to black South Africans, five of them situated in the Bantustans where they purported to be beacons of ethnic nation-building. By the 1970s, several of these institutions derisively referred to as 'bush colleges' – including Fort Hare, Turfloop, Western Cape, and Durban-Westville – became centres of opposition to apartheid.[34]

In the late-apartheid period, and especially following the Soweto uprising of 1976–7, state resources were disproportionately devoted to supporting weapons production and, notoriously, to developing a nuclear capability. The technopolitics of apartheid power involved public demonstrations of national prestige and prowess, often underpinned by clandestine military and strategic cooperation with Israel, the United States, France, Germany and Britain. A radio astronomy installation at Hartebeesthoek provided tracking facilities for American satellites. After the withdrawal of such cooperation, in part because of anti-apartheid protests, this plant was remodelled for research as the HartRAO radio observatory. Foreign-funded

33 B. Freund, *Twentieth Century South Africa: A Developmental History* (Cambridge: Cambridge University Press, 2019), 83.

34 S. Dubow, *Apartheid 1948–1994* (Oxford: Oxford University Press, 2014), 165–6, 186.

observatories had proliferated in South Africa and the CSIR oversaw a programme of consolidation and nationalisation in the 1970s, leading to the South African Astronomical Observatory at Sutherland. As a highly networked science, astronomy was dependent on international collaboration and some of these connections were sustained notwithstanding boycotts and isolation during the apartheid era.

South Africa's ecological diversity and environmental challenges were a focus both of local expertise and international attention. Research in botany and pasture sciences informed a sustained attempt at conserving the natural veld which had been battered by a century of intensive livestock farming. Interventions on white-owned farms and in African homelands were framed and legislated before 1948 but implemented in the apartheid era. At their core were systems of rotational grazing in fenced camps, with devolved water supplies, designed to save the veld from overgrazing.[35] On white-owned farms, conservation was incentivised. But these ideas informed a coercive system of agricultural 'Betterment' in the Bantustans and triggered widespread opposition. The state saw such measures as essential for conservation and development, supplemented by irrigation projects, in the hope of winning allies in the Bantustans and spurring them to the elusive goal of economic self-sufficiency.

Scientific work in some spheres directly justified apartheid and white domination. Exponents of *volkekunde*, a distinct form of anthropology entrenched in some Afrikaans-speaking universities, worked with the state in order to conceptualise and implement tribal forms of governance as the integral core of 'separate development'. Keith Breckenridge has shown how computerised biometrical schemes were designed to implement the apartheid Population Registration Act and to support the implementation of the hated pass laws.[36] Armament programmes launched in the 1970s with the help of the CSIR and its

[35] W. Beinart, 'Soil Erosion, Conservationism and Ideas about Development: A Southern African Exploration, 1900–1960', *Journal of Southern Africa Studies (JSAS)*, 11, 1 (1984), 52–83; P. Delius and S. Schirmer, 'Soil Conservation in a Racially Ordered Society: South Africa 1930–1970', *JSAS*, 26, 4 (2000), 719–42.

[36] K. Breckenridge, *Biometric State: The Global Politics of Identification and Surveillance in South Africa, 1850 to the Present* (Cambridge: Cambridge University Press, 2014).

spin-off companies, resulted in the development of armoured vehicles, long-range artillery, aircraft, and nuclear weaponry.

The natural sciences, however, took many other directions, exploring and protecting biodiversity, researching invasive plants, and analysing the ecological significance of wildfire.[37] National and provincial parks expanded in the apartheid era, providing sites for pioneering research in wildlife conservation.[38] Translocation techniques, while not invented in South Africa, were adapted for use with rhinos in order to extend the range of this globally threatened species. Such efforts brought favourable publicity.

While black South Africans were drawn into wildlife conservation as rangers and workers, wildlife tourism in the 'great outdoors' was largely but not entirely a white preserve. As Jacob Dlamini reminds us in *Safari Nation*, there was considerable historic African interest in the Kruger National Park. Conservationist successes and the apartheid government's devotion to wild animals were sometimes contrasted with its mistreatment of black South Africans. Yet displacement did not entirely efface a black 'presence'. There were many black South Africans, including leading African nationalists like Selope Thema, who nurtured memories of, and ongoing connections to, the Park. And even at the height of apartheid, limited numbers of African, Indian and Coloured tourists were able to access parts of Kruger.[39] One of apartheid's many hubristic miscalculations was to exclude black communities from participation in national strategies and campaigns for conservation.

Following the political transition in 1994, often referred to as a 'miracle', the ANC took decisive steps to reorient the focus of science (Chapter 7). In decommissioning its nuclear weapons capacity, South Africa became the first and only country to engage in voluntary nuclear disarmament. This decision was the consequence of an unlikely convergence of mutual interests: a white supremacist regime that did not want to entrust nuclear weapons to an incoming black government; and a new leadership, headed by Nelson Mandela, which made a

37 S. Pooley, *Burning Table Mountain: An Environmental History of Fire on the Cape Peninsula* (Cape Town: University of Cape Town Press, 2015).

38 J. Carruthers, *National Park Science: A Century of Research in South Africa* (Cambridge: Cambridge University Press, 2017).

39 J. S. T. Dlamini, *Safari Nation: A Traveling History of the Kruger National Park* (Athens: Ohio University Press, 2020), 148–50, 187, 260.

virtue of global cooperation and sought to repurpose the state to advance collective human and societal well-being. The ANC took pride in remaking the country as a modern, democratic state and it promoted a vision of science and technology working for the common good. It also backed theoretical research unconnected to immediate utilitarian needs, not least in the field of astronomy which was a particular beneficiary of the new dispensation. Existing infrastructure was massively expanded from the 1990s with the construction of new optical portals into the southern skies such as the Southern African Large Telescope (SALT) at Sutherland and a major new expansion into radio astronomy, MeerKAT. The descriptor 'Southern' in SALT's name signalled a desire to encourage regional African cooperation as part of the post-apartheid democratic dividend.

Thabo Mbeki, who succeeded Mandela as president (1999–2008), advocated an 'African Renaissance' as the basis for national renewal that included ambitious scientific work in Africa. In launching SALT, Mbeki predicted that 'this giant eye in the Karoo' will 'tell us as yet unknown and exciting things about ourselves' and 'provide all humanity with the largest and most modern single optical-infrared telescope in the southern hemisphere'.[40] Like Smuts, Mbeki was strongly attracted to the possibilities of human palaeontology; yet, by contrast with Smuts, Mbeki espoused a common humanity sharing a political future. In the early twenty-first century, a rapid succession of exciting new palaeoanthropological finds in the Cradle of Humankind near Johannesburg gave scientific substance to South Africa's singular claims to harbour the vestigial traces of humankind, and this was used to support the idea of a 'rainbow nation'.

At the same time, Mbeki became deeply suspicious of the authority of Western science. Following on from Mandela, he took a personal interest in the repatriation from the Musée de l'Homme in Paris of the human remains of Sara Baartman for reburial in the Eastern Cape. Mbeki considered Baartman's public degradation in early nineteenth-century London and Paris, where she was exhibited as an exotic specimen, to epitomise European power and scientific arrogance. His renewed Africanist affinities, coupled with a mistrust of aspects of

[40] 'Speech of the President of South Africa, Thabo Mbeki, at the official inauguration ceremony of the Southern Africa Large Telescope (SALT): Sutherland, Northern Cape', 10 November 2005, available at www.thepresidency.gov.za

Western science, impelled him to intervene in the controversy surrounding HIV/AIDS. Hostility to multinational pharmaceutical companies and sensitivity to African dignity led Mbeki to support AIDS denialism. Key figures in his government advocated health foods or traditional medicines rather than antiretrovirals (Chapter 7).[41]

In his intervention into the HIV/AIDS controversy, it is often said that Mbeki found himself caught between 'indigenous' and 'Western' knowledge systems. We explore how both of these dynamic and hybrid knowledge traditions were invoked and promoted in support of post-apartheid nation-building. In contrast to the polarised debates around the causes and cures of AIDS, botanists have generally seen them as complementary. The demanding process of developing a new spirit of 'South Africanism' in the post-apartheid era has resulted in divergent invocations of science and knowledge. The ANC government, however, has largely positioned South Africa as an African nation that associates science with development and the provision of leadership in Africa through 'big science' initiatives.

The Scientific Imagination and the Circulation of Knowledge

South Africa has produced eleven Nobel laureates. Those awarded for contributions to literature (Gordimer and Coetzee) and for peace (Luthuli, Tutu, Mandela and de Klerk) are perhaps the best known – an indication of the global prominence of apartheid and its opponents. Five South African-born scientists were awarded Nobel prizes: Max Theiler (1951), Allan Cormack (1979) and Sydney Brenner (2002) in physiology/medicine; Aaron Klug (1982) and Michael Levitt (2013) in chemistry. In contrast to peace prize winners, all these recipients largely worked in the United States and the United Kingdom.

Many more South African-educated scientists migrated. Capetonian Solly Zuckerman, who trained in medicine, rose to prominence in the field of primatology in the 1920s, worked in the British wartime scientific establishment, and capped his career as chief scientific adviser to the British government (1964–71).[42] Michael Thackeray, who

41 S. Robins, '"Long Live Zackie, Long Live": AIDS Activism, Science and Citizenship after Apartheid', *JSAS*, 30, 3 (2004).

42 J. Burt, 'Solly Zuckerman: The Making of a Primatological Career in Britain, 1925–1945', *Studies in History and Philosophy of Biological and Biomedical Sciences* 37, 2 (2006), 295–310.

worked on lithium battery technology in South Africa for two decades from 1973, achieved his greatest success after 1994 at the Argonne National Laboratory in the United States. In this book we deal little with those who left the country and concentrate instead on those who migrated to, and made their reputations in, South Africa.

These patterns of migration are clues to the globality of South African science. They remind us that science does not observe national boundaries, though it can at the same time be profoundly national in orientation. The encounters of 'skilled communities and individuals', which Kapil Raj has identified in South Asia, are also clearly evident in South Africa.[43] The resulting mixture of perspectives, inward and outward, collective and individual, adds complexity to our sense of the scientific imagination, particularly as these inform the constitution of national identity and pride. Scientific linkages were mostly, but not exclusively, north–south by way of geographical orientation. Yet in the case of South African relations with Australia, there was substantial interchange, and also intense competition on a south–south axis in spheres such as astronomy and botany.[44]

Historians of science have proposed a diverse range of models to capture the ways in which knowledge circulates in the extra-European world. In a widely cited article, Roy Macleod challenged George Basalla's uni-directional, radial hub-and-spoke model of the diffusion of knowledge outwards from metropolitan centres.[45] Macleod argued instead for a polycentric approach, while rejecting the idea that 'colonial science' was derivative, inferior, or valuable only for its fact-gathering function. Following W. K. Hancock, biographer of Smuts, Macleod conceived of a 'moving metropolis'. This influential, decentred, approach has been taken further by David Livingstone,

[43] K. Raj, *Relocating Modern Science: Circulation and the Construction of Knowledge in South Asia and Europe, 1650–1900* (Basingstoke: Palgrave Macmillan, 2007).

[44] J. Carruthers, 'Trouble in the Garden: South African Botanical Politics ca. 1870–1950', *South African Journal of Botany*, 77, 2 (2011), 258–67; Brett Bennett, 'Model Invasions and the Development of National Concerns over Invasive Introduced Trees: Insights from South African History', *Biological Invasions*, 16, 3 (2014), 499–512.

[45] G. Basalla, 'The Spread of Western Science', *Science*, 156, 3775, 611–22; R. Macleod, 'On Visiting the "Moving Metropolis": Reflections on the Architecture of Imperial Science', *Historical Records of Australian Science*, 5, 3 (1980), 1–16.

who argues that knowledge is always produced in particular locales and has to be understood contextually: 'as it moves it is modified; as it travels it is transformed'. In this sense, science is always also a form of 'local knowledge'.[46]

James Poskett takes this further in his study of worldwide phrenology, arguing that it is no longer meaningful to distinguish between 'the local and the global' since what is described as local may be globally constituted as well.[47] This insight is borne out in the history of proponents and opponents of racial science where South Africa was both producer and consumer. Key figures, such as Robert Knox and Francis Galton, the inventor of eugenics, developed their ideas and careers in the mid-nineteenth century in part from their southern African experiences (Chapter 2). The leading Nazi racial anthropologist, Eugen Fischer, wrote an early twentieth-century doctoral dissertation on racial 'hybridisation' in Namibia. Amongst anti-racist campaigners, Lancelot Hogben, a notable British critic of eugenics, was sensitised to these issues while working at the University of Cape Town in the 1920s; in the high apartheid era, the Wits comparative anatomist Philip Tobias gained international attention for his challenge to racial science (Chapter 6).

How do we characterise these ever-shifting relationships between different sites of knowledge collection, assimilation, and production? The concept of 'pericentricism', used by international relations theorist Tony Smith to understand the Cold War, may be apposite too in the global politics of knowledge, for the concept suggests that developments on, and contributions from the periphery, can impact disproportionately on centres of power.[48] A related view has been advanced by Jean and John Comaroff whose idea of 'theory from the South' proposes that 'modernity was, almost from the start, a North-South

[46] D. N. Livingstone, *Putting Science in its Place: Geographies of Scientific Knowledge* (University of Chicago Press, 2003), 4, 81.

[47] J. Poskett, *Materials of the Mind: Phrenology, Race, and the Global History of Science 1815–1920* (Chicago University Press, 2019), 251; J. M. Kuukkanen, 'I am knowledge. Get me out of here! On localism and the universality of science', *Studies in History and Philosophy of Science Part A*, 42, 4 (2011), 590–601.

[48] T. Smith, 'New Bottles for New Wine: A Pericentric Framework for the Study of the Cold War', *Diplomatic History*, 24, 4 (2000), 567–91.

collaboration – indeed, a world-historical production – albeit a sharply asymmetrical one'.[49]

Uneven development, sustained by mutually reinforcing disparities in power, remains key. This was a theme common to many disciplines in the social sciences in the 1960s and 1970s. From the 1980s, the promise of globalism and a web-driven knowledge society gave rise to the prospect of a 'flat earth' in which ideas and commodities would spread freely, frictionlessly, and to the universal benefit of humanity. New variants of postcolonial relativism sustained this positive view (perhaps inadvertently) in celebrations of 'hybridity', 'fluidity', 'co-production' and 'interpenetration' – and in so doing moved away from earlier, often mechanistic conceptions of postcoloniality that fixed on knowledge as power and on binary processes of 'othering'. This book is concerned with fluidity and hybridity where this can be demonstrated, yet we remain alert to power differentials in knowledge systems and to blockages in exchange.

Our approach to the history of science seeks to examine knowledge and networks in their socio-economic, cultural and historical contexts. There is no shortage of examples in which South Africa has been treated as a 'reservoir of raw fact', destined for processing or refining in Europe or North America.[50] But these examples are not necessarily the rule. From the early nineteenth century the trend has been towards barter and exchange from a position of relative advantage or disadvantage, with locally based institutions and scientists eager and able to play an active role in generating knowledge, either through informal networks or meditated by purpose-built 'centres of calculation'.[51] By the twentieth century, organised scientific knowledge, geared towards observation or experimentation, was increasingly generated and produced in South Africa.

Although we decentre the history of science by virtue of focusing on a single country in the Global South and its transnational connections, we also *recentre* science by paying close attention to multiple constructions of South African nationhood. We argue that scientific and applied technical knowledge have been key to the country's modernity

[49] J. and J. Comaroff, 'Theory from the South: Or, how Euro-America Is Evolving toward Africa', *Anthropological Forum*, 22, 2 (2012), 116.
[50] Comaroff and Comaroff, 'Theory from the South', 114.
[51] Bruno Latour, *Science in Action: How to Follow Scientists and Engineers through Society* (Cambridge, MA: Harvard University Press, 1987).

and development and that it is impossible to understand complex patterns of change and competing ethnic and racial claims without an appreciation of the impact of science and technology.

Science and African Studies

Historians of science and technology increasingly recognise the need for approaches that include discussion of politics and power along with innovation and discovery. In our earlier work, we have illustrated the significance of scientific ideas in the formation of interventionist colonial policies as well as in racial segregation and apartheid.[52] For those Africanists and advocates of decoloniality who are strongly critical of the role of science in colonialism, we hope to show that scientific thinking is enormously diverse, often relatively autonomous of colonial institutions and power, and even exciting and inventive. Our project challenges assumptions in southern African historiography, and in the interdisciplinary field of African studies, which treat science with suspicion as an external, foreign, and hostile force. Such ideas surface periodically and sometimes stridently, for example in academic literature about medical and environmental interventions, in politicised debates about HIV/AIDS, and in the Rhodes Must Fall movement since 2015.[53]

This book is in part an attempt to get beyond the politics of rhetoric in understanding the ubiquity of science. We have been interested in exploring examples where scientific knowledge, while generated within colonial contexts, was not reflexively responsive to the specific imperatives of colonial rule or white dominance in South Africa.[54] Even at the height of apartheid there were socially progressive and dissident academics who chose their own paths. We therefore discuss science in South Africa's colonial environment as well as colonial science.

[52] Beinart, 'Soil Erosion, Conservationism'; Dubow, *Scientific Racism.*

[53] W. Beinart, 'African History and Environmental History', *African Affairs*, 99, 395 (2000), 269–302; W. Beinart, K. Brown and D. Gilfoyle, 'Experts and Expertise in Colonial Africa Reconsidered: Science and the Interpenetration of Knowledge', *African Affairs* 108, 432 (2009), 413–33; S. Dubow, 'Introduction' in S. Dubow (ed.), *The Rise and Fall of Modern Empires, Volume II Colonial Knowledges* (Farnham: Routledge, 2013).

[54] Harries *Butterflies and Barbarians*; Dubow, *A Commonwealth of Knowledge*; Beinart, *The Rise of Conservation.*

Similarly, we suggest that it is possible to distinguish between scientific research conducted in the apartheid period and 'apartheid science'.

In the late-apartheid years some scientific work was explicitly aimed at the transformation of apartheid society, and many scientists engaged in fierce debates within their disciplines. By this time, too, scientific and medical training was beginning to be more open to black South Africans and Africans from other countries. This further diversified the social location of practitioners and the sites of scientific work. Jan Todd has argued, with respect to Australia, for a complex model that embeds scientific work in political and institutional developments as well as allowing analytical scope for the unpredictability and diversity of technological innovation and scientific ideas.[55] This book offers evidence to support such a non-instrumentalised view of science and technology.

In African and South Asian contexts, historians have explored the way that colonised people acquired and adapted new forms of knowledge under colonialism.[56] In an influential collection on health and healing in Africa, Steven Feierman and John Janzen explored the idea of 'therapeutic diversity'.[57] Local knowledge was itself dynamic and increasingly hybrid.[58] While we do not develop a detailed treatment of these processes, our perspectives are informed by an attempt to explore 'polycentric communication networks', 'reciprocal influences and multidirectional flows', and 'circulating knowledge'.[59] In the South African context, we have argued for the terms 'vernacular' and 'local'

55 J. Todd, *Colonial Technology: Science and the Transfer of Innovation to Australia* (Cambridge University Press, 1995); Beinart, Brown and Gilfoyle, 'Experts and Expertise'; Tilley, *Africa as a Living Laboratory*.

56 G. Prakash, *Another Reason: Science and the Imagination of Modern India* (Princeton, NJ: Princeton University Press, 1999); T. E. Huff, *Intellectual Curiosity and the Scientific Revolution: A Global Perspective* (Cambridge: Cambridge University Press, 2010).

57 S. Feierman and J. Janzen (eds.), *The Social Basis of Health and Healing in Africa* (Berkeley: University of California Press, 1992); K. E. Flint, *Healing Traditions: African Medicine, Cultural Exchange, and Competition in South Africa, 1820–1948* (Athens: Ohio University Press, 2008); Anne Digby, *Diversity and Division in Medicine: Health Care in South Africa from the 1800s* (Oxford: Peter Lang, 2006).

58 W. Beinart and K. Brown, *African Local Knowledge and Livestock Health: Diseases and Treatments in South Africa* (Oxford: James Currey, 2013).

59 D. Chambers and R. Gillespie, 'Locality in the History of Science: Colonial Science, Technoscience, and Indigenous Knowledge', *Osiris*, 15, 1 (2000), 221–40; K. Patiniotis et al., 'Science and Technology in the European Periphery: Some Historiographical Reflections', *History of Science*, 46, 2 (2008), 153–75.

rather than 'indigenous' knowledge, in order to recognise such interaction.[60] For, as Peter Burke points out, there exists a plurality of knowledges in every culture. Europe, too, had its margins and marginalised people. The distinctions between different kinds or levels of knowledge are therefore always relative, as well as bound up in contestation and conflict.[61] Patrick Harries explores such processes in writing about Junod, and Nancy Jacobs illustrates the relationships between European ornithologists and African assistants.[62] Yet an awareness of boundaries in knowledge and differentials in power means that it is important not to overstate connections and reciprocity or to compensate for histories of exclusion by selective appropriation of the past in order to make the present more palatable.

Consider the example of Christiaan Barnard, who performed the first human heart transplant at Groote Schuur hospital in 1967, thereby bringing prestige to the University of Cape Town (UCT) and, by extension, to apartheid South Africa. After 1994, reports circulated that Hamilton Naki, a self-taught black technician, was a key contributor behind-the-scenes to the transplant programme, and in 2002 Naki was awarded South Africa's Order of Mapungubwe for his outstanding contribution to medical science. On Naki's death in 2005, the *New York Times* quoted Barnard (who died in 2001) to the effect that Naki was 'one of the great researchers of all time in the field of heart transplants'. A correction soon followed. Naki was indeed a skilled technician, as were black cleaning staff at Wits like Alfred Nemudzivhadi who helped to operate on baboons as part of research into kidney transplantation. Yet Naki was not at Groote Schuur Hospital when the first human heart transplant was undertaken, and he was therefore not in a position to have extracted the heart of the donor as claimed.[63]

[60] Beinart, *Rise of Conservation*;Beinart and Brown, *African Local Knowledge*.

[61] Dubow, 'Introduction', xxii; A. Cooper, *Inventing the Indigenous: Local Knowledge and Natural History in Early Modern Europe* (Cambridge: Cambridge University Press, 2007).

[62] N. Jacobs, *Birders of Africa. History of a Network* (New Haven, CT: Yale University Press, 2016).

[63] M. Fox, 'Hamilton Naki, 78, Self-Taught Surgeon, Dies', *New York Times*, 11 June 2005; M. Wines, 'Accounts of South African's Career Now Seen as Overstated', *New York Times*, 27 August 2005; S. Horwitz, 'Race, Kidney Transplants, Immunosuppression Research, and White Supremacy under Apartheid, 1960–80', *Canadian Bulletin of Medical History* (online, 2020).

Another example comes from big science astronomy in South Africa and Australia, where the 'Shared Sky' art exhibit, which toured Europe in 2018, suggested that indigenous San and Wajarri aboriginal people in South Africa and Australia have traditionally viewed the heavens in ways that anticipate or complement the SKA project. The romanticised claim is misleading because it implies that there is a homogeneous 'first nation' cosmological system spanning Australia and South Africa. Moreover, it overlooks the distinct possibility that mega-projects like the SKA, rather than fulfilling a shared human desire to comprehend our place in the universe, might in fact impact negatively on the rights of local indigenous communities.[64]

We seek in this book to bridge divides, albeit with full awareness of their existence and how they have evolved. The scientific imagination in South Africa was multifaceted: it drew on global innovation, but was often shaped to local imperatives; it was an instrument of racial power in some contexts but autonomous and inventive in others; most of all, the imagination was shaped by the specific environments in which it developed. By attempting to find some alternative ground, we hope also to contribute to public discourse and policy by suggesting that local knowledge alone is insufficient to address complex medical, environmental and developmental issues.

In bringing together diverse examples over a long period, we hope that our book will provide a revised sense of the country's contribution to science. This is the core of our project: an argument for the importance of scientific research in the region; an argument for the importance of the region in some branches of scientific work; an argument for the inescapable necessity of understanding the impact of science and technology in South African history; and also an account that addresses the hybridity of knowledge. This explains our wish to explore over the long term, and in a wide range of disciplines and areas, the imbrication of the local and the global in the generation and understanding of knowledge about South Africa.

[64] C. Walker, D. Chinigò, and S. Dubow, Editorial introduction to 'Karoo Futures: Astronomy in Space and Time', *JSAS* special issue, 45, 4 (2019). 653–4.

1 *Scientific Imagination and Local Knowledge at the Cape in the Eighteenth Century*

During the eighteenth century, a spate of scientifically curious visitors arrived at the Cape, looked at southern Africa with new eyes, and recorded what they saw in great detail. The scale of reporting escalated from the 1770s, by which time the southern tip of Africa had been a frequent stop-off point for Europeans for more than two centuries: from the early sixteenth century, Portuguese coastal settlements were established in Mozambique while the Dutch East India Company (VOC) set up its station at Cape Town in 1652. By the 1770s, the European settlers who gradually spread out from Cape Town, chafing against Company authority, had penetrated into the interior to establish farms and grazing outposts 600 km to the east and 400 km northwards in the dry interior. They were in the final stages of conquering the indigenous Khoekhoen and had reached the boundaries of Xhosa and Tswana settlements. The VOC recognised that the original purpose of the settlement, to supply maritime traffic, was being superseded. In 1780 the colonial boundary was extended to the Great Fish River, absorbing some Xhosa land, and in 1786 an administrative centre was established in the eastern interior at Graaff-Reinet.

The new visitors largely travelled the routes opened up by colonial expansion. They were increasingly products of the Enlightenment and of the multiple shifts in thinking among educated Europeans that encouraged a more systematic desire to observe, document and classify nature and people. Multinational in origin and cosmopolitan in outlook, they brought with them a comparative capacity, born of reading and training, as well as the ambition to publish and establish their place in the literary record. Their writings served a rapidly expanding audience and their books were often translated into different languages. During the late eighteenth and early nineteenth centuries, a stellar multilingual reading list developed around the subject of the Cape. Of particular interest to us are Peter Kolb and Hinrich Lichtenstein (German), Anders Sparrman and Carl Thunberg (Swedish), Nicolas-Louis de Lacaille and François Le Vaillant (French), John Barrow and William Burchell

(British). We also make reference to some Dutch authors, particularly Robert Gordon, a Dutch military officer of Scottish descent at the Cape, whose writing was not published at the time, but whose knowledge was available to a restricted group of officials and visitors.

Their published descriptions came to constitute what they saw as an emerging scientific understanding of the region, especially its plants, animals and people. They did not generate this knowledge alone and depended on conversations with local people as well as on observation. The information and ideas they recorded were derived in part from their engagement with the Cape's inhabitants; local knowledge was absorbed into their texts and provided an important source for their attempts at scientific understanding. Knowledge and skills to some degree flowed in the other direction, too, so that some of the observations and perceptions of the travellers fed back into local discussions at the Cape. Together with Khoekhoe and African guides, settler farmers and government agents, these travellers were instrumental in conceptualising the region, and imprinting onto it natural and social features that helped to shape the subsequent scientific imagination about the expanding colony.

Plants and Marvels before 1770

After the dramatic voyages of the Portuguese navigators Dias and da Gama in the late fifteenth century, and the confirmation of a sea route for European mariners to India, shipping around the Cape gradually increased; all ships bound for India and China stopped at some point in southern Africa. Table Mountain and its immediate environs became a familiar landscape rendered in strikingly different ways in contemporary drawings and sketches. The early navigators were above all looking for water and food for their journeys. But sailors were also alert to the possibility of finding riches, their imagination and desire fuelled by the gold, silver, cochineal and plants of the Americas and East Asia.

The rounding of the Cape coincided with the early phases of book printing in Europe, which spread knowledge slowly but on a new scale. Early accounts of the Cape drew not only on observation but on a tradition of reporting marvels in far-off lands.[1] Some authors described features of a natural world that would be familiar to a modern reader,

[1] S. Greenblatt, *Marvelous Possessions: The Wonder of the New World* (Chicago: University of Chicago Press, 1991).

adding to it imaginary elements that populated their symbolic world. These included fanciful creatures like mermaids and unicorns. In 1555, Antonio Galvão described 'certain fishes', seen near Mozambique, that 'commonly went upright in the water, having the faces and natures of women'.[2] Jean Moquet, writing in the early seventeenth century, observed on the coast of Mozambique 'the peixe-mulher, that is to say the woman fish because this fish is like a woman being possessed of her nature. It carries its young under its fins that, on two sides, serve as arms, and often goes on land where it even produces its young. The black fishermen are lectured on the perils of sleeping with these fish-women.'[3] According to Moquet, mermaids' teeth were used as a protection against haemorrhoids and fevers. Perhaps they were embellishing reports of seals with ideas about mermaids that lay deep in European folklore; myths of human–animal hybrids were universal. Material objects from rare creatures were collected and housed in cabinets for display by the wealthy. Ostrich feathers, elephant tusks, hippopotamus teeth, rhinoceros horns, shells, seeds, civet, amber and geological specimens were all curiosities with possible scientific, medical or monetary value.

Had Portuguese authority been stronger, the area that became South Africa may have been colonised most intensively from Mozambique rather than Cape Town. Portuguese traders reached the Zimbabwean highveld and described the Mutapa Shona kingdoms – an early travel literature on the region prefiguring that on the Cape. João dos Santos published an account of the natural history of Mozambique in *Ethiopia Oriental* (1609).[4] This work, partly translated into English in 1625, contained chapters on fruits and cereals as well as descriptions of animals such as crocodiles, lions, leopards, wild dogs, monkeys and baboons.

But the major momentum for colonisation came from the VOC, founded in 1602 in a Dutch society that was already well-attuned to recording and painting nature.[5] In the seventeenth century it was the largest trading company in the world, focusing initially on Asian spices.

[2] A. Galvano, *The Discoveries of the World, from Their First Original unto the Year of Our Lord 1555, Corrected, Quoted and Published in England by Richard Hakluyt (1601)* (London: Hakluyt Society, 1862), 43.

[3] X. de Castro (ed.), *Voyage à Mozambique et Goa: la relation de Jean Mocquet (1607–1610)* (first published 1617; Paris: Chandeigne, 1996), 78–80, translated by P. Harries.

[4] J. dos Santos, *Ethiopia Oriental, e varia historia de cousas notaveis do Oriente* (first published Evora, 1609, excerpts in English 1625).

[5] Cook, *Matters of Exchange*.

The transfer and cultivation of plants, as well as botanical knowledge, was intimately bound up with colonial expansion. The Cape was an intriguing collecting point both because of the singularity and variety of its species – features that were recognised very early on. A Dutch fleet stopping at the Cape in 1597, well before settlement, brought back what was probably the first example of a protea in Europe.[6] A pelargonium (then, as now, called geranium) also probably found its way to the Netherlands before 1600. Such plants became the basis for an extraordinary diversity of cultivars that would come to fill the window boxes and greenhouses of European cities. Many species had value as food, medicine, commodities and flowers. The Dutch tulip mania of the seventeenth century demonstrated their potential worth.[7] Tulips, originally from Turkey, inflated massively in price in the mid-1630s, peaking during 1637, fifteen years before the VOC started the Cape settlement.

When Jan van Riebeeck arrived in 1652 to establish the VOC station, plants were an important feature of the project; the Company immediately founded a garden both to grow food and to experiment with imported and local plant species. Van Riebeeck, the first governor, was a product of empire. At thirty-five years old, he had already travelled widely to Dutch colonies in Asia and after his ten years in South Africa he left for Batavia (Indonesia). Simon Pooley has scoured van Riebeeck's daily diary to trace his keen interest in plants and his personal role in their importation.[8] His party brought vegetables and herbs from Europe and by 1653 had also imported pumpkin, melon and watermelon seed, as well as sweet potato roots from Pernambuco, Brazil, which the Dutch had briefly colonised. Needing rich soil for these plants, the Cape settlement used manure from the *kraals* (protected gathering points) vacated by Khoekhoe cattle. Vines and tobacco were soon cultivated, and the first wine was produced in 1658.

The colonists tried indigenous plants such as wild garlic (*Tulbaghia* species), their value learnt from the Khoekhoen. In 1654 van Riebeeck noted that 'bitter almonds' (*Brabejum stellatifolium*) were 'commonly collected and eaten by the natives, who first of all peel them and, after drying them in the sun for a few days, roast them on the fire.'[9] Perhaps

[6] J. P. Rourke, *The Proteas of Southern Africa* (Cape Town: Purnell, 1980), 5.
[7] A. Pavord, *The Tulip* (London: Bloomsbury, 1999).
[8] S. Pooley, 'Jan van Riebeeck as Pioneering Explorer and Conservator of Natural Resources at the Cape of Good Hope (1652–62)', *Environment and History*, 15, 1 (2009), 3–33.
[9] Pooley, 'Jan van Riebeeck', 18.

because they were poisonous unless carefully treated in this way, these nuts did not become a significant food source for the colonists. But by 1660 the tree's value as hedging was recognised, and it was planted as part of a boundary between the colonial settlement and the Khoekhoen. Small sections of this hedge survive – carrying symbolic weight to the present day as the first, unsuccessful attempt to segregate whites from Africans – and remain an attraction in Kirstenbosch Botanical Garden.

Experimentation and collection continued under successive governors, notably the two van der Stels, Simon and Willem Adriaan, from 1679 to 1707. In addition, minerals were a major source of interest for exploration, and eleven expeditions to the north were mounted between 1659 and 1686, stimulated initially by rumours of the gold-rich kingdom of Monomotapa, and then by the promise of copper.[10] Under the patronage of VOC director Nicolaes Witsen, later mayor of Amsterdam, Simon van der Stel led an exploratory expedition north of the settlement in 1685 looking both for minerals and plants. Those on the expedition drew knowledge from indigenous people encountered en route and made detailed drawings.[11]

The most comprehensive and influential attempt to record nature and society in the early colony was Peter Kolb's book *Caput Bonae Spei Hodiernum* (published in 1719 in German, 1727 in Dutch and 1731 in English).[12] Born in 1675 in Germany, Kolb studied astronomy, with a thesis on comets as natural phenomena. He was sent by a German patron, with the support of Witsen, to gather astronomical data at the Cape. While he did not realise his astronomical ambitions, he broadened his interests to include climate, geology, flora and fauna – and, increasingly, the people of the Cape too. In 1707 he became a VOC official based at Stellenbosch, from where he travelled, drew maps, and

[10] S. Huigen, 'Travellers to Monomotapa: the Representation of Southern Africa by the Dutch in the Seventeenth Century', *History and Anthropology*, 9, 2–3 (1996), 207–30.

[11] R. P. Brienen, 'Nicolas Witsen and His Circle: Globalisation, Art Patronage, and Collecting in Amsterdam circa 1700' in Nigel Worden (ed.) *Contingent Lives: Social Identity and Material Culture in the VOC World* (Cape Town: University of Cape Town Press, 2007).

[12] N. Penn, 'Written Culture and the Cape Khoikhoi: From Travel Writing to Kolb's "Full Description"', in A. Delmas and N. Penn (eds.), *Written Culture in a Colonial Context: Africa and the Americas 1500–1900* (Leiden: Brill, 2014), 171–93; A. Good, 'The Construction of an Authentic Text: Peter Kolb's Description of the Khoi', *Journal of Early Modern History*, 10, 1 (2006), 61–94.

met Khokhoe people as well as settlers. Kolb was obliged to return to Amsterdam in 1713, when his eyesight and relationship with the governor both began to fail. He wrote his German text in three parts on natural history, the Khoekhoen, and colonial politics. Kolb spent more time at the Cape than most of the other European visitors who published in the eighteenth century. His natural history constituted the most ambitious attempt to list species of plants and animals at that time. He did not write a travel narrative in the manner of later texts but rather presented his material as a compendium: 'Kolb's book is, in many ways, a printed version of a wonderful collection of facts and curiosities from the end of Africa' and, as Nigel Penn puts it, 'a hybrid of pre- and post-Enlightenment discourses'.[13]

By this time, attempts were being made by natural historians, for example John Ray in *Historia Plantarum* (1686–1704), to divide plants found in Britain, Europe and 'new worlds' into species and groups according to their morphology, distribution and habitat. Kolb did not attempt to classify species in this way but instead listed plants alphabetically. He divided them into local, or indigenous (*inlandschen* in Dutch), and exotic (or *uitlandschen*). He clearly paid close attention to the indigenous plants, mentioning over 400 species, and he included a special section on indigenous figs. He knew that plants such as agaves (which became known as American aloes in South Africa), *Cannabis sativa* and edible figs, already common at the Cape, were exotics.

With respect to animals, Kolb combined observation with occasional hearsay and reportage from earlier texts. He claimed to have seen, and to have been charged by, a rhino and his description of its movements does suggest first-hand experience. However, his German publisher included only the usual European representation of the rhinoceros with armour-plating and one horn, probably influenced by Dürer's imaginative woodcut of an Indian rhino made in 1515.[14] Thus Kolb's textual description was not matched by the illustration, which may seem to raise doubts about the authenticity of his report. However, the Dutch

[13] Penn, 'Written Culture', 183–4.

[14] K. Rookmaaker, 'The Slow Recognition of the African Rhinoceros from Hondius to Camper', in S. Huigen and J. Kommers (eds.), *Interpretations of Colonial Representations: Reflections on Alterity, Colonial History, and Intercultural Contact* (Saarbrucken: Verlag für Entwicklungspolitik, 2004), 35–54.

publisher (1727) included two detailed drawings of a rhinoceros.[15] The first was labelled as an outdated depiction of rhinos with one horn. (In fact, two of the five rhino species, the Indian and Javan rhinos, do have one horn.) The second drawing, captioned 'Rhinoceros volgens deze Beschryving' shows a smoother skin and two horns of different sizes. Kolb's description in the text indeed emphasised two horns and skin without scales or armour-plating. The second drawing in the Dutch edition has been called 'the first naturalistic [European] representation of the African rhinoceros, which remains easily recognizable today'.[16]

Figure 1.1 Kolb 1727: Rhinoceros volgens deze beschryving (following this description). Possibly the first printed image that did not use a version of Dürer's drawing of 1515 as pictured on left. Source: Peter Kolb, *Naaukeurige en uitvoerige beschryving van de Kaap de Goede Hoop*, 2 volumes (Amsterdam: Balthazar Lakeman, 1727)

15 P. Kolb, *Naaukeurige en uitvoerige beschryving van de Kaap de Goede Hoop*, 2 volumes (Amsterdam: Balthazar Lakeman, 1727), between pages 188 and 189, and 190 and 191.

16 Rookmaaker, 'The Slow Recognition of the African Rhinoceros'.

Kolb recorded that the rhino had small eyes and ears, and a sensitive capacity for smell. But he seemed to drift into the realms of hearsay when he added that when the rhino killed a man, by throwing the body with its horn, it 'feeds upon him by licking, with his Rough and prickly Tongue, his Flesh from his Bones'.[17] He also mentioned that rhinos were in perpetual enmity with elephants, which they would charge and pierce with their horn, ripping out the entrails. The illustration of such an attack in the German edition was dropped in the Dutch version, but the textual description remained. Kolb also reported that poison could be detected by pouring it into a cup fashioned from rhino horn, which would break on contact with the toxic substance. Rhino horn chips were said to be used as a medicine to fight convulsions, fainting and 'other illnesses', while the blood of the rhino healed 'inward sores' and 'opened obstructions'.[18]

Such stories are interesting in themselves and may, as in the case of those about unicorns or ostriches, reflect local knowledge. But they are a reminder that scientific observation could be imbricated with rumour or marvel. Kolb said he saw a sea lion in 1707 and described it in detail, comparing its head to a lion's without hair. He believed that when ostriches were scared, they did not run away, but instead hid their heads in the sand. Yet his writing about other species is more directly based on observation, and he used many local Dutch names that were already in common currency for Cape species. He also published realistic representations of fish.

Aside from its descriptions of the natural world, Kolb's text constituted the fullest and most sympathetic report thus far published on Khoekhoe society – a proto-anthropology.[19] Anne Good argues that 'it is not just veracity that he was aiming for; Kolb also crafted his description of the Khoikhoi to talk back to European culture'.[20] His narrative was part of the emerging notion of the noble savage, which drew especially on North American examples. At the time, before the terrible smallpox epidemic of 1713, there were still a number of relatively independent Khoekhoe groups living close to the colonial settlements at Stellenbosch and Drakenstein. Kolb spoke to them, gave their names and recounted what they said. He employed them as

[17] Kolb, *Naaukeurige,* II, 129.
[18] Kolb, *The Present State,* II, 104–5.
[19] Penn, 'Written Culture'.
[20] Good, 'The Construction', 85.

interpreters and guides – and he recounted how they saved him from lions.

Kolb described the Khoekhoen as a people with religion, laws, chiefs and customs. He demonstrated appreciation of their material culture, noting, for example, their use of musical instruments. A detailed drawing of 'hoe de hottentotten haar huysen bouwen' (how the Khoekhoen build their houses) showed the ordered arrangement of the homestead in a circle and the construction of dwellings with a symmetrical framework of branches overlaid by matting.[21] Khoekhoe people were depicted as devoted to their animals, and he illustrated their practice of riding oxen.[22] They showed ingenuity by digging large holes, with upright wooden stakes, to trap wild animals. In contrast with most earlier commentators, Kolb could explain such practices as body-smearing with fat and the buchu herb – even though he did not find this attractive. He could understand their dancing, singing and shouting as an expression of religion – and contrasted their 'zeal' on this front with the staid practices of Christianity.[23] In part this helped him to understand why they were reluctant converts; he recounted a story in which Khoekhoen raised in European households rejected European clothing, dwellings and religious beliefs and returned to their settlements.

Some of Kolb's most intriguing observations concerned Khoekhoe healing. Plants still lay at the heart of European medicine, and travellers were alert to their potential. The Dutch merchant and author Willem Bosman, who worked in West Africa for fourteen years, wrote in 1704 of the 'wonderful efficacy' of the 'green herbs' employed by healers there and 'deplored that no European Physician has yet applied himself to the discovery of their nature and virtue'.[24] Medical plant knowledge was also culled from the slaves brought to the Americas from Africa.[25] Kolb followed suit in suggesting that 'Hottentot Physicians' possessed 'great skill in the Vertues of their Herbs', knew

[21] Kolb, *Naaukeurige* II, 72–3, 114–15. [22] Kolb, *Naaukeurige* II, 38–9.

[23] Good, 'The Construction', 89.

[24] W. Bosman, *A New and Accurate Description of Guinea, Divided into the Gold, the Slave and the Ivory Coast* (London: James Knapton, 1705), 225. First published in Dutch, 1704.

[25] W. Churchill, 'Sloane's Perspectives on the Medical Knowledge and Health Practices of Non-Europeans' in A. Walker, A. MacGregor, M. Hunter (eds.), *From Books to Bezoars: Sir Hans Sloane and His Collections* (London: The British Library, 2012), 91–4.

how to bleed and cup patients, 'handle a lancet well', perform amputations and restore dislocations.[26] 'Great Cures are performed', he wrote, 'such as, perhaps, could not be performed by the ablest Physicians or Surgeons in *Europe*. The *Hottentots* who give themselves to the Study of Medicine are generally well skilled in the *Hottentot* Botany. They have some excellent Notions of the Virtues of the Multitude of Herbs and Roots that are produced in the Hottentot Countries; and often apply Herbs and Roots in very difficult and dangerous Cases with wonderful Success.'[27] Although they did not dissect bodies, they had 'pretty good notions of the human anatomy' and, despite their implements consisting of little more than a knife, horn and bird's bones, he thought that they practised their profession 'with much surprising dexterity'. There does not seem to be evidence of post-mortems in precolonial southern Africa with respect to human corpses, but animals were examined after slaughter.[28]

Because of its originality and novelty, Kolb's book was widely circulated. In addition to the translations, sections were included in compilations of travel narratives in English, French and German. Kolb's book served as a reference point for those who visited and wrote about the Cape later in the century, both to criticise and – especially in his approach to the Khoekhoen – to emulate. His book was one source, although not the main one, that fed into Enlightenment conceptions of the noble savage developed by Rousseau. He found ways to conceptualise Khoekhoe society as a totality with a sensitive and, to a degree, empathetic ethnographic eye.

Although the VOC did not specifically fund scientific research of the kind that Kolb had pursued, the Cape remained a major subject of study and research for scientific circles in the Netherlands. Amsterdam-based plant expert Johannes Burman, who studied and named the pelargonium, published on Cape plants in the 1730s; he also mentored the young Swedish botanist Linnaeus. The latter was later in touch with Rijk Tulbagh, Cape governor from 1751 to 1771, who

26 P. Kolb, *The Present State of the Cape of Good Hope* (London: W. Innys, 1731), I, 87–8, 302–11.

27 Kolb, *The Present State*, I, 302. C. Low, 'Understanding of Khoisan Healing, Past and Present'. Unpublished DPhil, University of Oxford (2004) and 'Finding and Foregrounding Massage in Khoisan Ethnography', *JSAS*, 33, 4 (2007), 783–99.

28 Beinart and Brown, *African Local Knowledge*.

was particularly attentive to the Company garden and the value of plants. Linnaeus wrote Tulbagh a flattering letter thanking the governor for the receipt of 'so great a number of rare Cape plants, which I had never before seen'.[29]

Kolb did not contribute much to astronomy, but the potential remained of a southern hemisphere vantage point on the heavens. The French astronomer Lacaille made an independent trip to the Cape via Rio de Janeiro in 1751–2.[30] A follower of Newton, whose ideas, especially on gravitational theory, were not yet well established in France, he was particularly interested in the movement of planets around the sun and sought to map the stars visible from the southern skies. Already established as an astronomer and mathematician, on which subjects he produced key textbooks, Lacaille taught at the Collège Mazarin in Paris, a leading scientific institution. In 1742, he was responsible for erecting a new observatory, where he made detailed observations that modified understandings of gravitational impacts on the planetary orbits.

Lacaille became convinced that measurements in the southern hemisphere would enable testing and improvement of calculations for the distances of the moon, planets and sun. Governor Tulbagh had recently taken office when Lacaille arrived at the Cape and supported his research, providing resources to build a small wooden observatory. He was extraordinarily industrious, listing about 10,000 stars and giving scientific names to the constellations that he observed. When this task was completed, he attempted the first measurement of the earth's radius in the southern hemisphere. Astronomers already understood that the earth was not a perfect sphere. They had detected a flattening at the pole in the northern hemisphere and the question was whether the same oblate shape could be found at the South Pole. Understanding this was not only important for defining the earth's shape, it also had implications for analysis of earth-bound and astronomical distances, gravity and the earth's rotation.

The astronomer Ian Glass, biographer of Lacaille, describes how he fixed the latitude of points on land and measured the distance between them. The Swartland area north of Cape Town provided a good site

[29] J. H. Verduyn den Boer, *Botanists at the Cape 1. Biographical Notices of Cape Botanists with 21 Portraits* (Cape Town and Stellenbosch, 1929), 64.

[30] I. S. Glass, *Nicolas-Louis De La Caille, Astronomer and Geodesist* (Oxford: Oxford University Press, 2012).

because there were peaks separated by flat land, allowing uninterrupted views for measurement. Lacaille's measurements showed that the oblate shape at the South Pole seemed to be flatter than at the north so that the earth was, logically, a little pear-shaped. This finding was anomalous and so, seventy years later, George Everest, chief assistant of the geodetic Survey of India, after whom the Himalayan mountain was named, tried to retrace Lacaille's steps while on leave to the Cape in 1820.[31] Everest pointed to possible errors and questioned the result which 'would assign to the southern meridians a different ellipticity from that found to obtain in the northern hemisphere'.[32] As shown in Chapter 2, the issue was only resolved when Thomas Maclear established that the oblate curve of the southern hemisphere was not significantly different from that in the north. Lacaille had failed to take adequate account of the gravitational effects on his plumb line of the mountains close to his key sites.

Lacaille stayed longer at the Cape than he had initially intended and made notes on plants, fish and animals. An enthusiastic walker, he scaled Table Mountain, via Platteklip Gorge, at a rapid pace. Familiar with Kolb's writing, he became curious about the Cape's people, including some old French speakers among the Huguenot settlers who had arrived about sixty years before. He also noted that younger Huguenots no longer spoke French and was told, correctly as it turned out, that the language would largely die out at the Cape. On his return to France in 1754, Lacaille became a celebrity, bringing with him 'not the spoils of the orient but ... those of the southern sky'.[33] A major astronomical paper and a map were published. The Cape, as a portal to the stars and planets, acquired a firmer place in the emerging global scientific imagination. Lacaille's notes were published in 1763, after his death.[34] If his astronomical observations proved stellar, his insights into Cape society were more prosaic. He was critical of Kolb, whom he accused of exaggerating the size and number of 'hottentot nations'.[35] His observations do not rival Kolb's, nor does he appear to have evinced the social openness that suffuses Kolb's writings. For example, he was aware of the importance of honey beer (*karee*), made with a

[31] Glass, *De la Caille*, 152. [32] Glass, *De la Caille*, 157.

[33] Glass, *De La Caille*, 115.

[34] N-L. de La Caille, *Journal historique du voyage fait au Cap de Bonne-Esperance* (Paris: Guillyn, 1763).

[35] Glass, *De La Caille*, 106–7.

root, to the Khoekhoen, but saw only its negative effects, rendering them 'absolutely incapable' for long periods.[36] He did, however, recognise that Khoekhoe society had been destroyed by smallpox in 1713.

Enlightenment Scientific Travellers

Compared to Kolb, those who visited the Cape from about 1770 were more deeply influenced by Enlightenment ideas and a scientific, secular view of nature and society. His late-eighteenth-century successors enhanced his picture of Cape society as a totality, though none wrote as fully on Khoekhoe society. Religious ideas, stories, even marvels, did not disappear from their texts. But the new travellers had a clearer idea of themselves as observers and scientists, attentive to classification and recording the diversity of species. Some, such as Sparrman, explored ideas that Carl Linnaeus promoted of a balance and order in the natural world, as a route to understanding interactions in nature.[37] Others were especially curious about the potential economic value of natural resources or expressed their aesthetic pleasure in nature.

From the mid-eighteenth century a new scientific momentum was generated in a number of countries, particularly Sweden, Britain, France and Germany. Linnaeus, professor at Uppsala in Sweden from 1741, who had wide connections in Europe, developed an overarching system for classifying and recording species and was able to impart a sense of urgency about the need to populate it on a global scale. One of the major problems then facing scientists was the proliferation of names given to the same plant or animal. In his landmark *Systema Naturae* (1758), Linnaeus aimed to standardise naming with the binomial system of classification in Latin. Every plant or animal was placed within a genus and assigned a unique species so that they were both grouped and distinct. With respect to plants, Linnaeus chose reproductive characteristics rather than morphology (form or structure) in order to establish the genus. As the 1758 volume listed only twenty

[36] Beinart and Wotshela, *Prickly Pear*.

[37] F. N. Egerton, 'Changing Concepts of the Balance of Nature', *Quarterly Review of Biology*, 48, 2 (1973), 322–50; D. Worster, *Nature's Economy: A History of Ecological Ideas* (Cambridge: Cambridge University Press, 1994); J. Kricher, *The Balance of Nature: Ecology's Enduring Myth* (Princeton, NJ: Princeton University Press, 2009).

mammals and thirty-five birds from Africa, fauna became a particularly important area for inquiry.

For Linnaeus, science was a crusade against ignorance. He wrote in 1759 that 'without the sciences, our economy would be run by charlatans ... Demons of the forest would hide in every bush, spectres haunt every dark corner. Imps, gnomes, river spirits and others in Lucifer's gang would live with us as if gray cats, and Superstition, Witchcraft and Black Magic, swarm around us like mosquitoes.'[38] Linnaeus stimulated loyalty and commitment among his students and followers, the most favoured of whom he called 'apostles'. Although wealthy amateurs and collectors still sponsored expeditions, and his own students travelled with VOC ships, expeditions were increasingly controlled by men of science themselves who were devoted to learning rather than to serving companies. Linnaeus himself was offered a trip to the Cape but turned it down; Thunberg and Sparrman, both inspired by him, went instead.[39] Vernon Forbes, editor of modern editions of both of these authors, thought that 'incontestably the two best books on the Cape in the 18th century were written by Sparrman and Thunberg'.[40]

Joseph Banks, Oxford-educated and from a landowning background, travelled around the world with James Cook on the *Endeavour* between 1768 and 1771, prioritising the study of botany. He took with him several specialists, including Daniel Solander, another student of Linnaeus, and the illustrator Sydney Parkinson, who drew 800 plants. Banks maintained contact with numerous botanists and travellers, becoming the centre of a rapidly growing network of knowledge, and was a key figure in the development of London's Kew Gardens. Francis Masson, working at Kew, was sent to the Cape

[38] L. Koerner, 'Purposes of Linnaean Travel: A Preliminary Research Report' in David Philip Miller and Peter Hanns Reill (eds.), *Visions of Empire: Voyages, Botany, and Representations of Nature* (New York: Cambridge University Press, 1996), 117–52, 126.

[39] A. Sparrman, *A Voyage to the Cape of Good Hope towards the Antarctic Polar Circle Round the World and to the Country of the Hottentots and the Caffres from the Year 1772–1776* (Cape Town: Van Riebeeck Society, 1976, 1977), 2 vols., edited by V. S. Forbes based on 1785–6 English editions, first published 1783; V. S. Forbes (ed.), *Carl Peter Thunberg, Travels at the Cape of Good Hope 1772–1775* (Cape Town: Van Riebeeck Society, 1986), based on the 1793–5 English edition, first published 1788.

[40] Forbes, *Thunberg Travels*, ix.

where he joined Thunberg on two exploratory trips. He collected over 1,000 plants between 1772 and 1775, including various pelargoniums and strelitzia. Masson's report, published in 1776 in the Royal Society's journal, was at that time the fullest botanical account of Cape plants in English by an eyewitness. Sparrman, already in the Cape, joined Cook's second expedition in 1772 and then, on his return, made further Cape journeys.

The expansion of scientific knowledge involved a network of people and institutions who collected and disseminated information. Most registered, published and compared their findings, and encouraged collaborators whom they encountered on their travels. A number of the key visitors published short pieces on Cape animals and plants in the emerging range of scientific journals before they went on to produce authoritative volumes. The changing European context, in which scientific publications and institutions were expanding, enabled them to benefit from their travels and to secure funds, or posts in museums and medical faculties (where botanical knowledge was still highly valued). Sparrman became a professor of natural history and medicine in Stockholm; Thunberg took over Linnaeus's professorship at Uppsala. Joseph Banks brought back a huge collection of seeds, plants, shells, insects and bottled specimens as well as examples of the material culture of the people he encountered. He appointed Solander to sort them out. They were displayed at Banks's house in Soho Square, which later became the base of the British Linnean Society. His collections, supplementing those of Hans Sloane, went to the British Museum.

Scientific travels in southern Africa were thus a route not only for scientific endeavour but to long-term employment in key European institutions. They enabled a continuous store of knowledge to evolve – with improving records of where and when specimens were found, how they reproduced themselves, what variants there were, and how they were placed into the new taxonomy. Increasingly, animals as well as plants were preserved. While the Cape was by no means the only point of reference for the new knowledge, the region was disproportionately significant as a node in the emerging European scientific world, even though it remained relatively insignificant as a colony. One reason for this was its location on the sea routes, another the diversity of its people, plants and animals. The Cape captured European scientific imaginations, enhanced them and was at the same time advertised by them.

Dutch Company officials continued to assist the enterprise. Joachim van Plettenberg, governor (after Tulbagh) from 1771 to 1785, was a young man with experience in Batavia and a keen interest in exploration. Robert Jacob Gordon, who first visited in 1773, travelling briefly with Thunberg and Masson, returned as a soldier and became commander of the Cape Garrison from 1780 to 1795. He explored the interior at least four times in the 1770s, once with van Plettenberg in 1778. He probably developed a deeper knowledge of Cape society, languages and environment than any of the visitors we have mentioned, some of whom he assisted. However, he committed suicide in 1795 at the age of fifty-two, before he was able to publish his notes, which appeared for the first time in 1982. Other key Dutch texts also remained unpublished: van Plettenberg's journal first appeared in 1896, and the substantial manuscript by Hendrik Swellengrebel, horticulturist and son of a Cape governor, about his travels in 1776–7, in 1982.

In all of these texts, the travellers and officials commented to varying degrees on the economic value of natural resources at the Cape: this interest was intrinsic to their imperial and scientific vision. They often noted the characteristics and productive potential of farms. Some of the expeditions to the interior had political purposes, although this was not necessarily made explicit in travel narratives. Public policy was the primary purpose of van Plettenberg's journey and also of John Barrow's. From relatively humble origins, Barrow attracted patrons and joined the British diplomatic service through his academic abilities. On a mission to China, under George Macartney, he learnt Mandarin and then published on Chinese astronomy. In 1797, aged thirty-three, he went to the Cape as secretary to Macartney, appointed as first British governor. Barrow married Anna Maria Truter, from an elite Cape Dutch family, in 1799, and planned to settle. His wife had established herself as a botanical illustrator and they shared an interest in plants. The retrocession to Dutch control in 1803 compelled him to move back to Britain. Barrow's book on the Cape (1801) was a particularly influential social and scientific travelogue as well as an audit of the Colony's resources.[41] He too became prominent, as the second secretary of the Admiralty from 1804 to 1845, giving state

41 J. Barrow, *An Account of Travels into the Interior of Southern Africa in the Years 1797 and 1798* (London: T. Cadell and W. Davies, 1801).

support to expeditions in many parts of the world, including a hydrographic survey of the southern and eastern coast of Africa (1822–5). He contributed to the formation of the Royal Geographical Society and was influential in the decision to build a Royal Observatory at the Cape. In 1831 he published *Mutiny on the Bounty*.[42]

Hinrich Lichtenstein, who arrived at the Cape from Germany in 1802, recently qualified as a doctor, had read Kolb, Sparrman, Thunberg, Le Vaillant and Barrow's very new *Account* before he joined Dutch Commissioner-General De Mist on a three-month tour of the northern and eastern frontier.[43] Lichtenstein became tutor to the son of the Dutch governor and was absorbed into the small intellectual community at the Cape including military officers such as Paravicini di Capelli and Lodewyk Alberti, who also wrote books. He recorded natural resources that might be of economic benefit but was especially interested in zoology. In 1805 he made a second journey northwards, reaching Tswana territory, and returned there to help with inoculation against smallpox before leaving the Cape when the British took over control again in 1806.

To a greater extent than Barrow and Lichtenstein, William Burchell travelled independently, driven by his scientific curiosity. Son of a successful plant nursery owner in Fulham, London, he studied botany and drawing, formed links with Kew and became a member of the Linnean Society in 1802. He lived on the island of St Helena in his twenties (1805–10) as a trader, schoolteacher and botanist, before travelling in the Cape from 1810 to 1815, during which time he collected botanical and zoological specimens. Burchell's *Travels*, published in 1822 and 1824, like Barrow's, made a significant impact on the increasingly anglophone coterie of scientists and travellers who wrote about South Africa – the culmination of a rich genre of books, circulating at the time, accompanied by a profusion of articles, reviews and comments.[44]

[42] J. Barrow, *The Eventful History of the Mutiny and Piratical Seizure of H.M.S. Bounty: Its Cause and Consequence* (London, 1831).

[43] H. Lichtenstein, *Travels in Southern Africa in the Years 1803, 1804, 1805 and 1806*, 2 volumes (Cape Town: Van Riebeeck Society, 1928, 1930). First published, 1811–12; reprint of the original English translation by A. Plumptre in 1812 and 1815.

[44] Texts produced at the time included Masson (1776); George Forster (1777, as part of his journey around the world with Cook); Hop (in Dutch, French and German, 1778); Johann Forster (1778–81, on Cape animals); Sparrman

Through such publications, readers at the time had access to a wide-ranging textual and visual record of the Cape. These outsiders aimed to produce authoritative works of observation and commentary: they frequently criticised their predecessors either for fictional excesses or inaccuracy. Kolb ended his book with a declaration that, unlike his predecessors, his own reportage was first-hand and truthful. Paterson wrote that his book was not 'a novel under the guise of a travel book'.[45] Thunberg described Paterson as 'a mere gardener'; Lacaille, as noted, was a critic of Kolb.

Sparrman asserted to his readers that he was a man of science who knew Linnaeus. During his exploration, he was 'fired with the love of science and the truth' and had 'a passionate regard for the truth'. He was particularly critical of Kolb whom he found too willing to accept the existence of 'prodigies and uncommon appearances' in Africa. Sparrman assured his readers that he would tell no tales of 'men with one foot'; nevertheless, he relished a good story, was careless with orthography, susceptible to reporting hearsay where it accorded with his views, and intentionally speculated on natural history and animal behaviour. While Thunberg collected more plants, Sparrman is nevertheless the most insightful when it comes to understanding the Cape's environment: he had a dynamic, proto-ecological view of nature.

Le Vaillant in turn wished to distance himself from the Swedish naturalists, whom he felt were insufficiently detached from the colonists with their 'chit-chat and tall stories'.[46] Le Vaillant also claimed to be scientific in approach but he shared more with Sparrman than he admitted, including disapproval of the settlers. Gordon wrote critical

(published in Swedish, German and English, 1783–5); Thunberg (1788 in Swedish and 1793–5 in English); Mentzel (1785–7 in German, based on an earlier stay); Paterson (1789); Le Vaillant (1790, with French, English and German translations and a second volume in 1795–6); Barrow (1801–4); Percival (1804); van Bouchenroeder (1806, in Dutch only); Alberti (1810); Lichtenstein (German in 1811, English in 1812, and Dutch in 1813) and Burchell. Subsequent publications of texts or notes written at this time include Gordon, van Plettenberg, Swellengrebel, Somerville, Barnard, De Mist, van Ryneveld and Paravicini di Capelli.

45 W. Paterson, *Narrative of Four Voyages in the Land of the Hottentots and Caffraria* (London: J. Johnson, 1789).

46 I. Glenn, *François Le Vaillant's Travels in the Interior of Africa via the Cape of Good Hope* (Cape Town: Van Riebeeck Society, 2007), vol. 1, 70 (first published 1790).

comments on Sparrman which were first published in Forbes's van Riebeeck Society edition.[47] Gordon had greater opportunity to observe and was also less discursive in his writing, but he was perhaps disturbed by Sparrman's censure of company rule, slavery and the Dutch colonists. Barrow started his account with a similar strategy in order to present his work as original and reliable. In turn, Burchell was particularly scathing about Barrow's map. Lichtenstein, as a self-aware young man with scholarly ambitions, analysed the practice of criticising precursors: 'Le Caille and Mentzel are severe upon Kolb, Sparrman criticizes Le Caille and Mentzel, and Le Vaillant comes under the censures of Mr. Barrow. For myself I must confess that the descriptions of the two latter, partly given without sufficient impartiality, partly too much loaded with ornament, have had a considerable influence upon the form which I have given to my work, earnestly desirous as I was of avoiding former failures.'[48] Lichtenstein explicitly recognised, in the manner of many modern scholars, that every traveller was influenced by 'his peculiar turn of thinking', by experience and by the chance often involved in obtaining information.[49]

For the modern reader, the striking feature of all of these books, notwithstanding some differences in voice, focus and attitude, is their similarity. Most used a chronological, narrative form, often based on their diaries, interspersed with observations and discussion. They also tended to use the first person and described incidents and adventures, such as hunting and encounters with wild animals. Although this strategy of presentation is seldom now used in scholarly texts, they were part of the foundation of modern academic practice in presenting evidence, critical commentary, and reasoning. Academics can still be competitive in seeking scientific authority.

Local Knowledge in Travellers' Texts

A distinctive aspect of some of these texts is the degree of empathy the travellers evinced with the indigenous people they encountered, and in this sense they followed Kolb. Sparrman, Le Vaillant and Burchell were especially forthright in their views and, together with Barrow, were

47 Sparrman, *Voyage to the Cape of Good Hope*.
48 Lichtenstein, *Travels*, 1, xiii.
49 Burchell, *Travels in the Interior of South Africa*, vol. 1, 1822, 577–8.

explicitly critical of the colonists, or at least of their excesses and cruelty in dealing with indigenous people. Most of the journeys were undertaken in the company of Khoekhoe guides and wagoners, with stops at settler homesteads. Travellers depended on the geographical knowledge of local people as well as their accumulated experience of the landscape. Their guides too were in the process of developing a local, hybrid knowledge of the Cape that helped to define its social geography. Writers gleaned many insights from conversations with those they encountered, both black and white.

References made by the travellers to their Khoekhoe auxiliaries are generally sporadic in their published texts, but sometimes they are named and given credit for information. Burchell wrote about them in some detail and it is possible to reconstruct a little of the role and personality of his key assistant, Stoffel Speelman – of whom he also painted a portrait.[50] The latter was particularly skilled in managing animals and in hunting – central pursuits on the road for which he took responsibility. For example, he informed Burchell about the best strategy for hunting rhinos, approaching them downwind so that they could not use their acute senses of smell and hearing. Burchell acknowledged that Speelman had a particularly wide knowledge of birds and their names, and he was a key resource in collecting them.

Sparrman's text, especially, can be read as an example of interactive knowledge in the production of Western natural science. When he visited the Cape, ideas about the acquisition of scientific knowledge were loosely defined, especially in regard to the natural sciences. He was open to the ideas of both indigenous people and the colonists – with their different languages, modes of living, and views of nature. Travellers were to a great extent dependent on the colonial infrastructure – and this included the availability of bilingual Khoekhoe guides whose subservient role was taken for granted. But their very dependency necessitated intensive interaction with people at all levels of colonial society. The travellers often located and named African people in the environment, which is why the texts remain valuable sources for African historians. Sparrman gave Khoekhoe names for places and especially rivers, even though these had sometimes been given Dutch

[50] T. Hammel, 'Shaping Science and Society at the Cape: A Relational History of Mary Elizabeth Barber'. PhD thesis, University of Basel (2017), 74ff; W. Beinart, 'Men, Science, Travel and Nature in the Eighteenth and Nineteenth Century Cape', *Journal of Southern African Studies*, 24, 4 (1998), 775–99.

names, which were displacing the aboriginal in written descriptions.[51] Gordon, by contrast, named the Gariep River the Orange, after William V, Prince of Orange, and this seeped into official texts and maps.

Sparrman and others also noted Khoekhoe words for animals, plants and objects. Kolb recorded the Khoekhoe understanding of the difficult terrain of the Cape and how they found the most fertile land.[52] Sparrman commented favourably on the Khoekhoe capacity to find water and denied that this 'talent, which is universally acknowledged', had anything to do with a 'particular acuteness in the organ of smell'. As in the case of tracking, it resulted from their 'faculty of observation, and judgement'.[53] In the semi-arid environment of the Cape, knowledge of water sources was fundamental to travel as well as settlement; in the eighteenth century, roughly half of farm names included mention of some form of water.[54]

Local tracking skills also impressed sympathetic observers. Le Vaillant affirmed that 'the African can detect the passage of the animal, can see the path it has taken, and what path he must follow to catch up with it. . . . The most experienced European hunter would be at a total loss in such circumstances.'[55] Burchell compared hunting and tracking skills to complex branches of European knowledge. 'These Africans pay an extraordinary degree of attention to every little circumstance connected with the habits and mode of life of the wild animals ... There are a multitude of ... circumstances from which they deduce information... it is therefore a subject deserving of attention.'[56]

Sparrman recorded how he learnt from a 'Hottentot-Boshiesman ... that the root of the da-t'kai, a shrub of the *mesembryanthemum* kind pretty common here, eaten raw, was, in fact, very well-tasted'. His servants knew where to find honey in trees because they could see where ratels had gnawed at the trunks. He made regular, appreciative references to Khoekhoe techniques of making useful products from

51 Sparrman, *Voyage to the Cape of Good Hope*, vol. 2, 26, 31, 65, 148.
52 Kolb, *Description*, 36.
53 Sparrman, *Voyage to the Cape of Good Hope*, vol. 2, 223.
54 L. Guelke and R. Shell, 'Landscapes of Conquest: Frontier Water Alienation and Khoikhoi Strategies of Survival', *JSAS*, 18, 4 (1992).
55 Glenn, *François le Vaillant's Travels*, 107
56 W. J. Burchell, *Travels in the Interior of Southern Africa* (London: Batchworth Press, 1953), vol. 2, 66. Reprint of 1822–84 edition.

plants and animals, such as the secretion of a hartebeest gland for medicine, eland horns for tobacco pipes, sinews for cloak-ties, and grey rhebok horns for awls. Barrow described the Damara technique of smelting copper by making charcoal and layering it with ore, using a gemsbok skin as bellows. 'The metal thus obtained is then manufactured into chains, rings, and bracelets, by means of two pieces of stone, that serve as a hammer and anvil, and the workmanship would be no disgrace to an artisan furnished with much better tools.'[57] He did note, however, that they did not solder the links.

Sparrman thought, perhaps a little unfairly in view of the incorporation of wild plants into Afrikaner folk remedies, that 'the African colonists are not near so forward to investigate the virtues of the plants of this country as by encroachments to increase their property'.[58] He criticised the violence involved in capturing slaves and, with a clarity startling to the modern reader, offered a justification of the value of indigenous knowledge on a global scale. Surely, he argued, it was in the interests of 'Christians' to gain such knowledge by protecting rather than killing its bearers. 'The Peruvian Bark, senega, ophiorrhiza, sarsaparilla, quassia, with many other useful remedies, calculated for preserving millions of our species, have we not learned them all from those we call savages?'[59]

Sparrman added that while colonisation and firearms made the Khoekhoen less exposed to the ravages of predators, nevertheless 'I could not but agree with them, that the colonists themselves were a much greater scourge to them than all the wild beasts of their country put together'. He added a more subtle discussion of the impact of colonisation. He commented on Khoekhoe hospitality and that in 'well-governed Hottentot craals . . . no member of society is abandoned to any considerable degree of indigence and misery'. But the quantity of glass beads and other commodities which were penetrating their society was leading to a change: 'I think I can plainly foresee a speedy and not inconsiderable revolution in the turn of mind and manners of this society.'[60]

57 Barrow, *An Account*, 374.
58 Sparrman, *Voyage to the Cape of Good Hope*, vol. 2, 78.
59 Sparrman, *Voyage to the Cape of Good Hope*, vol. 1, 155; N. Ulrich, 'Dr Anders Sparrman: Travelling with the Labouring Poor in the Late Eighteenth-Century Cape', *South African Historical Journal* 61 (4) (2009), 731–49.
60 Sparrman, *Voyage to the Cape of Good Hope*, vol. 2, 235.

Plant remedies were frequently mentioned by travellers. Lichtenstein described the use by Xhosa doctors of bleeding for fevers and rheumatic pains: 'the diseased limb is pierced in several places with a sharp pointed iron, when the end of a cow's horn is applied to the wounded part, in the manner of a cupping glass, by which the blood is drawn out'.[61] He noted that Kolb had seen this among Khoekhoen and clearly found it resonated with techniques that he knew. He was struck also that when a diseased Xhosa man returned to his community, he was banished and disallowed entry till he was cured. This suggested a conception of infection and quarantine.

Barrow was highly critical of the frontier Boers, seeing in them not only cruelty but a 'degeneration' of European civilisation.[62] In doing so, he had an immediate political agenda in justifying what he saw as progressive, enlightened British colonialism by contrast to the Dutch. Yet Barrow married into the Truter family and went on an expedition led by his future brother-in-law. Such choices were not unusual at the start of British colonisation since the new English and established Dutch land-owning elites found advantage in working together. His expedition to the north informed a narrative by William Somerville and also part of Barrow's *Account of Travels*. Samuel Daniell, travelling with them as an artist, produced strikingly sympathetic drawings of Cape nature and indigenous people, which were published as aquatints.

Barrow's text reads like a charter written by a modern, efficient servant of empire, distinguishing British views of appropriate governance from those of their Dutch predecessors. Barrow thought that slavery at the Cape was unnecessary, largely the result of the laziness of the colonists, who would not have needed such captive labour if they showed the same 'spirit of industry' as their forebears in Europe. The fact that Boer indolence was underwritten by dependence on unfree labour was an added outrage. Without slavery, 'native Hottentots' might 'feel they had a place and a value in society, which the miserable policy of the Dutch government denied to them'.[63] Barrow supplied future commentators and historians with a well-stocked repertoire of stereotypes about the inertia and cruelty of the Boers. Such ideas were later taken up, for somewhat different

[61] Lichtenstein, *Travels*, vol. 1, 310–11.
[62] Huigen, *Knowledge and Colonialism*.
[63] Barrow, *Travels*, vol. 1, 43.

purposes, by opinion-forming philanthropists and abolitionists like John Philip.[64]

Le Vaillant displayed even stronger sympathy with the Khoekhoen. As a contemporary reviewer noted, 'exempt from European prejudice, exempt especially from the commercial self-interest that usually takes people to these far-off regions, he denounces and attacks openly vices and abuses'.[65] Another reviewer ascribed to him 'a sort of passion for the savages and a profound aversion for the whites'.[66] In all this, Le Vaillant drew on Rousseau, who in turn had absorbed Kolb. Despite his impatience with Sparrman, Le Vaillant echoed a number of his points such as the value of indigenous skills. He was particularly impressed with Khoekhoe weaving, so tight that they could keep milk in their baskets – 'an invention that reveals real craft.' He rarely used scientific names, preferring local names in Dutch or indigenous languages.

Le Vaillant explained approvingly how the Khoekhoen thought of the jackal as the spy for other animals, which warned them of trouble.[67] He may well have been taking folktales too literally. He described in detail the use of the buchu plant with fat, as an unguent. His lengthy description of a flirtation with a young Khoekhoe woman, whom he named Nerina, after a flower, has often been noted.[68] Whatever the nature of the relationship – and it may largely have been in his fertile imagination – he spoke of Khoekhoe beauty with a degree of sensitivity and made a literary gesture towards reciprocity in his account of their interaction.[69]

This romantic sensibility did not extend to Le Vaillant's hunting activities. The modern reader might warm to his careful observation of browsing elephants, but he noted: 'When I had sufficiently observed their feeding method, I shot the one nearest to me in the head and in

64 Coetzee, *White Writing*; M. L. Pratt, *Imperial Eyes: Travel Writing and Transculturation* (London: Routledge, 1992); N. Penn, 'Mapping the Cape: John Barrow and the First British Occupation of the Colony, 1795–1803', *Pretexts*, 4, 2 (1993), 20–43.

65 Glenn, *François Le Vaillant*, vol. 1, xlii.

66 Glenn, *François Le Vaillant*, vol. 1, xliii.

67 Glenn, *François Le Vaillant*, vol. 1, 157.

68 Although it seems that the genus of South African bulbs/flowers, Nerine, was only named in 1820 after Le Vaillant visited, the term was already used in Italian for a flower.

69 Glenn, *François Le Vaillant*, vol. 1, 164–8.

less than ten minutes did the same with the other three.' He unfairly blamed 'my rash Hottentot' for shooting a fifth – a small male. Even more uncomfortable is Le Vaillant's share in the prurient recording and drawing of the elongated labia of some Khoekhoe women, their steatopygia, and the ritual excision of a testicle by Khoekhoe men. Ian Glenn argues, however, for Le Vaillant's status as a scientist: despite errors and later criticism he made a major contribution to a more systematic African ornithology.[70]

Figure 1.2 Le Vaillant at Kok's kraal, Great Fish River, receiving birds from Khoekhoe assistants.
Credit: Ian Glenn; Library of Parliament, South Africa

Sparrman also hunted, as did all travellers, but he was a little more self-critical in reporting his attempt to shoot a honey-guide bird, highly valued by the Khoekhoen because it directed them to bees' nests, so that he could examine it more closely. 'This plan', he wrote, 'offended my Boshies-men not a little: and though I had previously promised an ample reward, consisting of glass beads and tobacco, to my *Zwellendam* Hottentots, on condition that they would assist me in catching and shooting a *honing-wyzer*, yet I found them too much the

[70] I. Glenn, *The First Safari: Searching for Francois Levaillant* (Auckland Park: Jacana, 2018); Jacobs, *Birders of Africa*.

bird's friend to betray it; a circumstance that gave me great pleasure, as it shewed that these people were in general possessed of good and grateful hearts.'[71]

Le Vaillant talked warmly of nightly discussions with his guides. It was perhaps this that made him feel he was absorbing a more authentic indigenous experience than those travellers who stayed largely with Boer farmers. Burchell echoed him. On the road, he found many of the Boers taciturn, albeit hospitable, but he clearly enjoyed the company of his guides and servants whose loquacity, especially around the fireside, he described sympathetically. He specifically noted, as if to affirm the authenticity of his experience, that he travelled for four years with 'no companion or assistant, nor other attendants than a few Hottentots, the number of whom never exceeded ten'. 'To me', he wrote, 'almost everything truly African was interesting; and nothing gratified these feelings more than an opportunity of observing and conversing with the Hottentots.'[72] Not all were so generous, and Thunberg, for example, was more attuned to the worldview of the settler farmers.

Sparrman appreciated the skills of the Khoekhoen as pastoralists. He noted how they moved with their animals as soon as grazing became exhausted or the cattle showed signs of sickness: this he thought was one reason why the 'cattle of the Hottentots, in some measure, keep up to their original standard whilst, on the contrary, those of Christians degenerate to a smaller race'.[73] He also reported that the Xhosa 'fondled and talked to their cattle a good deal as they stood in the craal; doubtless, in the same manner as the Arabians do to their horses ... making them thrive and rendering them brisk and lively, and at the same time more intelligent and tractable'.[74]

It would be a mistake to seek consistency in the approach of all scientific travellers. Lichtenstein, in keeping with his desire to distance himself from those he judged too emotional and critical, was disparaging about the Bushmen, whom he felt had 'no government, no secure dwelling-place, no social compact, nay, were even without individual property'; 'one might almost call this extraordinary race without customs and without language'. They were 'murderous' to all others because they had nothing to lose and were 'always in enmity with their

71 Sparrman, *Voyage to the Cape of Good Hope*, vol. 2, 148.
72 Burchell, *Travels in the Interior of South Africa*, vol. 1, 50.
73 Sparrman, *Voyage to the Cape of Good Hope*, vol. 1, 238.
74 Sparrman, *Voyage to the Cape of Good Hope*, vol. 2, 14.

nearest neighbours'.[75] He associated this trait with their physical appearance. To assist in his studies, Lichtenstein acquired the skull of a 'Bosjesman' who had died in prison, with facial skin still in place. It was placed in a collection in Göttingen and he published an article on this topic.[76]

Lichtenstein did not have the capacity of Sparrman, Le Vaillant, Barrow or Burchell to see that the overwhelmingly predominant routes of violence were by colonisers against these most vulnerable of the colonised people. Linked with the brief Dutch regime of 1803–6, he also defended the frontier colonists, whom he thought Barrow had unfairly criticized.[77] He appears to have thought Barrow prejudiced against the Dutch although he did agree that the livestock farmers on the colonial frontier were lazy, a common trope of nineteenth-century British settlers: 'all the trouble that the colonist has is to see that his Hottentots go out with the flocks in the morning, and that the sheep are brought home safe at night.'[78]

Barrow's qualified sympathy towards the Khoesan stands in counterpoint to his hostility towards Xhosa-speaking Africans, whose demographic strength represented more of a challenge to nineteenth-century British settlers. By contrast, Lichtenstein admired the Xhosa people that he met. He commented on their social and physical discipline in that he never saw them 'sneeze, yawn, cough, or hawk' and they did not rub or scratch themselves in public. Lichtenstein also regarded Tswana people he met on a journey to the north with interest and respect, as revealed in a long description of his first encounter.[79] Interestingly, he recognised them as slaveholders.

Seen collectively, scientific travellers were gleaners and recorders of a Cape vernacular – a term often used to describe an architectural style, but which can be applied also to a recognisable cultural repertoire of knowledge and practice.[80] Travellers' accounts also offer insights into the process by which information and understanding of the social and

75 Lichtenstein, *Travels*, vol. 1, 142–3.

76 P. Grogan, 'German Natural History Collectors and the Appropriation of Human Skulls and Skeletons in Early Nineteenth Century Southern Africa: Towards a Discursive Analysis of Collecting' in V. Arlt, S. Bishop and P. Schmidt (eds.), *Explorations in African History: Reading Patrick Harries* (Basel: Basler Afrika Bibliographien, 2015).

77 Lichtenstein, *Travels*, vol. 1, 443.

78 Lichtenstein, *Travels*, vol. 1, 446–7.

79 Lichtenstein, *Travels*, vol. 2, 363, 379.

80 Beinart, *Rise of Conservation*, 8, 46.

natural environment were accumulated and transferred. At their most curious, they were interested in society and environment, and how these interacted. Most were open to local knowledge but organised and used it in the context of their own training and experience. Their accounts were also helpful to those who worked to intensify colonial settlement.

Scientific Understandings

Taken together, the major contribution of this generation of scientific travellers was to record and describe, in publications accessible to Europeans and those literate in European languages, a large number of plant, animal and insect species that had not been formally noted. Some of this was reported in detailed specific articles, some as part of their travel narratives, and some as lists. Most writers noted and named species within Linnaeus's system and thus greatly expanded its content and authority. This enabled scientists to think even more systematically across the global range of plant and animal species.

Linnaeus's student Carl Thunberg joined the VOC as a surgeon in 1771 in order to get to the Cape, where he spent three years on the road. In addition to his travel book (1788), he published a *Flora Capensis* that listed 3,100 plant species. Although there was already some sense of the special and unusual diversity of Cape plants, this was the first systematic attempt to record, quantify, and provide an initial base for scientific understandings of the distinctiveness of the region. In the first edition of *Species Plantarum* (1753) Linnaeus had recorded 5,940 plant species and estimated 10,000 globally. Thunberg's influence was especially great because in 1775 he travelled on to Indonesia for two months before reaching the VOC station in Japan where he again collected many plants and produced a pioneering work on *Flora Japonica*. He became a long-serving professor at Uppsala, well connected globally, and supervised 294 dissertations.

The Cape, as well as other regions of botanical exploration such as Latin America, soon changed conceptions of diversity. Barrow, at the start of the nineteenth century, knew that Cape plants were among the most interesting in the world, 'more numerous, varied and elegant' than elsewhere.[81] Johann Franz Drège collected about 7,000 species in

[81] Barrow, *Travels*, 23.

the 1820s and 1830s; William Harvey, who spent time as an official in the Cape from 1834 to 1838, recorded 1,086 genera and 8,500 species. Ludwig Pappe, a physician who had collected plants since he arrived in Cape Town in 1831 and served as the first colonial botanist from 1858 to 1861, reckoned by this time that South Africa might have 18,000 plant species. (It is now estimated at about 22,000, with 9,600 in the Western Cape alone. Britain has about 1,500 native plant species.) Awareness of this 'richness' and 'almost incredible variety', as Pappe called it, developed early; it became a strong point of scientific interest in the first half of the nineteenth century and a focus of nascent national pride by the beginning of the twentieth.[82]

Sparrman was less significant than Thunberg in recording species but he offered innovative environmental insights that were picked up by later writers as representing the origins of environmental thinking at the Cape. T. D. Hall, one of the most interesting writers on conservation in the interwar years of the twentieth century, quoted Sparrman extensively and saw in his eighteenth-century text the first sustained discussion of the problem of overgrazing.[83] Dutch officials and farmers had already registered this point, but Sparrman was perhaps the first to publish the idea that heavy grazing by cattle and sheep tended to diminish the grass and thus facilitate the spread of unpalatable bushes. He believed that wildlife should be preserved, not only to avoid unnecessary extinction, but because by browsing and spreading their dung, they might help to extirpate unwanted bush and fertilise the veld. To this end he also suggested domestication of wild antelope species, as the Swedes had tamed the reindeer and the elk, so that they could help prevent both bush encroachment and desertification. Similar ideas were canvassed from the mid-twentieth century to support wildlife farming.

Sparrman was also an advocate of transhumance in resolving this problem. Colonists, he thought, not only kept too many cattle but were too immobile. The Khoekhoen, who frequently moved their dwellings, were more sensitive to environmental degradation. This defence of transhumance was echoed in the late twentieth century by environmentalists with Africanist sympathies. While the problems of

[82] Beinart, *Rise of Conservation*, 74.
[83] T. D. Hall, 'South African Pastures: Retrospective and Prospective', *South African Journal of Science (SAJS)*, 31 (1934), 59–97.

overgrazing were of obsessive concern to natural scientists and veld conservationists in South Africa from the late nineteenth century, they did not support transhumance but instead recommended rotation of animals in fenced camps or paddocks within the fixed boundaries of the farm.[84] Yet they recognised Sparrman's foresight in respect of 'the cumulative effects of man's destructive hand upon the environment'.[85]

Drawing on local ideas, Sparrman wrote of the need to 'keep in a just equilibrium the increase of the animal kingdom ... so that it may not exceed the supplies afforded it by the vegetable part of creation, and by this means prevent the necessary renewal of the latter by seeds, etc. and thus, by desolating it and laying it waste, in the end impoverish and destroy themselves'.[86] Locusts and fire, he suggested, played a role in the natural regeneration of valuable plants. 'The ground is ... stripped quite bare; but merely in order that it may shortly afterwards appear in a much more beautiful dress, being, in this case, decked with many kinds of annual grasses, herbs, and superb lilies, which had been choaked up before.'[87] The role of fire in maintaining biodiversity was only fully accepted by scientists studying fynbos in the 1970s.[88] Similar understandings of the balance needed between animals and vegetation were articulated by Aldo Leopold in the 1940s and are now seen as one of the foundation stones of modern environmentalism. Although Sparrman is not, as he might be, cited as one of the founders of global environmental thinking, he became a reference point for later South African authors.

Burchell wrote less on this theme, but when commenting on the paucity of grazing for his oxen around a farm in the Roggeveld, he spoke with feeling that 'the sheep of the place consumed, like locusts, every blade of grass and leafy twig within a moderate compass'. (The comparison between sheep and locusts was a metaphor that would be frequently used by environmentalists in later years.) Burchell suggested that some plants developed thorns or acrid sap to preserve themselves from 'annihilation' and he also associated overgrazing with the spread of unpalatable and poisonous plants. For him, this resulted from a disturbance of what Sparrman called a 'just equilibrium', or the

84 Beinart, *Rise of Conservation.*
85 Sparrman, *Voyage to the Cape of Good Hope*, vol. 1, 3. Comment by V. Forbes.
86 Sparrman, *Voyage to the Cape of Good Hope*, vol. 1, 170.
87 Sparrman, *Voyage to the Cape of Good Hope*, vol. 1, 330–1.
88 Pooley, *Burning Table Mountain.*

'balance of nature' – a proto-ecological concept, developed in Enlightenment thinking, with wide currency at the time.

Robert Gordon probably developed the deepest knowledge of Cape zoology in the late eighteenth century and contributed to articles by Jean-Nicolas-Sébastien Allamand, professor at Leiden. Following his first trip, Gordon 'gave to Allamand the skin of the head of a black Wildebeest'.[89] His observations of this species as well as the hippopotamus informed descriptions by Allamand in Buffon's encyclopaedic, multi-volume *Histoire Naturelle*. This compilation aimed at a complete global record of animal species. Gordon himself kept detailed records of the diverse range of animals that he encountered and had drawings made of the animals, birds and plants he sighted or collected; 362 survive in the Netherlands, with some scattered elsewhere.

Allamand contributed an entry on the giraffe in Buffon's work, informed by a skin he received in the 1760s, and Gordon was keen to supplement the information. 'It was a prime ambition of his', Cullinan writes, 'to bag one of these animals as the giraffe was a great rarity in the annals of eighteenth-century zoology.'[90] It was not till his fourth journey in 1779 that his party killed a 'cameelpaard', which he described with unusual excitement: 'this handsome and extraordinary animal, one of the most beautiful formed by nature'.[91] A painting survives of this hunt, showing eight giraffes. Gordon spent the next day examining the corpse and recorded seventy-eight separate measurements, with many anatomical notes; Allamand published the details in 1781, acknowledging Gordon. The skeleton and hide of another animal was sent to the Netherlands and mounted for display at the residence of the Prince of Orange. (Le Vaillant, for his part, returned to France with a giraffe for display.) Gordon corresponded with other leading European scientists and was acknowledged as 'a man of science of an active and well-cultivated genius'.[92]

As a government official Barrow was not only interested in the political context, but on the lookout for economic assets and resources.

[89] L. C. Rookmaaker, *The Zoological Exploration of Southern Africa 1650–1790* (Rotterdam: A. A. Balkema, 1989), 61, 75.

[90] P. Cullinan, *Robert Jacob Gordon 1743–1795: The Man and His Travels at the Cape* (Cape Town: Struik Winchester, 1992), 101.

[91] Cullinan, *Robert Jacob Gordon*, 101.

[92] Cullinan, *Robert Jacob Gordon*, 165. Quote from John White, official physician for New South Wales.

Coal deposits could be important, because wood for fuel was short – a point he underlined by recording that some households at the Cape kept a slave for the purpose of collecting wood. He was alert to the potential of commercial crops and noted possibilities for many fruits, cotton, tea, indigo, coffee, hemp, as well as olive, sesame and other oil-producing plants. Cochineal was still valued as a colouring and he observed that 'different species of the cactus, the plant on which the cochineal insect feeds, grow just as well here as on the opposite continent'.[93] Prickly pear, from central America, was well established at the Cape by this time. The British government appointed an agriculturalist, William Duckitt, as early as 1800 in order to improve farming production and experiment with new crops. Barrow thought that Algoa Bay, where Port Elizabeth was later established in the eastern part of the Colony, looked especially promising for a settlement: it had water, timber, fish for food, and the land appeared open and cultivable. There were also potentially good livestock supplies from farmers in the interior, who otherwise had to send most of their animals a long way to Cape Town.

Alongside his economic concerns, Barrow was guided by 'curiosity, science and botanical research'. He carried on his travels books to identify plants, a compass, a sextant, a surveyor's chain, a chronometer and a thermometer. He often recorded the temperature and published these measurements in his book. He was immersed in botany, could identify many indigenous plants, and noted their distribution – for example, that silver trees were restricted to particular slopes of Table Mountain whereas other proteas were more widespread. Travelling a few days beyond Graaff-Reinet, he was excited to find what he thought was a new species of strelitzia with a smaller, thinner leaf and a violet blue flower.[94]

Barrow was sufficiently immersed in the Linnaean system of classification to discuss its modification. For example, in an extended section on the gnu – or wildebeest – he recognised that, despite its appearance, it was a species of antelope (as Linnaeus had written in his *Systema Naturae*) but was part of a distinctive genus.[95] (Antelopes were later classified into thirty genera.) He also mused on the classification of humans in relation to their facial angles, rather than their skin

93 Barrow, *Travels*, vol. 1, 17.
94 Barrow, *Travels*, vol. 2, 115.
95 Barrow, *Travels*, vol. 1, 264.

colour, as part of a widespread concern to anatomise presumed racial hierarchies. Finally, as an indication of his wide scientific interests, his observations of rock formations led him to question the biblical conception of the formation of the earth in 6,000 years and to suggest that the landscape must be the result of a very long period of 'gradual decay' or mutation.[96]

Ironically, it was partly because of his scientific interests that Barrow was open to the search for the unicorn and reluctant to dismiss evidence of its existence. The term had been used scientifically by Linnaeus with reference to *Rhinoceros unicornis* – though this was later applied only to the Indian rhino (with one horn). Sparrman had earlier noted that the animal featured in Bushman rock art, 'usually represented like a horse with a horn on its forehead'.[97] He thought it might have existed in the past, if not in the present. Gordon dismissed the unicorn as a fable. But Anne Barnard, whose stay at the Cape coincided with Barrow's, briefly touched on a story about unicorns in letters to her family.[98] Intriguingly, this arose because African people recognized it in the British royal coat of arms. The white unicorn, depicted as a rearing horse with one horn, had been a Scottish symbol since medieval times – probably of danger, purity and power. It had earlier roots in classical texts.

While travelling at the edges of white settlement in the Tarka valley and Bamboesberg, Barrow's party visited at least two caves with rock art depicting animals.[99] He copied an image of a possible unicorn and published a version in his book.[100] Although Barrow accepted that the unicorn in Europe was a 'fancy', because it had never been seen, this did not mean it did not exist in Africa. He adopted a post-Enlightenment scientific approach. The greatest number of depictions of this animal, he said, were in the Bamboesberg, remote to settlers, and 'this part of Africa is as yet untrodden ground'. He also knew that 'certain animals, as well as plants, confine themselves to certain districts'.[101] He had commented on this idea of endemic species with a narrow range, or phytogeographical specificity, with respect to the silver tree in Cape Town. Barrow remained open-minded about the

[96] Barrow, *Travels,* vol. 1, 363.
[97] Sparrman, *Voyage to the Cape of Good Hope*, vol. 2, 116.
[98] A. Barnard, *South Africa a Century Ago: Letters Written from the Cape of Good Hope (1797–1801)*, edited by W. H. Wilkins (London: Smith Elder, 1910).
[99] Barrow, *Travels*, vol. 1, 291. [100] Barrow, *Travels*, vol. 1, 294.
[101] Barrow, *Travels*, vol. 1, 296–7.

unicorn for the kind of scientific reasons that emerged out of Enlightenment rationality rather than a susceptibility to hearsay. His openness would resonate with a more modern scientific approach and the possibility that new species could be found. Lichtenstein also offered money for a complete skin of a unicorn, with horn and skull, and even Burchell did not rule out, at least humorously, the possibility of their existence – remarking on the potential value of a unicorn skin.[102]

On his travels Lichtenstein, like Barrow, also equipped himself with the technology of the scientist, including a case of anatomical instruments, ointment and brandy for preserving species, and a thousand needles for fastening insects. He was fascinated by some animal species that he considered to be inadequately understood such as the hyena and the ostrich. Surprisingly, he thought the spotted hyena 'by far the most abundant of any among the beasts of prey in the colony', even near Table Mountain.[103] He understood that it largely ate carrion but could be a predator of livestock and was therefore the target of hunting parties and traps. He was quick to dismiss the well-established, half-joking popular conception, that hyenas were hermaphrodites – a belief that outlasted his text. Male and female, he confirmed, were distinct although difficult to tell apart by observation, especially when young. He did, however, accept another popular story: 'A circumstance with regard to these animals, held by many to be a fable, I can from my own knowledge aver to be a fact, that they appear by night to be much larger, and of a brighter colour than they really are; they even appear wholly white.'[104]

Lichtenstein wrote a separate paper on ostriches in 1809 and was keen to establish his authority on this species – claiming that 'the habits of the ostrich are so remarkable, and have been so imperfectly described by travellers in general'. He recorded their behaviour in groups and their egg-laying habits in a round cavity in the soil with the eggs upright to fit more in. He also made an interesting observation that as a 'prudent, wary animal', they were valuable to quagga herds in providing warning of danger, and in turn the ostriches benefited from the beetles that were attracted to the quagga's dung.[105] This type of complementary behaviour between species was quite widespread, but

102 J. Pickering, 'William J. Burchell's South African Mammal Collection, 1810–1815', *Archives of Natural History* 24, 3 (1997), 312.
103 Lichtenstein, *Travels*, vol. 2, 19.
104 Lichtenstein, *Travels*, vol. 2, 31–5.
105 Lichtenstein, *Travels*, vol. 2, 262.

in this case it did not save the quagga from extinction by humans. Ostrich eggs were not only eaten by all of the rural Cape's inhabitants but were brought to Cape Town for sale and the feathers used as gifts by colonists. In a striking sentence, Lichtenstein noted another example of complementarity between species. Remarking on the numerous and large 'Danish' dogs kept by the Boers, he wrote: 'it is not too much to assert, that without dogs, and such dogs, this country would be absolutely uninhabitable'.[106]

Lichtenstein's travels established his standing as a scientist and on his return to Berlin he became a local celebrity. Naturalists pored over his specimens of insects, animals, plants and rocks. About half of the 600–700 species of insects in his collection were 'new to science'. Among early articles, he published on the Tswana in 1808 and on the speech of 'Südafrikanischen wilden Völkerstämme'. He also wrote a 'History of South Africa', of which the first part appeared in a journal in 1810 but the rest remained unpublished. His *Travels* were published in German in 1811, when he was appointed as the first professor of natural history at the University of Berlin (later Humboldt University). The Museum of Natural History, which he established in 1813, became an important site for the collection and analysis of zoological and botanical data. A successful field trip could be the basis for an academic or institutional career. The presence of German and northern European scholars is not only a reflection of strong academic traditions in these countries; it is also a reminder that, prior to national unification, Germans posed little threat either to the Dutch or British occupiers of the Cape. They could function as sojourners and settlers with relative freedom as fellow Protestants, benefiting from the security and sense of opportunity offered by the Dutch and British maritime empires. Eighteenth- and early nineteenth- century science in South Africa was thus, rather like in South Asia, 'a shared European project'.[107]

Conclusion

These travellers were engaged collectively in an increasingly systematic production of knowledge about South Africa. They drew extensively

[106] Lichtenstein, *Travels*, vol. 2, 14.

[107] M. Von Brescius, *German Science in the Age of Empire* (Cambridge: Cambridge University Press, 2019), 5.

on local knowledge, from both settlers and Khoesan, but also developed new ways of writing about nature and its inhabitants. Influenced strongly by Linnaeus, they were curious, driven, and keen to establish themselves – sometimes in contradistinction to their predecessors – as the most knowledgeable authorities. Thus, although they largely wrote their major books in the form of personalised narratives, recalling their own adventures and experiences, they were simultaneously developing a form of critical writing and practice which they saw as truth-telling. They based their findings on observation and collection, and they brought a comparative framework that provided a means of testing and comparing. They dispensed with some, but not all, hearsay, and also sought to assess fables through a system of logic and observation. They were in a transition from storytellers to scientists, and they absorbed the expanding frontiers of knowledge into a system of universal recording and analysis.

This was, in Linnaeus's terms, an escape from superstition and magic, but it was not, in Weber's, a disenchantment because many were excited and energised by their new visual and intellectual experience and understanding. Nor was it simply an exercise in categorisation. Some of the Cape travellers, such as Sparrman, began to observe and describe dynamic environmental relations. Burchell especially expressed ideas that were influenced by Romanticism – a form of thinking and artistic expression that also incorporated scientific ideas and evinced positive views about landscape and the natural world.[108] We can see in observers such as Sparrman a keen sense not only of the interrelationship of species, but of changing environmental conditions. Such observations were expanded and developed theoretically in the early decades of the nineteenth century by Humboldt and others, into a holistic view that underpinned modern ecology.[109]

These texts were all created in a conquered land, a slave-holding society where indigenous people were actively being dispossessed and suppressed. But some authors were critical of slavery and Dutch colonial rule. Some attempted to incorporate the knowledge of local people. In a few cases, such as Kolb, Sparrman, Le Vaillant and Burchell, authors on occasion explicitly attempted to express the

108 J. Bate, *Romantic Ecology: Wordsworth and the Environmental Tradition* (London: Routledge, 1991).

109 A. Wulf, *The Invention of Nature: The Adventures of Alexander von Humboldt, the Lost Hero of Science* (London: John Murray, 2015).

humanity and individuality of colonised subjects, rather than simply mirror or defend conquest, dispossession and racial hierarchy. While they were not consistent in this, and their writing also asserted asymmetrical relationships of race, class and knowledge, there was sufficient openness for the ideas of local people to emerge. Most were clearly excited by travelling in this new environment, writing about its diversity and breaking new ground in scientific collection and categorisation. Some of their interests would remain central to scientific concerns and also to the way that South Africa was understood and imagined.

2 *Scientific Governance and Colonial Institutions, c. 1800–70*

In 1806 colonial possession of the Cape, which had shifted uncertainly between the Dutch and British from 1795, was finally resolved and the tip of Africa became a permanent fixture of the British Empire. Under the Batavian Republic and the brief tenure at the Cape of the Dutch statesman J. A. de Mist (1803–4) there were suggestions of reforms in governance which may have given a boost to scientific inquiry and bureaucratic modernisation. But these were scarcely realised amid the disruption of the Napoleonic Wars.

Britain's major motive for claiming the Cape was the control of sea routes and its naval presence there became disproportionately large. The difficulties of conquest on the expanding eastern frontier of the colony necessitated substantial military forces. And, partly for this reason, the new British administrations sought to coax the Cape into an economically productive colony of settlement. Changes in the law, in administration and to currency, were all designed to improve governance. Yet policies were often inconsistent and the language of reform and progress belied authoritarian and illiberal practices.

The British remained beholden in key respects to the established Dutch elite, especially in the countryside. While British officials and military officers took over the key positions of power, the need for political compromise perpetuated some aspects of the 'old corruption' that Barrow and others had condemned. This included slavery, which remained intrinsic to the second British Empire. British rule thus brought opportunities for these established Dutch families. Some were also accommodating to the policy of anglicisation, first promoted by Governor John Cradock (1811–14). The first Chief Justice appointed in 1812 was a local specialist in Roman-Dutch law, Johannes Truter. He was also the first South African to receive a knighthood. Lord Charles Somerset, longest-serving Cape governor in the nineteenth century (1814–26), approved some modification of slave conditions, but he found a strong coincidence of interests with the landed

Afrikaner elite, the major slaveholders and wine producers. Between 1809 and 1825, when protection for Cape wines in Britain was ended, the number of vines doubled, and wine production increased by 80 per cent.[1]

Somerset, Barry and Pringle: The Foundation of New Institutions at the Cape in the 1810s and 1820s

The Cape presented a military and political challenge for this new colonial government, especially Somerset with his visceral fear of revolutionary republicanism or 'Jacobinism'. For their part, some Boers, especially livestock farmers in the eastern part of the Colony, resented government interference in their control over Khoekhoe workers and the extension of the centralised court system. The major indigenous threat to the Colony was presented by the Xhosa chiefdoms on the eastern frontier. Somerset's attempt to stabilise conflict in the eastern zone by completing a string of forts and recognising Ngqika as Xhosa paramount proved unsuccessful. Ngqika's uncle and regent until his accession, Ndlambe, challenged colonial authority in 1818–19, besieging Grahamstown, the major British military centre in the east. He was ruthlessly suppressed, and the colonial government annexed Xhosa land up to the Keiskamma River as the 'Ceded' territory. This included Ngqika's great place, which was confiscated despite his attempt to forge an alliance with the British.

The first couple of decades of British rule were striking not only for the assertion of control but also for the formation of new scientific and medical institutions and associations. In part this resulted from measures by the new state, which adopted a wider sense of reform-minded governmentality than the VOC; they also developed as reactions to what was perceived as autocratic, gubernatorial rule. New associational and scientific networks, mainly British in form but remarkably cosmopolitan, flourished as the pace of economic growth increased and as Cape Town's population grew from about 15,000 in 1800 to 25,000 in 1850.

Medical practitioners were by far the most numerous group with specialist scientific knowledge at the Cape, and in 1807 the principal

[1] T. Keegan, *Colonial South Africa and the Origins of the Racial Order* (London: Leicester University Press, 1996), 52.

medical officer of the Army, Alexander Baillie, formed a commission to regulate them. A proclamation established a Supreme Medical Committee that would examine the credentials of, and license, all physicians, apothecaries and retailers of medicine. As some were not professionally qualified, this constituted a major intervention on the part of the state. Harriet Deacon argues that monopolisation of medical practice by licensed practitioners, excluding both unlicensed settler providers and African practitioners, was unusually early.[2] While rigorous legislative control of medical practice preceded that in Britain, the law was not systematically enforced. There was also tension between physicians and apothecaries – the latter attempting to restrict the supply of drugs to themselves. Some well-established practitioners such as Dr Friedrich Liesching, who came to the Cape as surgeon-general in a company of mercenaries sent to support the Dutch by the Duke of Württemberg, acted as both: he was a physician who also ran one of the largest wholesale and retail apothecaries in Cape Town.

During Somerset's governorship medical institutions were further transformed. One key figure was his close friend James Barry, who has attracted a number of biographies and fictional treatments: he sustained a successful career as a British Army physician for over forty years, rose to the senior position of inspector-general of military hospitals, and, when laid out on his deathbed, was identified as biologically female. A detailed recent treatment by Du Preez and Dronfield seems to establish firmly that Barry was born Margaret Ann Bulkley, probably in 1789 in Cork, Ireland.[3] A confident young woman with great intellectual potential, Bulkley moved to London with her mother and was promoted by well-connected friends of her artist uncle, James Barry. These included Francisco de Miranda, an exiled leader of the Venezuelan revolution.[4] With this support, Bulkley was able to prepare for a medical degree at Edinburgh University (1810–13); entry was barred to women, and so she adopted her dead uncle's name, with the

[2] H. Deacon, 'Introduction: the Cape Doctor in the Nineteenth Century' in H. Deacon, H. Phillips and E. van Heyningen (eds.), *The Cape Doctor in the Nineteenth Century: A Social History* (Leiden: Brill, 2004), 17–43.

[3] M. du Preez and J. Dronfield, *Dr James Barry: A Woman ahead of Her Time* (London: One World, 2016).

[4] R. Holmes, *Scanty Particulars: The Strange Life and Astonishing Secret of Victorian Adventurer and Pioneering Surgeon James Barry* (London: Viking, 2002).

middle name Miranda. For the rest of his life, Barry sustained a medical career in the armed forces as a man.

After qualification, further work with a London surgeon, and a few short-term posts as a junior military doctor, Barry arrived in the Cape in 1816 with a letter of introduction to Somerset. By chance the governor's daughter was ill and Somerset gave Barry the responsibility of restoring her to health. Barry was by this time outgoing, well-dressed, and equipped with the metropolitan manners that attracted Somerset. He also presented himself as a 'romantic, progressive gentleman scientist', which chimed with Somerset's mindset.[5] The governor had been part of the entourage of the Prince of Wales (later George IV) and the fashionable Brighton set. Barry was 'welcomed into the bosom of the Somerset family', invited to important dinners and social events, and attended the theatre in the governor's party.[6] The new doctor also attended 'rainbow balls' described by the colonial secretary Christopher Bird as 'composed of each different hue in this many coloured town'.[7]

In 1820, when Somerset visited Britain for an extended period, his temporary replacement, Rufane Donkin, established a new post of colonial medical inspector, effectively taking over from the Medical Committee to head medical services in the Colony. When Somerset returned at the end of 1821, and the post became vacant, he appointed Barry, despite his inexperience. Barry benefited from the famously rigorous training in Edinburgh and his experience in London; between 1765 and 1825, only 20 per cent of Edinburgh medical students graduated with an MD. Scottish medical education provided many of the Empire's physicians and influenced both practice and, later, training in the anglophone colonial world. Up to this point, most licensed physicians at the Cape had trained in Germany or the Netherlands. Barry was disdainful of much of the medical practice that he had seen in the Cape and set about reform with enormous energy: 'sweeping through each of the public institutions of the Cape – hospital, leper colony, prisons, asylums, apothecaries, commerce in medicines – he ferreted out venality, criticised bureaucratic corruption,

[5] Holmes, *Scanty Particulars*, 65.
[6] Du Preez and Dronfield, *Dr James Barry*, 123.
[7] Holmes, *Scanty Particulars*, 68.

challenged inefficiency, created detailed policies of medical reform and relieved suffering'.[8]

Leprosy was reported in South Africa in the eighteenth century, and early British officials became aware that it had spread amongst the Khoekhoe population.[9] In 1813 it was made the second notifiable disease after smallpox. Initially it was left to the attentions of local officials, but by 1817 a more coordinated response seemed necessary and Somerset approved a leper colony in the Hemel en Aarde valley, between Caledon and present-day Hermanus. Together with the Somerset Hospital for the poor and outcasts of society, this became the first public health institution in the Colony and new buildings were completed in 1820. Little medical support was given, however, and when Barry visited in 1822, he was appalled at the prevailing conditions.[10] The Moravian missionaries at Genadendal were asked to take over and managed the establishment until it was disbanded in 1846.[11] Barry visited regularly to ensure that adequate standards were maintained. Hemel en Aarde was replaced by Robben Island in Table Bay, which served as a more isolated leper and 'lunatic' asylum but was also sufficiently accessible to Cape Town to facilitate medical care.

In 1824 Barry revealed dire conditions in Cape Town's gaol. He inspected and wrote an extensive report on Somerset Hospital, which was also the government facility for those declared insane. Again, he was highly critical: 'lunatics were kept in filthy cells better calculated for the confinement of brute animals...'.[12] He took on the Liesching medical family over the licensing of one of the sons, who had not been to Europe to qualify formally as apothecary. Rather than demonstrating flexibility, as others advised, Barry was uncompromising. Somerset's protection was no longer sufficient and in 1825 the post of colonial medical inspector was terminated. One of Barry's last acts as inspector was to begin the replacement of Cape Town's open sewers with iron pipes. He imagined a rigorous public health regime in Cape Town and went some way towards establishing it. There had certainly been initiatives in the British Army to ensure adequate ventilation,

8 Holmes, *Scanty Particulars*, 94.
9 E. H. Burrows, *A History of Medicine in South Africa* (Cape Town: A. A. Balkema, 1958), 103.
10 Burrows, *History of Medicine*, 105.
11 Du Preez and Dronfield, *Dr James Barry*, 175.
12 Du Preez and Dronfield, *Dr James Barry*, 177.

cleanliness and sanitation in military institutions, but the scale of change that Barry envisaged was at the cutting edge of scientific thinking.

Barry continued to practise as an army doctor with some private work, usually unpaid. When in 1827 the sister of his landlady, Wilhelmina Munnik, experienced acute distress in the early phases of childbirth, Barry was called. He could not get the foetus to pass through the cervix, even with forceps. Caesarean operations were known, but no British case had been reported of mother and child surviving. Barry's biographers suggest that he had focused on obstetrics (or midwifery) in Edinburgh and read about a successful operation in Switzerland. With a midwife and two servants assisting, the patient biting on a leather strap, Barry operated. He clearly knew sufficient anatomy from his training and practice, and he was already experienced in dealing with wounds in military patients (from fights, as much as wars). Mother and child survived, perhaps because of his fastidious concern with hygiene as much as his surgical skills. The procedure is now generally accepted as the first successful Caesarean operation in the anglophone world. Barry refused payment, suggesting instead that the child be named James Barry Munnik in thanks. The family retained this practice and a cousin of a descendant named their son, born in 1866, James Barry Munnik Hertzog. In this unlikely way, the first Afrikaner nationalist prime minister of South Africa (1924–39) carried the name of a radical British doctor from the early nineteenth century.

Among his other scientific interests, Barry also wrote on botany and pharmacology and was convinced that by examining 'the Natural History of the Cape, its Vegetable Kingdom will be discovered to abound in many rare and valuable medicinal productions'.[13] He experimented with *Arctopus echinatus*, which was used in African healing, and believed it could help to treat venereal disease, the curse of the garrison. Existing treatments were complicated and not very effective.

While Barry's international reputation has grown with successive biographies and studies, Somerset has long featured in the historiography as an aristocratic, illiberal governor, who ruled by patronage and suppressed the first stirrings of a liberal press. Yet Somerset actively promoted economic progress in the Colony. Theal, a

[13] Du Preez and Dronfield, *Dr James Barry*, 221.

conservative historian sympathetic to the settler view, judged that Somerset was 'as desirous as anyone could be for the improvement of the Colony and the prosperity of its inhabitants ... only everything tending to improvement and prosperity must emanate from himself'.[14] At the instigation of military officers in 1814, Somerset appointed the eccentric botanist and licensed apothecary, Dr Mackrill, to seek out a favourable place in the eastern Cape for the establishment of a government farm. Mackrill chose a site in the Agterbruntjieshoogte, a favourable spot east of Graaff-Reinet, where earlier travellers such as Sparrman had been hosted by Boer frontiersmen. One aim was to experiment with tobacco and so establish a source other than the United States. A further purpose was to develop agriculture in order to provision troops and horses on this troubled frontier; supply lines from Cape Town were very long and Port Elizabeth was not yet established.

Under Robert Hart from 1817, Somerset farm became a significant centre for intensive food and fodder production and also a depot for produce purchased from the local Boers and then distributed to military posts. The farm turned into the village of Somerset East. Hart started his own farm, Glen Avon, nearby. He was one of the first successful merino breeders in the Cape at the very beginning of the wool boom, which finally promised an export commodity to make the Colony solvent. Groote Post, near Darling in the western Cape, played a similar role, specialising in livestock breeding.

From the very start of his governorship, Somerset promoted the idea of a major British migration to South Africa. Many of the 4,000 new British colonists who arrived in 1820 found it difficult to establish themselves and felt that they had inadequate support from the government. One of their number, Thomas Pringle, headed a Scottish party that settled on a tributary of the Fish River, closer to Somerset farm (about 30 km away) than to the major concentrations around Grahamstown (about 150 km). Pringle visited the farm, which greatly impressed him, and met Hart, the head of the project.[15]

Pringle and Barry were born in the same year, 1789, and both overlapped in Edinburgh. In 1817, when Pringle, developing his

[14] A. K. Millar, *Plantagenet in South Africa: Lord Charles Somerset* (Cape Town: Oxford University Press, 1965), 6.

[15] T. Pringle, *Narrative of a Residence in South Africa*, vol. 1, 52. (Brentwood: Empire Book Association, 1986), first published 1834.

literary ambitions, helped to found the *Edinburgh Monthly Magazine*, Barry had arrived at the Cape. In 1822 Pringle moved to Cape Town to become librarian at the South African Public Library, newly established with Somerset's approval and drawing on the proceeds of a tax on wine and brandy. Soon after its foundation the library benefited from the addition of 4,500 books – including a first edition of the works of Descartes – assembled by the German VOC official Joachim von Dessin (d. 1761), whom Lacaille considered to be one of the most influential men at the Cape in his day.

Around the time that Pringle became librarian, Barry donated some of his medical textbooks, and advised the library committee about the purchase of further medical books. It is very likely that they met through this connection – and they shared an antipathy to slavery. Pringle was initially impressed by Somerset and later noted in his *Narrative of a Residence in South Africa* (1834) that the governor had approved capital punishment for a white man for the murder of a slave 'for the first time in the history of the colony'.[16] But Pringle crossed swords with the government over the plight of the 1820 settlers, for whom he acted in Cape Town. He soon found Somerset and key officials both uninterested and intolerant – leading what Pringle termed 'the Cape "Reign of Terror"'.[17]

British settlers, not least those in Cape Town, sought to assert themselves as citizens with rights as well as responsibilities. In the eastern Cape, the majority developed strong anti-emancipationist and racist views, fuelled by a sense of vulnerability and resentment at their indifferent treatment by officials. In Cape Town a small group of British colonists articulated specifically liberal, emancipatory ideas. Missionaries like John Read and John Philip took a lead in defending some of the legal rights of the colony's indigenous subjects. Although initially representing the settlers, Pringle was also an advocate of progressive colonial freedoms for black as well as white. Together with his Scottish compatriot John Fairbairn, Pringle started a school in Cape Town and founded the *South African Journal*. They claimed the right to publish without censorship and the necessity for a 'general diffusion of knowledge, Piety and civilisation throughout this extensive Colony'.

16 Pringle, *Narrative of a Residence in South Africa*, vol. 1, 148.
17 Pringle, *Narrative of a Residence in South Africa*, vol. 1, 157.

The *Journal* lasted for two issues only. A new venture followed in 1824 when Fairbairn and Pringle, with George Greig, founded the *South African Commercial Advertiser*. This was soon quashed by Somerset. The governor's fight against a free press was certainly rooted in reactionary sensibilities and fear of political opposition and Jacobinism. However, his immediate motivation was also to suppress the full record of a trial in which Somerset claimed he was libelled; one of those charged was also suspected of displaying a pamphlet suggesting a homosexual relationship between Somerset and Barry.

The *Commercial Advertiser* sought to root radical principles of the Scottish Enlightenment in African soil. Among the prominent themes were calls for social and economic progress, improvement and reason – ideas and values that also helped to underpin claims to colonial respectability and patriotism. The irony is that Barry was pursuing some of these improving aims in the sphere of medicine and public health and that he was immobilised not by Somerset, but by the Cape functionaries and medical men with whom he had clashed. Somerset remained adamantly opposed to a free press and seems to have been personally outraged by Pringle. Progress, as Theal suggested, had to be controlled by the governor.

The campaign to publish free of official censorship was accompanied by efforts to expand associational life by forming clubs and societies. From the mid-1820s a range of new scientific and literary institutions emerged in order to give voice to embryonic forms of civic assertion. A South African Literary Society was revived, but suppressed soon after its creation in 1824.[18] Sir John Truter had decided to join the Society, 'as the promoter of science and literature in the colony', but Somerset was so strongly opposed to Pringle and Fairbairn that he 'called the Chief Justice to his presence, and gave him such a rating … that Sir John, almost frightened out of his wits', withdrew.[19] Somerset also warned off the lawyers and merchants who had supported the revival.

Zoology, Astronomy and Survey at the Cape: Smith, Herschel and Maclear

Somerset and Pringle departed to Britain in 1826, and Barry left the Cape soon after. Pringle became secretary to the Anti-Slavery Society

[18] Dubow, *A Commonwealth of Knowledge*, 29–34.
[19] Pringle, *Narrative of a Residence in South Africa*, vol. 1, 163.

in London. Ordinance 60 of 1829 established the right to publish at the Cape and, following Somerset's departure, the scope for civic organisation expanded rapidly. The South African College was founded in 1829, as was the South African Institution, which launched the country's first scientific publication, the *South African Quarterly Journal*, in 1830. This publication, which remained in existence until 1837, served as an outlet for nineteen scientific papers by Andrew Smith, a medically trained Scottish explorer, traveller and naturalist.

Smith qualified in medicine at Edinburgh and worked initially in the eastern Cape. He was responsible for the creation of a private museum in Cape Town in 1825 dedicated to 'the reception and classification of the various objects of the Animal, Vegetable and Mineral Kingdoms which are found in South Africa'. The museum benefited from donations and natural history specimens contributed by locally based naturalists, including the botanical collector Baron Carl von Ludwig and the taxidermist Jules Pierre Verreaux. The Paris-based firm Maison Verreaux maintained a branch in Cape Town and was a premier supplier of natural history specimens to museums in Europe and America.[20] The Verreaux brothers were also responsible for exhibiting the stuffed corpse of a Tswana man in Paris. It ended up on display as 'El Negro' in Banyoles, Spain, and was repatriated to Botswana amid some controversy in 2000.

In 1826 Smith published a *Descriptive Catalogue of the South African Museum*, which constituted a significant contribution to cataloguing the country's mammalia. His coverage was expanded in the four-volume *Illustrations of the Zoology of South Africa* (1838) – the fullest record so far. Smith abandoned the form of presentation used by earlier travellers, who weaved their writing about animals into their narrative accounts, and wrote instead a systematic list with illustrations. His sojourn at the Cape lasted from 1821 to 1837, when he left in order to mount an exhibition in London. This featured natural history specimens from the South African Museum as well as from a major expedition that he undertook to the 'interior of Africa' in 1834–6. His journey brought him into contact with such African

[20] R. Summers, *A History of the South African Museum 1825–1975* (Cape Town: A. A. Balkema, 1975), 5–8.

leaders as Andries Waterboer of the Griqua, Moshoeshoe of the Sotho, and Mzilikazi of the Ndebele. He had earlier also visited the Zulu king Dingane.

Smith was supported in this venture by the newly created South African Literary and Scientific Institution as well as a committee of prominent Cape citizens who issued subscription shares to raise money for his trip. As an explorer and man of science Andrew Smith was a transitional figure, for although he drew on local and official support to sponsor and promote his scientific ventures, he remained an individual traveller in the eighteenth-century sense and his ambitions were ultimately focused on the metropole rather than the Cape. After his return to Britain, he became an army surgeon, rising to the position of director-general of medical services during the Crimean War.[21]

The career of Edinburgh-trained anatomist and naturalist Robert Knox provides a parallel. A decade before his involvement in the infamous Burke and Hare murder scandal, Knox spent three years at the Cape as a military surgeon during which time he likely worked with James Barry. After his return to Edinburgh in 1820, Knox travelled to Paris where he met Cuvier (shortly after Cuvier's dissection of Sara Baartman's body) and Le Vaillant. In 1823 he delivered a talk on anatomical differences between the 'Native Races' of southern Africa. When a group of 'Bushmen' were exhibited in London in 1847, Knox addressed a large audience at Exeter Hall on the relationship between Bushmen and the Khoekhoen – taking issue with Smith's view that they were racially the same. Knox's book *The Races of Men* (1850), a landmark in racial science, contains a lengthy and characteristically opinionated chapter on 'The Dark Races of Men' in which he expounds on the relative inferiority of indigenous South Africans to the 'white races'.[22] For Scots as diverse in their political views as Pringle, Smith and Knox, their experience of the Cape in the early decades of British occupation proved formative in their later British careers.

[21] P. R. Kirby, *Sir Andrew Smith, M. D., K. C. B.: His Life, Letters and Works* (Cape Town: A. A. Balkema, 1965).

[22] P. R. Kirby, 'Robert Knox and his South African Research', *South African Medical Journal*, XIV, 13 (1940), 254–60; R. Knox, *The Races of Men: A Fragment* (Philadelphia: Lea and Blanchard, 1850), chapter 6.

Figure 2.1 Charles Piazzi Smyth, composite sketch of Aloe Ferox, Springboks, African Rhinoceros, and Hippopotamus, from his sketchbook, 'South African Microcosm 1835–1846'.

Credit: Charles Piazzi Smyth Collection, Royal Observatory Edinburgh/Royal Society of Edinburgh, with thanks to Andy Lawrence and Denis Pellerin.

Smith's short-lived private museum catered to a self-selected elite, but it housed the largest variety of animal specimens yet assembled in South Africa for study and comparison. It provided a model for collection and display, and in 1829–30 many of the specimens were transferred to the newly established South African Institution with Verreaux acting as curator. This brought together private collections belonging to Smith, Verreaux, and Ludwig, but it was broken up in 1837 after Smith's departure. The remaining zoological holdings were transferred on loan to the newly created South African College, forerunner of the University of Cape Town.[23] Institutional memories proved short, however, for the establishment of the Albany Museum and the South African Museum in 1855 was mistakenly assumed by many subsequent historians to be the start of museology in South Africa.[24]

The most important early specialist scientific institution, not least judged by its permanence and prominence, was the Royal Observatory, approved in 1820 and completed in 1827–8 under the supervision of the Simonstown naval dockyard. It was intended to be the equivalent in the southern hemisphere of the Observatory at Greenwich. John Barrow, now firmly ensconced at the Admiralty, was one of the principal figures responsible for the decision to locate it at the Cape, together with Joseph Banks of the Royal Society. Banks opined that 'nothing could more essentially promote the glory' of Britain than to take a lead in such an undertaking. King George IV gave it the imprimatur of royal authority as well as financial support.[25]

The Royal Observatory was a handsome building designed by the Scottish civil engineer John Rennie, according to the principles of the Greek Revival, and at considerable cost. Its classical lines and symmetrical mathematical proportioning symbolised both its scientific importance and its significance as an outpost of British astronomy and imperial power (the cover of this book is a somewhat imaginary representation of the building with Table Mountain as a backdrop). The establishment of the Cape Observatory affirmed the Colony's new importance in the post-Napoleonic 'second British empire'. Christopher Bayly's description of this moment as the 'imperial meridian' acquires new meaning given that one of the Observatory's premier scientific programmes was to determine an arc of the meridian, essential for accurate mapping and geodetic surveying as well as for determining the curvature of the earth.[26]

Her Majesty's Astronomers at the Cape, Fearon Fallows (1820–31), Thomas James Henderson (1832–3) and Thomas Maclear (1833–70),

23 Summers, *South African Museum*, 14–15, 18.
24 Summers, *South African Museum*, 5.
25 D. Gill, *A History and Description of the Royal Observatory, Cape of Good Hope* (London: HMSO, 1913), vi; S. Ruskin, *John Herschel's Cape Voyage: Private Science, Public Imagination and the Ambitions of Empire* (Aldershot: Routledge, 2004), 45–6.
26 Bayly, *Imperial Meridian*.

made distinctive scientific contributions. Henderson measured the distance of the Alpha Centauri stars closest to the sun, but he lost patience with the Cape (reportedly deriding the environs of the Observatory as a 'Dismal Swamp among Slaves and Savages') and resigned his post. He subsequently became Scotland's first Astronomer Royal.[27] Interest in the potential of southern hemisphere astronomical observation had attracted eighteenth-century astronomers Kolb and Lacaille (Chapter 1). The creation of a permanent Observatory made systematic, ongoing research a viable project. Many activities were practical and applied: cataloguing southern hemisphere star positions as an aid to navigation, observing tidal movements, conducting meteorology, and maintaining reliable weights and measures. These allied 'observatory sciences' and routines were intimately associated with a repertoire of precise measurement that reinforced the symbolic and practical authority of British governance at the Cape. [28]

One activity where theoretical and practical objectives could be combined was geodesy (measuring the shape and dimensions of the earth) and land surveying. Both involved the laborious technical task of laying out baselines as a standard of length and triangulating further. This initiative was undertaken at the Cape by Maclear and his assistants Charles Piazzi Smyth and William Mann. Maclear's triangulation project began in 1836 with the establishment of an accurate baseline on the Grand Parade, Cape Town, a key point of public assembly. Thereafter, beacons were placed on mountaintops all the way into the interior. The project involved extensive travel over difficult terrain by ox-wagon. Cumbersome precision instruments (sectors, heliostats and theodolites) had to be carried up and down mountains.[29] Maclear's endeavour, carried out over some twenty years, resulted in accurate trigonometrical land surveys from Cape Agulhas to the Orange River, and also allowed him to verify and extend the arc of the meridian. His work resolved Lacaille's earlier

[27] B. Warner, *Astronomers at the Royal Observatory, Cape of Good Hope* (Cape Town: A. A. Balkema, 1979), 34.

[28] S. Schaffer, 'Keeping the Books at Paramatta Observatory', in D. Aubin et al. (eds.), *The Heavens on Earth: Observatories and Astronomy in Nineteenth-Century Science and Culture* (North Carolina: Duke University Press, 2010).

[29] Warner, *Astronomers at the Royal Observatory*, 51–2, 56–7; P. Moore and P. Collins, *The Astronomy of Southern Africa* (London: Robert Hale, 1977), 57–9.

error and demonstrated that the earth was indeed ellipsoid, as the laws of Newtonian physics suggested.[30] Maclear was awarded a knighthood for these efforts.

There were close but informal connections between the Royal Observatory under Thomas Maclear and the private observatory established in 1834 by Sir John Herschel at Feldhausen, a Cape Dutch country homestead in the present-day suburb of Claremont. Herschel was already established as a leading British scientist. Educated at Cambridge, he published an overview of scientific disciplines and the theory and methods of science in 1830.[31] He made a major contribution to an undulatory theory of light waves and advanced innovative conceptions in astronomy about the scale of the universe and the relationships within it. Following on from Newton, as well as his father William Herschel and others at the cutting edge of the discipline, he conceptualised an almost infinite universe with different types of stars and nebulae. 'Within this beautiful structured encapsulation', a historian has argued, 'is great complexity of detail, and mystery as to the physical nature of many things – nebular matter, comets, our sun itself.' Herschel published his *Treatise on Astronomy* in 1833, 'stretching ... the English imagination'.[32]

Herschel and his family arrived at the Cape with the express purpose of cataloguing the clusters, nebulae and double stars of the southern hemisphere. To this purpose he installed a 20-foot reflector telescope in the garden of Feldhausen. In the course of regular nightly 'sweeps' of the southern skies, Herschel identified over 1,700 nebulae as well as 2,100 double stars, which had important implications for his development of theoretical understandings of the Milky Way. Together with Maclear, to whom he was personally close, he observed and recorded Halley's Comet in 1835.[33]

Herschel also became absorbed by the Cape's unique botanical riches, planting his garden with indigenous fynbos bulbs and drawing

[30] B. Warner, *Royal Observatory, Cape of Good Hope, 1820–31: The Founding of a Colonial Observatory* (Dordrecht: Springer, 1995); B. Warner, *Charles Piazzi Smyth. Astronomer-Artist. His Cape Years 1835–1845* (Cape Town: A. A. Balkema, 1983), chapter 4; D. S. Evans, *Lacaille: Astronomer, Traveller* (Tucson: Pachart, 1992).

[31] W. F. Cannon, 'John Herschel and the Idea of Science', *Journal of the History of Science*, 22, 2 (1961), 215–39.

[32] Cannon, 'John Herschel', 238.

[33] Dictionary of National Biography, 'John William Herschel'.

landscapes and mountain views with the aid of a camera lucida.[34] The expansion of understanding of the range of species in botany, zoology and entomology was coterminous with new conceptions of scale in astronomy. Related scientific interests pursued by this polymath included experiments in chemistry, physics and photography. Herschel served on the committee responsible for Andrew Smith's 1834–6 scientific expedition into the interior and he received birds and fossils which Smith had collected. In recognition of Herschel's scientific standing, he was elected in 1834 as president of the newly constituted South African Literary and Scientific Institution, which now stood at the centre of the Cape's emerging institutional and intellectual life.[35]

Working closely with his wife Margaret, an accomplished illustrator of plants, Herschel participated in scientific and cultural activities. They maintained a wide social circle, yet kept their distance from the conservative elite at Government House. In part, this was because Herschel believed that scientists should eschew patronage – and with his own private means he was able to work independently. There was also a political dimension to this choice, since the liberal-minded Herschels disapproved of the harsh realities of colonial rule, in particular the bitter wars on the expanding eastern Cape frontier and settler resistance to slave emancipation. Margaret Herschel recorded in her diary that they were 'delighted' by the Cape – while noting their disapproval of the 'tyrannical Spirit which Colonies are apt to engender among the Inhabitants towards the Natives'. She identified her husband as being on the 'side which humanity & justice dictate, & which is guarded by the mournful minority at the Cape'.[36]

Among the regular visitors to Feldhausen were the missionary campaigner for aboriginal freedoms, John Philip, and the liberal journalist John Fairbairn. Griqua leaders Adam Kok and Andries Waterboer visited Feldhausen and were shown around the Royal Observatory in 1834. When a further war with the Xhosa broke out that year, the Herschels and Maclear showed their disapproval of colonial aggression by absenting themselves from a public dinner in honour of Governor Benjamin D'Urban and his military campaign.[37]

34 B. Warner, *Cape Landscapes. Sir John Herschel's Sketches 1834–38* (Cape Town: Juta, 2006); B. Warner, *Astronomers at the Royal Observatory.*

35 Ruskin, *John Herschel's Cape Voyage*, 32.

36 Ruskin, *John Herschel's Cape Voyage*, 56, 54–5.

37 Warner, *Cape Landscapes*, 132.

On his return to England, Herschel was treated as a scientific celebrity. At a dinner attended by leading scientists to honour Herschel's achievements at the Cape, the marquess of Lansdowne, a leading Whig politician and supporter of causes such as abolition and popular education, portrayed Herschel's expansion of the 'empire of knowledge' as a positive outcome of the post-Napoleonic British acquisition of the Cape and, further, as a benefit to all countries.[38] For all that Herschel opposed militarist imperialism and aggressive colonial expansion, he supported the principle of the progress and extension of British 'civilisation'. He viewed astronomy as playing a role in the enlightened science of imperial governance and efficient fact-gathering.[39] Albeit sympathetic to the plight of the Xhosa, Herschel considered that Africans would have to adapt to colonial rule and their social systems needed to change accordingly. In this regard, he was reform-minded rather than radical. Herschel's social instincts were evident, too, in the field of education where he helped locally based activists like Fairbairn in persuading Governor Napier to institute, from 1839, an advanced form of state-funded education that was independent of church oversight and geared to the acquisition of secular and scientific knowledge.

A notable visitor to Feldhausen was the young Charles Darwin, who docked at Simonstown for nearly three weeks in 1836, on the *Beagle*'s way back from Mauritius. Darwin, who had read Herschel's *Preliminary Discourse on the Study of Natural Philosophy*, hastened to meet him as well as Andrew Smith, Thomas Maclear, and the colonial secretary John Bell. A dinner at Feldhausen on 15 June was one of the highlights of Darwin's visit.

Yet from a scientific point of view, Darwin found the Cape generally disappointing and, despite travelling through some of the most beautiful parts of the western Cape, with its biologically rich and diverse fynbos and soaring fold mountains, he found the Cape aesthetically unappealing. Darwin scholars have rather little to say about his sojourn there, other than his inspection of a granite-schist contact on a Sea Point beach, shown to him by Smith, which had important implications for the

38 Ruskin, *John Herschel's Cape Voyage*, 64–5; E. Green Musselman, 'Swords into Ploughshares: John Herschel's Progressive View of Astronomical and Imperial Governance', *British Journal for the History of Science*, 31 (1998), 420–1.

39 W. Ashworth, 'John Herschel, George Airy, and the Roaming Eye of the State', *History of Science*, 36, 2 (1998), 160, 171–2.

understanding of geological upheavals.[40] Darwin's lack of enthusiasm for the country may have had to do with the fact that he was on the last leg of his journey and keen to return home. The Cape was not an exotic location like the Galapagos or tropical Mauritius, nor was it cultivated and ordered like the English countryside. The very fact that it was by now relatively well studied scientifically may also have diminished its allure: 'the Cape was simply too ordinary'.[41] Yet that ordinariness or familiarity also made him comfortable in the community of intellectuals with whom he engaged. The Cape, he remarked, was one of those colonies around the world in which 'little embryo Englands are hatching'.[42] Darwin began the *Origin of Species* by referring to John Herschel as 'one of our greatest philosophers'.

Knowledge and Power in the Zulu Kingdom

Almost coterminous with Lord Charles Somerset's governorship at the Cape was the rise of the Zulu kingdom under Shaka (c. 1816–28). There were direct links between the two centres in that Cape-based traders, who first arrived at Port Natal in 1824, established regular contact with the Zulu. Traders and emissaries such as Henry Francis Fynn and Nathaniel Isaacs, who met Shaka, and Andrew Smith, who visited his successor Dingane, sent back information to the Cape. Additional reports were also gleaned from other intermediaries and from African refugees in the eastern Cape and the interior. In turn Shaka and Dingane were keen to understand and make contact with British colonial power in the Cape; they were also warned of the dangers of colonial encroachment.

Recent analyses tend to dismiss Zulu-centric approaches to early nineteenth-century developments in African societies in the region, and thus minimise the specificity, scale and innovative features of the Zulu state.[43] Norman Etherington has argued forcefully that little about the Zulu state was new: its regiments, weapons, and methods

[40] S. Master, 'Darwin as a Geologist in Africa: Dispelling the Myths and Unravelling a Confused Knot', *SAJS*, 108, 9–10 (2012), 1–5.

[41] W. S. Barnard, 'Darwin at the Cape', *SAJS*, 100 (May/June 2004), 245.

[42] A. Desmond and J. Moore, *Darwin* (London: Penguin, 1991), 184; S. Sivasundaram, *Waves Across the South. A New History of Revolution and Empire* (London: Collins, 2020), 306–7.

[43] N. Etherington, *The Great Treks: The Transformation of Southern Africa, 1815–1854* (London: Pearson Education, 2001); J. Wright, 'Turbulent Times: Political Transformations in the North and East, 1760s–1830s', in C. Hamilton,

of warfare had all been employed for some decades before.[44] He is also sceptical about exaggerated estimates of the size of the army and the frequent comparison made between Shaka and Napoleon.[45] Yet it is important to recognise that Shaka and Dingane developed an African state that probably had twice the population of, and a larger military than, others in South Africa at the time. They were for a period (and in a similar period to Napoleon) formidably successful in battle, laying the foundation for a powerful and distinct African linguistic and cultural identity.

The Zulu kingdom, consisting of perhaps 250,000 people by mid-century, was not on the scale of major West African states such as the Asante, nor did it develop so elaborate a bureaucratic structure. But state formation was difficult in many African contexts. Relatively limited changes in resources, trade, military organisation and weaponry could confer significant advantages. Innovations by Shaka and Dingane made a difference – including new, militarised forms of royal residence. Excavations at Mgungundlovu, built in 1829 for Dingane, which survived until 1838, reveal an estimated 1,700 houses for a population of about 7,000 people.[46]

Other innovations related to the scale of the *amabutho* (or regiments) and military technology. At that time, African armies in the subcontinent generally went into war with metal-tipped assegais which they threw, and usually also shorter stabbing spears with a broader blade. Most historians accept that Zulu military technique focused on the stabbing spear; they may even have dispensed with the throwing spear. Improved shields warded off assegais. They were mass-produced centrally from oxhide in larger, oval, standard sizes of about 54 x 30 inches (137 × 76 cm), sometimes with uniform colours for the regiments.[47] The wooden pole on which the oxhide was fixed was

B. K. Mbenga and R. Ross (eds.), *The Cambridge History of South Africa: From Early Times to 1885* (Cambridge: Cambridge University Press, 2010), 211–52.

44 N. Etherington, 'Were There Large States in the Coastal Regions of Southeast Africa before the Rise of the Zulu Kingdom?', *History in Africa*, 31 (2004),157–83.

45 Etherington, 'Were there Large States?', 181.

46 P. Mitchell, *The Archaeology of Southern Africa* (Cambridge: Cambridge University Press, 2002), 372.

47 I. Knight, *The Anatomy of the Zulu Army from Shaka to Cetshwayo 1818–1879* (Barnsley: Frontline Books, 2015), 101–2.

sharpened at the bottom to provide an extra weapon. Zulu prowess in close-quarter fighting, using shields and stabbing spears, was honed by experience. Going into battle, the regiments – 'understanding the psychological advantages of display' – beat their shields rhythmically.

Even if Zulu armies were no more than 15,000 strong, this must have created enormous pressure for the production of strong and large-bladed iron spears. Archaeologist Tim Maggs finds evidence of an intensification of metal production at the time, an issue that has attracted little attention in historical analysis. Metal- and especially iron-working were deeply embedded skills in African societies of southern Africa. There were important sites of metallurgical experimentation within the area that became the Zulu Kingdom; by 1992, when Maggs wrote on this topic, archaeologists had recorded over 120 Late Iron Age furnaces.[48]

Although the timing is uncertain, Maggs argues that iron was increasingly mined, smelted and exchanged by specialist smiths in the Late Iron Age and that the Zulu kings attempted to expand their control over its production. Oral traditions suggest that 'the rapidly growing Zulu kingdom, in the third decade of the nineteenth century, saw a surge in demand for metalwork to be distributed through the king's patronage'.[49] Production seems to have been most fully developed around Nkandla, where contingents of the Zulu army assisted specialists with the heavy work of mining and transporting iron-bearing material to the furnaces. Shaka and Dingane controlled the output of spears and distributed them at the royal residences to the *amabutho*.

Maggs excavated a site at Mabhija in the upper Tugela Basin where there were a number of smelting furnaces, some organised with three adjacent pits.[50] These seem to have been worked by people identifying as Dlamini, whose chiefdom was conquered by the Zulu (and some of whom fled to become part of the Mfengu in the eastern Cape). The

48 T. Maggs, '"My Father's Hammer Never Ceased Its Song Day and Night": The Zulu Ferrous Metalworking Industry', *Natal Museum Journal of Humanities*, 4 (1992), 65–87.

49 Maggs, 'My Father's Hammer', 84.

50 T. Maggs, 'Mabhija: Pre-colonial Industrial Development in the Tugela Basin', *Annals of the Natal Museum*, 25, 1 (1982), 123–41; G. Whitelaw, 'Precolonial Iron Production around Durban and in Southern Natal', *Natal Museum Journal of Humanities*, 3 (1991), 29–39.

traveller Adolphe Delagorgue recorded furnaces of a similar scale at the Mkhuze River, in territory formerly controlled by the Ndwandwe, whom the Zulu conquered in the 1820s. He was told in 1846 that at the time of conquest, 'Djaka [Shaka] had a continual need for quantities of iron for his frequent wars and by these means he was assured of an ample supply, forged by his own subjects.'[51]

It seems, then, that Zulu-controlled iron production was expanded by incorporating existing sites in conquered territory, and by intensifying production both at these mines and furnaces and in the heart of the kingdom at Nkandla. The furnaces at the Mkhuze were still operating in 1846, with four relays of six men constantly employed in manning the bellows from about eight in the morning to midnight. 'Their skill in working the iron is worthy of note', Delagorgue recorded; 'their methods, however imperfect and crude they may seem, nevertheless enable them to produce some fine pieces . . . their weapons are elegant'. He was highly impressed as he watched the iron-workers sharpening blades.[52] Maggs calls this intensification a form of 'precolonial industrial development'. It has been suggested that the process used by Zulu smiths allowed some charcoal to mix with the molten ore, thus hardening the iron blades.[53] Indigenous knowledge was a dynamic field at the time and Shaka, as much as Somerset, appreciated the significance of controlling innovation in the exercise of authoritarian power.

Shaka was also aware of the value of firearms. In 1826, the Port Natal traders, who visited him regularly, joined his forces, armed with muskets, in a major sortie against the Ndwandwe. Under Zwide, the Ndwandwe had been his most formidable rivals but they had retreated north of the Phongola River. Shaka's armies marched north and routed those of Sikunyana, Zwide's son, in close combat.[54] It is not clear how significant firearms were in the battle, which involved as many as 40,000 Ndwandwe, a large number of whom were reportedly killed.[55] Dingane seems to have developed strategies both to use firearms (with small specialist contingents) and to counter them.[56] These, however,

51 Maggs, 'My Father's Hammer', 69. 52 Maggs, 'My Father's Hammer', 69.
53 Knight, *The Anatomy of the Zulu Army*, 110.
54 J. Stuart and D. McK. Malcolm, *The Diary of Henry Francis Fynn* (Pietermaritzburg: Shuter and Shooter, 1969), 126–7.
55 Etherington, *The Great Treks*, 151.
56 Knight, *The Anatomy of the Zulu Army*.

failed at the Battle of Blood River with the Voortrekkers in 1838. It seems possible that Etherington may have underestimated the significance of new military organisation and strategies, coupled with more intensive iron-working – emphasising instead the contribution made by the traders to Shaka's reputation: 'accounts which made his name far more familiar to the world at large than any other king of the heartland'.[57]

Karen Flint takes a different approach. She is less concerned with cutting the Zulu down to size or puncturing the mythologies of 'terrific majesty'.[58] Rather, she accepts Zulu distinctiveness and explores some of the innovative aspects of African and hybrid knowledge that contributed to state-making. She suggests that 'African healers played an important role in upholding the authority and national identity of a new Zulu nation. Healers and therapeutics served not only the physical and social body, but the body politic. As the Zulu Kingdom emerged and consolidated its power, medicine helped defeat Zulu enemies, strengthen the king, and create a new sense of national pride and obligation.'[59] Flint uses the term 'nation', as much as 'kingdom', 'chiefdom', 'state', or 'polity', with the implication that these ideas made a significant impact in welding a new sense of identity and unity. She argues for the centrality of medical and ritual practices, as much as the *amabutho*, in supporting 'a diverse but centralized state' and in securing allegiance to the king rather than a local lineage or chief. Her analysis echoes discussions of the abolition of male circumcision in the Zulu Kingdom at the time.

The oral traditions put particular emphasis on this aspect of Zulu state-building and provide 'insight into the imagination of the Zulu nation state'.[60] Poisoning and supernatural manipulation were seen as part of bids for power and revenge. Chiefs themselves could derive power from ritual medicine (*muthi*) and could use them against others. Traditions emphasise the national scale of ceremonies such as rain-making, with recruitment of key ritual specialists from outside the kingdom. Similarly, Shaka developed the annual first fruits festival into a national event. At such festivals, the king, as healer of the land, accompanied by his doctors and regiments, performed rituals aimed

[57] Etherington, *The Great Treks*, 151

[58] C. Hamilton, *Terrific Majesty: The Powers of Shaka Zulu and the Limits of Historical Invention* (Cambridge MA: Harvard University Press, 1998).

[59] Flint, *Healing Traditions*.

[60] Flint, *Healing Traditions*, 71.

at ensuring 'the well-being of the nation and all who lived in it'. Soldiers going into battle were doctored at the king's homestead and those returning were cleansed. In addition to the shield, the Zulu headband, or *inkatha*, became a pivotal symbol protecting the nation and fostering national unity.

Anthropologists writing in the 1930s, such as Monica Hunter (later Wilson), saw medical and magical or supernatural ideas as entwined with and inseparable from African notions of healing.[61] Zulu medical thinking at a later stage also included an environmental understanding of disease, and there is evidence of it in the nineteenth century.[62] The layout of settlements on higher ground, as well as building styles, given the available technology, reflected local knowledge of environmental hazards. Low-lying malarial areas were avoided. Grass was regularly burned around homesteads to keep snakes and insects at bay. Europeans visitors remarked that dwellings were 'kept scrupulously clean', swept daily inside and out; 'neither dirt nor ash were to be seen'.[63] Tsetse fly zones were avoided. White settlers observed in the 1870s that Africans knew that sleeping sickness in both cattle and humans was transmitted via tsetse flies and sought to minimise contact. David Bruce found the same in the late nineteenth century (see Chapter 3).

Shaka was also interested in the healing skills of the traders and allowed Fynn to treat him for a stab wound. As it was the only material he had on hand, Fynn washed the wound with camomile tea and bound it with linen. After Shaka's doctor administered an emetic – central to African concepts of medicine – and administered healing roots, Fynn obtained further medicines and dressed the wound again.[64] Fynn also learnt African healing strategies and administered them to other traders and settlers. Medicine, Flint argues, was an important means of interchange and, as in the Cape in the eighteenth and nineteenth century, elements of hybrid medical practice were established early in encounters between the Zulu and people from Britain and the Cape.

61 M. Hunter, *Reaction to Conquest* (Oxford: Oxford University Press, 1964), first published in 1936.
62 H. Ngubane, *Body and Mind in Zulu Medicine* (London: Academic Press, 1977).
63 Flint, *Healing Traditions*, 73.
64 Stuart and Malcolm, *The Diary of Henry Francis Fynn*, 85.

Zulu chiefs were also concerned to oversee an element of social order that could be described as public health. They were keen to hear about sick people: 'the king would then most likely give the order to consult diviners so as to discover the nature and cause of his illness. A sick man in Zululand was always an object of great importance.'[65] Although the causes of disease were not generally attributed to infection, elements of the idea were latent; by the mid-nineteenth century, possibly because of white settler influence, there were reports of sick people being isolated. Mpande, the Zulu king from 1840 to 1872, attempted to ban cattle from Natal during the outbreak of lungsickness in the 1850s; and there are similar reports from Mpondoland and the eastern Cape. The notion of infection was probably very restricted at the time, as was also the case in Western conceptions. Nor were all Zulu practices conducive to health; Flint cites kraaling livestock, later believed to spread cattle disease by enclosing them nightly. This practice was also general among settler farmers and became the focus of attempts by early vets to improve livestock management from the 1870s.[66]

In these ways, hybrid ideas about naming and treating diseases began to evolve in southern Africa. Medical missionaries started to train African assistants, and, from the mid-nineteenth century, African knowledge of plant medicine was sometimes recorded in botanical publications (Chapter 7). More generally, social practices in nineteenth-century colonial Natal were to some degree shaped from below by the colonised.[67] Thus, from an early period in the colonial encounter with independent African chiefdoms, exchanges in ideas and practices, along with the development of hybrid forms of knowledge, began to take place throughout the region – often far away from the centres of British power and settlement at the Cape.

Museums and Scientific Innovation, 1840s to 1860s

The development of institutions at the Cape devoted to the propagation of knowledge in the 1820s was important, yet they resembled small clubs that were for the most part reliant on private subscription

[65] Flint, *Healing Traditions*, 83. [66] Beinart, *Rise of Conservation*, chapter 4.
[67] K. Atkins, *The Moon is Dead! Give Us Our Money! The Cultural Origins of an African Work Ethic, Natal, South Africa, 1843–1900* (London: James Currey, 1993).

with additional public subvention. Notably, boundaries between public and private, amateur and professional, and arts and sciences, remained highly permeable. The South African Public Library lay at the centre of these public-private institutions. Herschel called it 'the bright eye of the Cape', regarding it as vital for the wide diffusion of knowledge.[68]

There were many testimonies to the Library's literary and public value. But its precise purpose and function remained unclear and contested. Among the running debates was whether it should be for reference alone or should operate as a circulating library, the extent to which it should cater for working people as well as the middle class, and whether its remit included the stocking of novels and entertaining literature as well as serious instructional material. The leading liberal politician and attorney-general, William Porter, a great supporter of the Library, took pride in its being a socially inclusive colonial institution which could serve as a beacon of enlightenment. (We have found no records of black people making use of the facility). But there were other voices among its subscribers who favoured restricting it to the anglophone male elite.[69]

The South African Museum was established anew in 1855 with strong backing from Governor George Grey and local dignitaries like Charles Aken Fairbridge. Grey has been criticised for his role during and after the Xhosa Cattle-Killing (1856–7) when he used the opportunity to further incorporate the Xhosa chiefdoms into the Cape colonial order and reduce their people to colonial servants.[70] But an older liberal historiography recognised him as a leading patron of progressive institutions including schools, museums and libraries. It was on Grey's initiative that the German comparative philologist Wilhelm Bleek came to the Cape. Bleek was responsible for cataloguing Grey's personal library and went on to conduct pioneering and systematic research on Bushman languages, oral traditions and myths.[71]

68 Proceedings of the 25th Anniversary Meeting of the Subscribers to the Public Library, Cape Town, Saturday 6th May 1854 (Cape Town, 1854), 20.

69 Dubow, *A Commonwealth of Knowledge*, 48.

70 J. B. Peires, *The Dead Will Arise: Nongqawuse and the Great Xhosa Cattle-Killing of 1856–7* (Johannesburg: Ravan Press, 1989).

71 A. Bank, *Bushmen in a Victorian World: The Remarkable Story of the Bleek-Lloyd Collection of Bushman Folklore* (Cape Town: Double Storey, 2008).

Grey's official patronage helped to persuade more than 200 public figures to pledge an annual sum of one guinea to help with the museum's establishment. The first two trustees were the colonial secretary Rawson W. Rawson (replaced in 1860 by Fairbridge) and the colonial botanist Ludwig Pappe. The astronomer Thomas Maclear was elected as a further trustee.[72] Grey was personally responsible for recruiting the first curator of the South African Museum, Edgar Layard, son of a senior official in the East India Company and younger brother of the Assyrian archaeologist Austen Layard. Edgar Layard arrived at the Cape in 1854 to join Grey's civil service and was seconded to the museum on a part-time basis.

The Cape government also offered a grant to support the Albany Museum in Grahamstown, established in 1855 by a committee of doctors led by William Guybon Atherstone (1814–98). Atherstone was a polymath and a local celebrity with broad interests in medicine, geology, botany and astronomy. He arrived with his large family in 1820, was schooled in Uitenhage, and served as an assistant surgeon in the 1835 frontier war. After medical training in Europe, he returned to Grahamstown in 1839. Atherstone is known for administering the first anaesthetic in South Africa, during a leg amputation in 1847. In addition to the museum, he founded the Albany Hospital in 1858 and became its first medical officer. Geology became a major research interest of Atherstone's after he inspected Andrew Geddes Bain's fossil collection. Botany was another passion, and he helped to establish the Grahamstown Botanical Gardens, donating 100 specimens to start a herbarium. He also became involved in exchanging plants with overseas institutions and experimented with local medicinal plants, drawing on the deep knowledge of Xhosa informants to treat his own patients. In 1864, Atherstone was elected a fellow of the Geological Society of London. Three years later he identified a stone found at Hopetown as a diamond, thus sparking South Africa's mineral revolution.[73]

Atherstone's activities as a serial institution-builder included his help in founding the Grahamstown Public Library. In 1881 he was elected as a member of parliament representing Grahamstown. In the 1860s and

72 Summers, *History of the South African Museum*, 22–5.

73 W. G. Atherstone, entry by Alan Cohen in *Oxford Dictionary of National Biography*; Hammel, 'A Relational History of Mary Elizabeth Barber', 69.

1870s, when political tensions between the eastern and western parts of the Cape Colony were intense, 'Grahamstown men' like Atherstone were proud of their local achievements. The depth of civic pride associated with these public institutions reflected broader rivalries. Thus, the Albany Museum, whose immediate roots lay in the Grahamstown Medico-Chirurgical Society (of which Atherstone was a co-founder), strove to equal or exceed the South African Museum in importance. But the Cape Town-based South African Museum commanded far greater resources and was duly incorporated by legislation, based on the legal precedent applied to the Australian Museum in Sydney, which opened to the public in 1857.

Layard administered the South African Museum with energy and imagination. He had specialist interests in ornithology and shells and in 1867 he published *The Birds of South Africa*, at the time the most comprehensive distillation of the subject. But he was otherwise a generalist and amateur, reliant on ingenuity, hard work, and personal connections to build up collections. The mouldering remains of Andrew Smith's defunct museum provided a foundation, as did material objects deriving from Joachim von Dessin's collection. Layard's approach was populist, based on the view that the museum should offer something to everyone and so be a repository for all kinds of 'curiosities'. Animals, insects, geological specimens, coins and ethnographic materials were all put on display, as were figures said to be illustrative of the colony's 'four principal tribes' – Xhosa, Sotho, Khoekhoen and Bushmen.[74] One of the most significant acquisitions was a stuffed quagga foal, obtained around 1860, which remains the only example of this extinct species of zebra on display in Africa. Layard relied heavily on donations from private individuals, whose offerings were recorded in a regular column in the *Cape Monthly Magazine*, a mark of public commitment and recognition.

This distinctive local publication, established in 1857 on the model of British quarterlies, and edited for many years by the Scottish brothers Roderick and John Noble, published a rich variety of items of literary and scientific interest. The venture did much to stimulate the development of civic institutions and functioned as a key brokerage for the exchange of ideas. In 1860 the peripatetic South African Museum, which had been temporarily housed above a bookseller in St George's

[74] Summers, *History of the South African Museum*, 28.

Street, was moved to the wing of a new, well-designed, purpose-built building (modelled after the Fitzwilliam Museum in Cambridge) situated in a prime position at the bottom of the Botanical Gardens. The museum shared these quarters with the South African Public Library, and the dual-purpose building was opened with great ceremony by the young visiting British royal, Prince Alfred, in 1860. Four years later, a sculpture of Sir George Grey was erected by public subscription in front of the entrance.[75]

The combined South African Museum and Library building bookended the rectangular grid-space of the old Company Garden, which extended south-east along Government Avenue towards Table Mountain. The Garden, established by the VOC in 1652, was perhaps the oldest colonial institution with a continuous existence in the country. Over the course of Dutch and British rule, public buildings were sited around or within its precincts, cementing its position at the heart of civic life. An ornamental gateway was constructed in 1791 by the Dutch architect Louis Thibault; a zoo with gateways, designed by the German baroque sculptor Anton Anreith, was opened at the mountain end. By the beginning of the twentieth century, buildings around the Garden included the Old Supreme Court (formerly, the Slave Lodge), Parliament, an official government residence (the Tuynhuis), St George's Cathedral, the Great Synagogue, the Supreme Court (now the High Court), and the South African College (now a campus of the University of Cape Town).

The Garden had historically served as a site for cultivating indigenous species and imported plants. In 1848, after a period of neglect, prominent Cape citizens persuaded the government to proclaim part of the Garden as a botanical preserve and a site of scientific interest. Local critics such as 'Hortulanus' charged that the Garden was neither large enough to serve the interests of the country nor ambitious enough in its scope: its purpose should go beyond the receipt and acclimatisation of exotic imports, to include collecting and preserving indigenous botanical species such as the silver tree or *witteboom* on Table Mountain. These were 'fast disappearing from their native wilds'.

In 1856 a parliamentary committee concluded that the Botanical Garden, while attractive as a 'pleasurable resort', was a failure from a scientific point of view. It was vital to appoint a curator just as the

75 Summers, *History of the South African Museum*, 35–6.

museum had done.[76] In 1858 Ludwig Pappe, a German physician settled at the Cape, was appointed to the position of colonial botanist. He had been an enthusiastic collector of plants since his arrival in 1831 and collated his publications on indigenous plants and remedies in two editions of a book on *Florae Capensis Medicae Prodromus* (1850, 1857), which detailed the medicinal uses of local plants and drew distinctions between Boer and Khoekhoe usage.[77] Pappe was closely inserted into local scientific and professional networks and was well known as a public champion of a dedicated botanical garden as well as the preservation of indigenous trees and vegetation. He had strong connections to Kew, the centre of the British Empire's networks of botanical exchange and knowledge, to which he sent specimen plants. He also prepared a list of Cape trees and shrubs, *Silva Capensis* (1862), for a display of indigenous plants at the Paris Universal Exhibition of 1862 (though the exhibit was not included).[78] A friend of Baron Carl von Ludwig, a pharmacist and businessman who assembled a private garden at his home in Kloof Street, Pappe was able to stock the revived Botanic Garden with trees and shrubs from the deceased Ludwig's estate.

Pappe's scientific importance rests on his awareness of the extraordinary variety and uniqueness of the Cape's botanical kingdom, his appreciation of the need to conserve these resources, and his understanding of environmental change – in particular the dangers of degradation through the practice of veld burning. Unlike pioneer botanists who travelled through the Cape, collected specimens, and then returned home to Europe in order to publish their scientific findings, Pappe chose to build his reputation at the Cape. He successfully maintained local as well as international connections and helped, both at Ludwig's private garden and, subsequently, in the public sphere, to accord status and standing to Cape Town's new scientific centres of calculation. Like the museum and library, these gave scope to local collectors and enthusiasts, providing a repository for donations and a

[76] 'Hortulanus', 'A Cape Botanical Garden, as It Should Be', *Cape Monthly Magazine*, 2, 9 (1857); Report of the Select Committee on the Botanical Garden, House of Assembly, 1856, iii–iv.

[77] C. W. L. Pappe, *Florae Capensis Medicae Prodromus; or an Enumeration of South African Plants Used as Remedies by the Colonists of the Cape of Good Hope*, 2nd edition (Cape Town: W. Brittain, 1857).

[78] Beinart, *Rise of Conservation*, 74–5.

focus for talks, research and writing. Locally published journals like the *Cape Monthly Magazine* and its Grahamstown rival, the *Eastern Province Monthly Magazine*, became vital outlets for the exchange of scientific and literary ideas.

The field sciences were particularly well suited to the accumulation of local expertise by amateur scientists. Botany provides a good example because it was an enthusiasm that could be pursued both for aesthetic and scientific reasons. Notably it was a science that included women as well as men. In genteel Victorian society, flower arranging had traditionally been part of the female domain. The fashion for botanical painting drew in women artists and botanical 'picnics' provided occasions for family excursions. Following Anna Maria Truter (Barrow's wife) in the late eighteenth century, Margaret Herschel was an expert painter of Cape flora, as was Arabella Roupell, whose *Specimens of the Flora of South Africa, by a Lady* (London, 1849) was published to satisfy the exotic tastes of an exclusive group of English subscribers that included Queen Victoria. Katherine Saunders and Marianne North, both socially well-connected English botanical collectors and travellers, spent extended periods in South Africa in the second half of the nineteenth century.

With the establishment in 1854 of Representative Government in the Colony, parliament provided another vehicle for detailed inquiries and reports that fed into these scientific streams. For example, Andrew Wylie, first government geologist (1855–9), wrote in detail about the mineral resources of the Colony. Pappe and his successor, John Croumbie Brown, published annual reports on botanical matters. In 1860, a parliamentary select committee reported on *Xanthium spinosum*, spiny burrweed, a coloniser from South America that stuck in the wool of sheep, which was by then the Colony's most valuable export. It was the first plant to be declared a noxious weed. Committees and commissions sat to collect evidence on irrigation, drought and forestry. They drew on the knowledge of members of parliament and of farmers, as well as on early scientific experts where available.

Geology and Palaeontology

Geology, like botany, was a foundational science in South Africa, which drew freely on new approaches to the study of earth history in Britain especially. Foremost among the self-taught geologists at the Cape was Andrew Geddes Bain, who arrived from Scotland in 1816, settled in Graaff-Reinet, and embarked on an itinerant career as a saddler, hunter, ivory trader, explorer and road-builder. In the 1820s he travelled north to Tswana territories and east to Mpondoland, where he briefly met King Faku and encountered the trader Henry Fynn.[79] Bain kept journals of his journeys and wrote frequently for newspapers. He absorbed some of the eastern Cape prejudices, expressed in his dramatic comedy *Kaatje Kekkelbek; or Life among the Hottentots* (1838), which satirised the work of the emancipationist missionaries. It is often credited as an early example of creolised Afrikaans literature.

In Britain, William Smith made observations for his first geological map, and found fossils, while surveying and engineering canals.[80] In South Africa, Bain's introduction to geology came in the process of building roads and engineering extraordinary mountain passes. In 1837, while constructing the Queen's Road from Grahamstown to Fort Brown, Bain chanced on Charles Lyell's *Principles of Geology* (1830–3), whose uniformitarian account argued that the earth's existence went back much longer than the conventional 6,000 years since Creation. Lyell's ideas had an immediate influence on Bain. He pronounced himself 'smitten' and advanced his knowledge of the subject with further reading, including Gideon Mantell's *Wonders of Geology* (1838) which popularised the study of geology and fossil-hunting for an enthusiastic Victorian audience.

Together with William Borcherds (the civil commissioner of Fort Beaufort) and Atherstone, Bain hunted for fossils in the Karoo hinterland. His search was rewarded in 1838, when he chiselled a strange turtle-like skull with a beak and two canine tusks that corresponded to no known animal. He called it 'Bidental' and in 1844 sent it to the Geological Society of London. In 1845, the year in which he described his findings in the *Transactions of the Geological Society of London*, Bain's confidence was boosted by a grant from the Society. In England, the Bidental remains aroused the intense interest of T. H. Huxley and Richard Owen, who used the name *Dicynodont* (two dog teeth) to characterise the extraordinary mammal-like reptile (now known as a therapsid). In tribute, he named Bain's specimen *Oudenodon bainii* in 1860. Bain was also associated with the first fossil mammal to be

[79] M. H. Lister (ed.), *Journals of Andrew Geddes Bain* (Cape Town: Van Riebeeck Society, 1949).

[80] S. Winchester, *The Map That Changed the World: A Tale of Rocks, Ruin and Redemption* (London: Penguin, 2002).

collected in South Africa, called *Pelorovis bainii*. This buffalo-like creature was found around 1839 on the Modder River and sent to the South African Museum. Bain viewed the fossil remains, measuring and describing them in a letter to Andrew Smith. The information then went on to Charles Darwin, who reported it to the Geological Society of London.[81] These networks and associations created a critical mass of interest in Cape palaeontology.

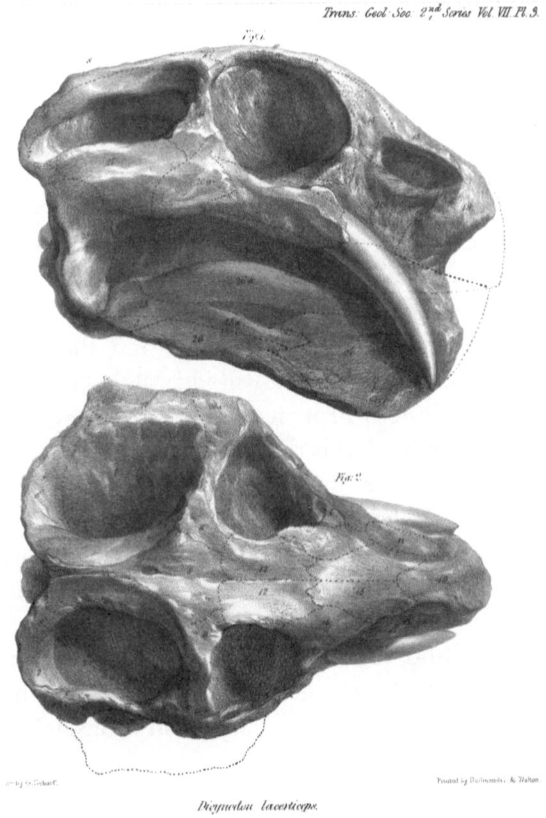

Figure 2.2 Side and upper views of the fossil skull of *Dicynodon lacerticeps*, a mammal-like reptile from South Africa, discovered by Andrew Geddes Bain and sent to Richard Owen. Drawing by George Scharf.
Credit: The Geological Society of London

[81] Summers, *History of the South African Museum*, 31.

Using his prestige and authority to control the collection and description of these dicynodont fossils from the Karoo, Owen also took receipt of a further collection sent to the British Museum by Governor Grey and reciprocated by naming one of the species *Oudendon greyii.*[82] Other South African collectors were not so well honoured. For example, Alfred 'Gogga' Brown sent a large collection of fossils from the Stormberg region to T. H. Huxley, who used the material without due acknowledgement or recompense.[83] Atherstone expressed disappointment in 1871 that the fossil bones unearthed by himself and Andrew Bain 'lie still in the vaults of the British Museum, unknown, unnamed, unexamined'.[84] Collectors on the far-flung periphery of the British Empire could just as easily be acknowledged – or taken for granted and ignored.

Bain was also responsible for producing the first comprehensive stratigraphic survey of South Africa, published in 1856, and acclaimed by the Geological Society of London.[85] He posited the remarkable idea that the interior bowl of the Cape, centred on the Karoo beds, had once been a vast freshwater lake, possibly double the size of all the Canadian lakes combined. Together with Atherstone and the Port Elizabeth doctor and palaeontologist, Richard Rubidge, Bain contributed articles on geology to the *Eastern Province Monthly Magazine* in a conspicuous act of identification with this region of the Cape Colony.

Geology was one of the first sciences in receipt of state support: the Cape Colony retained Andrew Wyley from 1854 to 1859 to conduct survey work on mineral deposits. P. C. Sutherland, who was medically trained and in 1850 sailed to Baffin Bay on an expedition to find survivors from Franklin's Arctic expedition, served Natal as its government geologist from 1854 and subsequently as surveyor-general. George Stow worked for a period as a geologist on behalf of the Orange Free State and Griqualand West administrations. The

82 Professor Owen, 'On Some Reptilian Fossils from South Africa', *Quarterly Journal of the Geological Society*, 16 (1860), 49–63.

83 M. R. Drennan, *Gogga Brown: The Life-Story of Alfred Brown, South African Hermit-Naturalist* (Cape Town: Maskew Miller, 1938).

84 W. G. Atherstone, 'From Graham's Town to the Gouph', *Cape Monthly Magazine*, April (1871), 307.

85 A. G. Bain, 'On the Geology of South Africa', *Transactions of the Geological Society of London*, series 2, 7 (1856), 175–92; A. G. Bain, 'Geology of South Africa', *Eastern Province Monthly Magazine*, 1 (Sept. 1856) and 1 (June and July 1857).

English-born Stow arrived in 1843 in the eastern Cape where he worked as a teacher, military officer, trader and farmer. In the midst of the frontier war of 1850–3, Stow discovered reptilian fossils that were described in the *Quarterly Journal of the Geological Society* (1858). He was elected a fellow of the Geological Society of London in 1872. Stow's 'Geological Notes upon Griqualand West', published in the *Quarterly Journal* (1874), were based on work commissioned by the governor of the Cape Henry Barkly. Stow is today best known for his descriptions of rock art, another of his enthusiasms.

Geology received a significant boost as a result of the discovery of diamonds and coal, and it was through commissions to search for such riches that self-taught geologists like Atherstone and Stow developed parallel interests in ethnology and palaeontology. The mineral revolution, diamonds and gold in particular, transformed South Africa's cultural, political and intellectual environment almost as profoundly as it affected its political economy.

Imagining the Deep Past: The Beginnings of Archaeology

Gender and racial boundaries restricted scientific training to white men in the first half of the nineteenth century. However, a limited number of white women in Britain were able to navigate patriarchal norms and establish themselves as scientists, often utilising family networks. Although medical training was barred to women, Mary Somerville forged a specialism in mathematics, largely self-taught, and was supported by her second husband, William Somerville, who had briefly worked at the Cape. Likewise, Caroline Herschel was able to become an innovative astronomer by working with her brother, William, father of John Herschel, who worked in close association with his wife, Margaret, a talented artist.

Mary Elizabeth Barber (1818–98), a member of the large Bowker family of eastern Cape British settlers, was one of the few who was able to pursue scientific interests systematically and to write about botany, entomology, zoology, ornithology, lepidoptery and archaeology. Initially she focused on botany, drawing and collecting plants and became an authority on the beautiful but foul-smelling *Stapelia* succulents. William Harvey, who worked at the Cape in the 1830s as treasurer-general, used the opportunity for botanical collection and published a short book in 1838, inviting collectors to correspond with him. This led to a sustained connection with Barber, who sent dried specimens, notes and illustrations. After his appointment to Trinity College Dublin in 1844, Harvey started a comprehensive herbarium and the publication of *Flora Capensis* (from 1859), drawing extensively on Barber's specimens. She also corresponded with Darwin and William Hooker at Kew. William Atherstone, a relative, interceded with the British scientists to ensure that Barber's achievements as a botanist and illustrator were recognised in spite of her gender.[86] That was not fully achieved, but she published in the prestigious journal of the Linnean Society in 1870 – the first South African to do so.

Barber was an inventive thinker and early adopter of Darwin's theory of natural selection. At a time when studies of animals were moving from the classificatory and anatomical to consider animal behaviour, she drew on Darwin to develop new insights into mating relations in birds. Hammel argues that Barber saw female birds as equally active in sexual selection and thus viewed mating as a 'space of gender equality'.[87] She implicitly questioned the way in which male ornithologists theorised bird behaviour by applying Victorian gender stereotypes. Barber also contributed to a range of other debates, from the environmental degradation caused by millions of sheep in the Cape to the mesmerising effects of snakes on their victims.[88]

On the question of race, however, Barber was anything but enlightened. Like many eastern Cape whites, she was firmly of the view that black people were intellectually inferior to white.[89] Although Barber learned about the medicinal and toxic properties of plants through direct or indirect association with Khoesan, Xhosa and Afrikaner medical practitioners and naturalists, she did little to credit their contribution to her own botanical knowledge.[90]

Mary Barber's brothers, known primarily for their military roles in wars against the Xhosa in the eastern Cape, were also associated with scientific advance. James Henry Bowker was a noted lepidopterist and plant collector. He and his sister fostered a close working relationship

86 Hammel, 'Shaping Science and Society', 109–10.
87 T. Hammel, 'Thinking with Birds: Mary Elizabeth Barber's Advocacy for Gender Equality in Ornithology', *Kronos*, 41, 1 (2015), 85–111.
88 A. Cohen, 'Mary Elizabeth Barber, the Bowkers, and South African Prehistory', *South African Archaeological Bulletin*, 54, 170 (1999), 120–7; Beinart, *Rise of Conservation*, 117.
89 Beinart, 'Men, Science'.
90 Hammel, 'Shaping Science and Society', 62–3.

with Ronald Trimen of the South African Museum, to which he donated a collection of butterflies in 1860.[91] Another collection of Bowker's butterflies and moths was presented to the Albany Museum in Grahamstown. Thomas Holden Bowker is credited with being the first to accumulate a systematic collection of stone artefacts in South Africa.

The big questions being raised by geology and fossil collection led to a recognition of stone tools as the product of ancient societies. In 1857–8 Thomas Bowker identified stone points as arrowheads crafted by humans, after being shown Danish equivalents by Edgar Layard. The similarities caused Bowker to believe that they must have been produced by white people. Mary Barber disagreed because, in line with her evolutionist views, she could not countenance the possibility that white inhabitants had somehow vanished from history. Samples from Bowker's collection were donated in 1866 to the Royal Artillery Museum in London, where they remained for a century until they were returned to South Africa for display in the Albany Museum.[92] Stone implements found by James Henry Bowker at Pniel near Kimberley, and in Lesotho, were sent to the British and the South African Museums.

In the 1860s the Cape educationist, Langham Dale, a frequent contributor to the *Cape Monthly Magazine* (he wrote as 'Delta'), found and recognised stone tools on the Cape Flats near his home at Mowbray. In an article published in 1870, Dale introduced his readers to the subject of archaeology, referring to it as a 'new science interposing itself between Geology and History'.[93] Dale was aware of related finds in Europe, Japan and Australia, and he followed with avid interest debates about the antiquity of palaeoliths arising from Boucher de Perthes's discoveries in the Somme Valley. Dale was also acquainted with the recent work of John Lubbock, originator of the terms 'palaeolithic' and 'neolithic', who was a strong supporter of

91 Hammel, 'Shaping Science and Society', 102–3.
92 H. J. and J. Deacon, *Human Beginnings in South Africa: Uncovering the Secrets of the Stone Age* (Cape Town: New Africa Books, 1999), 2–3; S2A3 Biographical Data Base: T. H. Bowker, www.s2a3.org.za/bio. Barber later changed her mind and came to believe that perforated stones may have been made by a lost tribe of Israel.
93 Delta, 'Stone Implements in South Africa', *Cape Monthly Magazine* (New Series, October 1870), 236.

Darwin's still controversial theory of evolution. No supporter of Darwin, Langham Dale's views carried weight at the Cape on account of his key position in the Cape educational establishment. He served as superintendent-general of education at the Cape from 1859 to 1892 and also played a major role in the creation in 1873 of the University of the Cape of Good Hope, which he went on to lead as vice-chancellor and chancellor.

In his essay on archaeology, Dale was clear that the 'rude' specimens of pottery, arrowheads and scrapers he inspected were fashioned by 'native races'. Yet, in his opinion there was as yet nothing to suggest that they were 'prehistoric'.[94] Stimulated by Dale's intervention, Alfred Brown, fossil collector in Aliwal North, drew attention to ancient mounds, stone implements, and 'innumerable paintings representing the history and habits' of cave dwellers, probably Bushmen, in the vicinity of the Stormberg. Some of the bones associated with these finds were 'in a sub-fossil condition', but Brown could not be sure that the stone tools were of the same age as the 'mammalian remains' found with them.[95] In 1871 Atherstone, now deeply engaged in geology and palaeontology, reported on caves in Mossel Bay where significant accumulations of shells were to be found. The same year, Henry Piers of the South African Museum found shell mounds in caves far above sea level at sites around the False Bay side of the Cape Peninsula. In the columns of the *Cape Monthly* there was vigorous debate as to what these discoveries meant. Might they be the result of geological upheaval rather than human habitation? Were birds or inanimate forces such as the action of the sea responsible for such deposits?

Piers, a frequent contributor on the geology of the Cape Peninsula, thought it 'inconceivable' that shell middens might be the work of ancient human hands, notwithstanding evidence of entombed human skeletons in caves to the north.[96] A much bolder explanation was offered by Dr Peter Comrie, naval surgeon and naturalist, who visited the Cape en route to New Zealand on the naval ship *Basilisk*. Comrie, after examining arrowheads, pottery, perforated stones and human remains at Cape Point, dismissed various theories including the notion

94 Delta, 'Stone Implements in South Africa', 237.

95 Note from 'AB' in *Cape Monthly Magazine*, December (1870), 366–7, appended to an article by Delta, 'Stone Implements in South Africa', 365–6.

96 H. W. Piers, 'On the Geology of the Cape Peninsula', *Cape Monthly Magazine*, September (1871), 172.

that shells found in elevated caves might have been brought there by birds. He was also impressed by samples of pottery, reticulated on the inside, which were shown to him by P. D. Martin of Simonstown. In an article entitled 'The Stone-Age at the Cape of Good Hope', Comrie declared in 1874 that these finds were evidence of prehistoric peoples, indicating that 'our present civilization is the result of simple and rude beginnings. In the language of the first writer on this subject, De Perthes, "Let us not, then, disdain these few essays of our forefathers."'[97]

Meanwhile, research undertaken by the linguist Wilhelm Bleek was revealing the 'astonishing' traditional literature – or 'folklore' – that might have been 'first told in those so-called pre-historic times when the art of smelting any metal was still unknown, and when stone implements were the only ones in use'. It was as yet too early to speculate on their origins, but he thought it was clear that tales with similar structures reappeared in the 'remotest parts of the earth' and 'there is clear evidence that a literature of fables existed among the Hottentots before the advent of Europeans'.[98] In 1874, Bleek commented enthusiastically on newly described and reproduced Bushman paintings from the Maloti Mountains, sent to the *Cape Monthly Magazine* by Joseph Millard Orpen, a leading colonial official and expert on Basutoland. In Bleek's view, the rock paintings cast 'a good deal of new light upon the subject of Bushman mythology, and upon the vastness of the legendary lore of this highly curious people'.[99] Bleek's pronouncement was amply borne out, for Orpen's article relating his encounters with a Bushman guide named Qing, remains one of the foundations of contemporary understandings of Bushman mythology and rock art.[100]

In the view of William Bisset Berry, a physician based in Queenstown who later became a prominent politician and speaker of the House of Assembly, evolutionary theory was soundly based. Berry

97 Dr Comrie, 'The Stone-Age at the Cape of Good Hope', *Cape Monthly Magazine*, November (1874), 290.

98 W. H. I. Bleek, 'African Folk Lore', *Cape Monthly Magazine*, September (1870), 168–9.

99 J. M. Orpen, 'A Glimpse into the Mythology of the Maluti Bushmen', *Cape Monthly Magazine*, July (1874), 1–10 with 'Remarks by Dr. Bleek', 10–13.

100 R. King, '"A Loyal Liking for Fair Play": Joseph Millerd Orpen and Knowledge Production in the Cape Colony', *South African Historical Journal*, 67, 4 (2015), 410–32.

took issue publicly with Langham Dale in the columns of the *Cape Monthly*, but his painstaking rebuttal of Dale's position was a minority one. Just a few years before, in 1867, Berry's support of evolution was sufficiently controversial to make him temporarily leave Queenstown to take up a medical practice in the small town of Burgersdorp. Langham Dale's circumlocutory prevarications on the matter of evolution, as evidenced in his 1874 public address at the South African Public Library, were evidently more in keeping with the mood of the anglophone intelligentsia.[101]

Darwinian ideas were nevertheless gaining currency, and debates in the *Cape Monthly* were conducted in a respectful spirit of positive, if prudent, inquiry. It was while on a government-sponsored expedition in 1870 to verify reports of gold found in a remote region of the Karoo between the Swartberg and Nieuweberg mountains that Atherstone declared the real 'nuggets' were not bullion but extraordinary rock formations and evidence of Silurian fossils. He wrote a facetious travelogue in which he joked about the lazy, ignorant farmers he encountered and wondered, in a gesture to degenerationist theories, what the Darwinians would say 'of this undoubted development of Boer into Hottentot by the mere solarisation of a dozen summers'.[102] This parodic treatment did not require Atherstone to declare his own view on the scientific veracity of natural selection, but it did indicate that such ideas were in the air and that there was a public eager to find out more.

Such responses – anxiety, willing acceptance, or jokes about primitive man and hints of polygenesis in the case of Atherstone – are indicative of the different ways in which human evolution was considered by Cape-based intellectuals in the 1870s, less than two decades since the first mention of shell middens was raised by way of a 'query' in the *Cape Monthly Magazine* by 'Barnacle'. Darwin's *Descent of Man* (1871) rather than the *Origin of Species* (1859) provided a further challenge. Langham Dale's intellectual and social conservatism led him to address, ponderously, the question of human origins and

[101] D. N. Livingstone, *Dealing with Darwin: Place, Politics, and Rhetoric in Religious Engagements with Evolution* (Baltimore: Johns Hopkins, 2014), 16–18; Dubow, *A Commonwealth of Knowledge*, 96–9.

[102] W. G. Atherstone, 'From Graham's Town to the Gouph', chapter III, *Cape Monthly Magazine*, July (1871), 50.

evolution as a compelling but unproven hypothesis. As well as scriptural objections to the theory of evolution, which were shared by investigators in Europe, Darwinian ideas raised particular problems in the context of the colonial Cape where conceptions of deep time and history brought to the fore alluring but troubling questions about sequences of human habitation and the increasingly vexing issue of racial origins and hierarchies.[103] Quite aside from theological issues, the implications of revealing that the country might have been inhabited for millennia by people who were not from Europe were profoundly disturbing. Yet by the 1870s, Comrie, Bleek, Orpen, Stow, and others were beginning to assemble a deepening understanding and imagination of just such societies, with stone tools, middens and rock art building a picture made of an ever-denser array of pixels.

New Connections

As is clear from the first two chapters, Cape society was not isolated from Europe; ideas and networks were carried by sea in both directions over 10,000 km. Shipping routes were the sinews of the maritime empires and until 1869, when the Suez Canal opened, every vessel going from Europe to the Indian Ocean, Asia or Australasia passed, and generally stopped at, a South African port. A mail steamship became a regular feature of the passage to England from the early 1850s; in 1857 the Union line won the mail contract and also started services between Cape Town and Durban. A sailing ship took around three months to reach the Cape in 1800; Thomas Pringle, 1820 settler, was at sea for about seventy-four days. By 1876, when the Castle line joined the Union as a mail carrier, steamships cut this to thirty days. Within the Colony a twice-weekly postal service, transported by horse, linked the main centres. The arrival of ships in Table Bay was closely watched and anticipated.

The first railway out of Cape Town reached Wellington in 1863, and in 1879 lines arrived at Graaff-Reinet and Grahamstown. Internal telegraphic communication was slow to take off. It was possible to

[103] Unsigned review of Darwin's 'The Descent of Man', *Cape Monthly Magazine*, June (1871), 321–30, which treats the 'highly speculative' theory of evolution with noncommittal respect; also Langham Dale 'Anthropology: A Review of Modern Theories', *Cape Monthly Magazine*, May (1874), 350–61.

send messages between Cape Town and Simonstown in 1860, and Grahamstown in 1863, but telegraphs remained a novelty. The utility of the medium was proved in 1878 when the colonial governments were warned by telegraph of the imminent war against the Zulu.[104] Over the course of this great conflict, military requirements and public hunger for information expressed itself in a need for more rapid links with Britain. This was partly achieved with the extension of an unreliable telegraph link to London, via Aden, at the end of 1879. A direct link with London was only securely established by the time of the South African War.

The increase in transoceanic maritime traffic required a major improvement in the capacity of the Cape Town harbour; work on a new breakwater and dock began in 1860. In the same year the acting colonial secretary, Richard Southey, an eastern Cape farmer and government administrator with a range of scientific interests, became inaugural chair of the Cape government's Meteorological Commission, effectively the first weather service in southern Africa and one of the oldest in the world.[105] From a small, isolated rural colony at the turn of the nineteenth century, with few institutions other than those directly linked to the government and the maintenance of order, the Cape was now a distinctly more connected society. The first parliament in 1854 and the grant of Responsible Government in 1872, with an elected Prime Minister, helped to expand the range of scientific initiatives sponsored by the government.

The scientific and literary societies that took root from the late 1820s did much to encourage intellectual curiosity and activity. By expanding the public sphere, institutions became important loci of colonial knowledge and self-awareness. Together with the battle to publish without censorship and to express civil and political autonomy, associations and societies – including journals, libraries, museums and botanical gardens – gave expression to an assertive sense of colonial pride and respectability. Many of the politicians who led the movement for self-government – John Fairbairn, William Porter, Saul Solomon among

[104] A. F. Hattersley, *An Illustrated Social History of South Africa* (Cape Town: A. A. Balkema, 1969), 208–10.

[105] P. R. de Kock and H. D. Moraal (eds.), 'Physics in South Africa', unpublished manuscript (2011), 417.

them – were intimately connected with the development of the particular brand of Cape liberalism that expressed itself through intellectual, scientific, and cultural as well as economic pursuits.

While the western Cape remained the fulcrum of institutional development and scientific knowledge, rival developments were evident in the eastern Cape, supported by a rapidly growing agrarian economy, and also in the neighbouring colony of Natal. Over this period of seventy years, independent African kingdoms such as the Zulu innovated in spheres of indigenous knowledge and incorporated colonial technology in attempts to centralise political power. While they were increasingly drawn into the colonial web, they remained significant alternative nodes of authority and power.

By this time, substantial scientific activity was evident in fields such as astronomy, botany, geology and medicine. Awareness of South Africa's deep history and rich ethnography was stimulated by growing interest in archaeology, linguistics and rock art. Theories of evolution helped to frame questions of deep time, patterns of human habitation, and progress. As well as the rapid assimilation of books and journals arriving from overseas, newspapers and journals like the *Cape Monthly Magazine* and the *Eastern Province Monthly Magazine* served as vital conduits for discussion and debate. Prior to the establishment of universities, these were the networks that built on an older, traveller-based scientific curiosity, coalescing in a new, locally based, largely anglophone scientific imagination. The institutions of this time, such as the museums, libraries and gardens, survive to the present day as national assets.

As yet, scientific specialisation and professionalisation were limited: there was no teaching university in existence, nor any system of state-funded research. But a wide range of research was being generated by accomplished amateurs with close links to government institutions who regarded the Cape Colony as their permanent place of residence. The establishment of a parliamentary system began to give scope to detailed inquiries on matters of applied science through government commissions. Although no party system had yet been established in parliament, political differences corresponding to humanitarians and anti-humanitarians, Dutch- and English-speakers, colonial nationalists and imperialists were taking clear form. Aside from interest in phrenology, there were as yet few sustained intimations of the brand of social Darwinism that came to express itself, from the 1870s, in efforts to classify and rank different races in terms of racial superiority and inferiority, and to impute corresponding political rights. Such ideas were about to coagulate as processes of colonial conquest and industrialisation gathered momentum. In such discussions and debates, intellectual ideas played no small part.

3 *Technological Innovation and the Scientific Imagination in Mining and Agriculture, 1870–1902*

Between the 1870s and 1890s, the remaining independent African kingdoms and communities were conquered and incorporated into the four settler states of the Cape, Natal, the Orange Free State and South African Republic (Transvaal). Britain went to war with the Boer republics in 1899 and by 1902 had asserted its authority over the whole of the area that was shortly to become South Africa. In 1910 the four regional colonial governments, guided by the imperial power, came together to establish a consolidated Union of South Africa. This was led by two generals who had spent the war fighting the British, Louis Botha and Jan Smuts. In this chapter, we will focus largely on the period from 1870 to 1902. Even by the standards of South Africa's violent experience of colonisation, these were decades of intense conflict and imperial assertion.

In the late nineteenth century South Africa also experienced breakneck economic growth. Agriculture was in many respects transformed in scale by innovation. In the Cape especially, new livestock breeds underpinned a flourishing commercial pastoral economy that brought in over £3 million a year in exports by the early 1870s, as well as a similar value in locally supplied meat and draught.[1] Much of the transport system still depended on ox-wagons. One of the very few successes, in a global context, in expanding the range of farmed animal species was the semi-domestication of ostriches for feathers. Plantation production of sugar and timber took root in Natal, assisted in no small measure by indentured Indian labour, at much the same time as the Zulu state was finally conquered and dismembered. Maize, already well established on African smallholdings and settler farms, became the country's staple. Although they were a minority of landowners, a

[1] Beinart, *Rise of Conservation*, 14. The figures for meat and draught are very rough estimates and are not included in secondary sources. The point is to emphasise the neglected contribution of livestock production to the economy.

significant number of self-consciously 'progressive' farmers, together with a sprinkling of agricultural experts employed by governments, increasingly adopted and advocated scientific prescriptions in their husbandry. To some degree these innovations bridged racial divisions.

In these decades, too, large-scale mining enterprises were established in the interior which shaped the economy and the society for the next century. By the late nineteenth century, Kimberley was exporting about £4 million worth of diamonds a year and the Witwatersrand about £15 million worth of gold.[2] After a lapse during the South African War, gold exports revived to £16 million in 1904 and £34 million in 1910. Diamond exports remained at £4–5 million while pastoral exports from the Cape alone were about £4 million at the end of the century and £7 million in 1910. From a region that had barely exported £1 million a year in 1860, South Africa was by the end of the century exporting about £23 million in minerals and agricultural produce, and this doubled by 1910. The country was now a significant destination for foreign investment and a focus of imperial intervention.

The four most valuable export commodities in 1910 – gold, wool, diamonds and ostrich feathers – were mainly produced for the international market: South Africa was in this sense fundamentally a colonial economy. However, the scale of domestic food production and manufacturing had also suddenly become more significant. Regional agriculture largely supported consumption in towns, which grew tenfold from perhaps 150,000 people in 1875 to 1.2 million in 1904 and 1.5 million in 1911.[3] Colonisation also led to increased consumption by blacks and whites in the rural areas.

This does not necessarily imply that standards of living rose for all; many lives were shattered by war, displacement and dispossession. For example, an estimated 8,000–10,000 Zulu people died in the British invasion of Zululand during 1878–9, which was followed by civil war in the eviscerated Zulu domain. Lives and lands were lost in many

[2] Cape Statistical Registers for Diamonds. Figures for gold from Jean Jacques van Helten, 'Empire and High Finance: South Africa and the International Gold Standard 1890–1914', *Journal of African History*, 23 (1982), 544. These figures are surprisingly difficult to find in the secondary literature and this represents the values imported into Britain from South Africa, which was overwhelmingly the most important market. Very little gold was processed or kept in South Africa until the mint was established in 1919.

[3] Figure for 1875 from Cape census with estimate for other colonies; 1904 and 1911 figures from censuses.

other colonial wars against African kingdoms from the Pedi and Ndzundza Ndebele in the Transvaal, to the Xhosa, Mpondomise and Sotho in the Cape. Perhaps 6,000 Afrikaner soldiers and 28,000 women and children as well as 20,000 Africans died in the South African War (1899–1902). British forces applied scorched earth policies in both the Zulu and Boer conflicts. These figures represented about 5 per cent of the population of independent Zululand and 10 per cent of the white population of the Boer republics. In addition, 22,000 British and colonial troops died from war and disease in the South African War, making a total, at a rough estimate, of at least 75,000 people.

In earlier generations, South Africa – its boundaries as yet not clearly defined – had been an important site for the expansion of European scientific horizons in fields such as botany, zoology, geology and astronomy. Local specialists and collectors contributed to the fund of European-based knowledge largely through sending materials, and some ideas, back to the major museums, gardens and universities. Some local institutions found their feet from the mid-nineteenth century; specialist volumes such as *The Birds of South Africa* (1867), by E. L. Layard, were published by enthusiasts and semi-professionals (Chapter 2). By the late nineteenth century, the momentum of local scientific interest quickened, spurred by government funding and industrial requirements. Scientists and technical practitioners based in South Africa developed innovative ideas and technologies focusing on specific local problems. Veterinary scourges and deep-level mining were perhaps the two most striking spheres – both directly related to some of the most important sectors of the South African economy. While the scientific metropolis remained in Europe, South Africa became a significant branch of the rapidly growing and interconnected global scientific world. Science in the colonies was increasingly part of a 'polycentric communications network', with multidirectional information flows.[4] The United States and Australia, as much as Europe, were important to South Africa for new thinking and experimentation in spheres from chemistry to parasitology and entomology. In South

[4] MacLeod, 'On Visiting the Moving Metropolis'; D. Wade-Chambers and R. Gillespie, 'Locality in the History of Colonial Science', *Osiris* 15 (2000), 221–40.

Africa professional scientific associations were established, providing a vehicle for ideas among local networks of innovators and specialists.

Crucially, the diversification of scientific endeavour received political backing. Historians addressing the late nineteenth-century Cape have been attracted to the Cape white liberals, their African allies, and the fading influence of both constituencies as mining interests, personified by Cecil Rhodes, became more dominant. Liberals tried to defend a more inclusive political system, subject to a qualified franchise; opportunities for black people in education and on the land were important dimensions of this process. There has, however, been less focus on the political influence of what may be broadly called 'progressivism', a term that intersected with but did not necessarily imply liberality, let alone assimilation. Anglophone interests, such as English-speaking farmers, urban merchants and capitalists mobilised themselves politically by the late nineteenth century into a Progressive Party. In his period as Cape prime minister, Rhodes allied himself with the Afrikaner Bond until his support of the Jameson Raid violently sundered such cooperation. Both parties – the Progressives and the Bond – shared an interest in restricting the expansion of the African franchise as demands for cheap labour trumped support for an improving, loyal African peasantry. The alliance of diamonds and wool in the late-nineteenth-century Cape therefore preceded the better-known alliance of 'gold and maize', which shaped the reconstruction of the country after the South African War.[5] The Progressive Party, which came to power in the Cape from 1900, independent of the Bond, was pro-imperial and antagonistic to the more racially inclusive constitutional civic liberalism of 'colonial' leaders like John X. Merriman and W. P. Schreiner.

Progressives inherited some of the strident and racialised discourses that had characterised eastern Cape settler interests since the 1840s. Anglophone progressives looked outwards, especially to similar movements in the United States which reached their apogee under Theodore Roosevelt's presidency. At the heart of their thinking was a broad-ranging commitment to economic growth, investment, improvement,

[5] S. Trapido, 'South Africa in a Comparative Study of Industrialization', *Journal of Development Studies*, 7, 3 (1971), 309–20; S. Marks and S. Trapido, 'Lord Milner and the South African State', *History Workshop Journal*, 8, 1 (1979), 50–81.

and the 'gospel of efficiency'.[6] Increasingly, they saw a central role for science and the state.[7] In the late nineteenth century, government departments of Health, Agriculture and Mines, all employing scientists and technical experts, took shape. Although cultural divisions within white society made the Afrikaner Bond reluctant to unite in a common political front, the educated leaders of the Bond shared many of these aims. Educated, Christian Africans also saw themselves self-consciously as progressive, though they did not share the hard-edged political ideas of anglophone white progressives and identified instead with the civilizational ideals associated with missionaries and liberals.

Technologies of Conquest: Firearms and the Anglo-Zulu War

'Historians of colonialism in nineteenth-century South Africa', William Storey argues, 'have produced important works of social, political, and cultural history without much reference to technology.'[8] His book on firearms aims to fill this gap. The conquest of South Africa, as in the case of many of Europe's colonies, would not have been possible without horses and firearms, yet Storey is rightly cautious in assigning overwhelming weight to technology. One significant complicating factor is that from about the 1820s to the 1870s many African societies managed to secure and breed horses, and to acquire a limited supply of firearms. This briefly rebalanced the asymmetries of power and altered the dynamics of conquest.

Firearms were only one element in the command of resources that enabled Dutch and British colonial powers, supported by settlers and sometimes by African allies, to assert military superiority. By the end of the nineteenth century, the capacity to mount professional standing armies of unparalleled scale in Africa facilitated British conquest of the most powerful African states and the Boer republics. Access to science and rapidly improving technologies kept British forces a step ahead

[6] S. P. Hays, *Conservation and the Gospel of Efficiency: The Progressive Conservation Movement, 1890–1920* (Cambridge MA: Harvard University Press, 1959).

[7] K. Brown, 'Agriculture in the Natural World: Progressivism, Conservation and the State. The Case of the Cape Colony in the late 19th and early 20th Centuries', *Kronos*, 29 (2003), 109–38.

[8] W. K. Storey, *Guns, Race and Power in Colonial South Africa* (Cambridge: Cambridge University Press, 2008), 5–6.

throughout the critical period of conquest from the 1860s to 1902. Although improved munitions, designed by European and American inventors, did not have colonisation as their primary purpose, breech-loader rifles, artillery and machine guns opened an arms gap that made it very difficult for Africans to defend themselves effectively. British financial, bureaucratic and organisational capacity as well as technological power ensured that their armies incorporated such weapons rapidly and that tactics were adapted to use them.

Given the centrality of military power in the battle for dominance in nineteenth-century Europe, from Napoleon at the beginning of the century to competing nationalist forces in the Franco-Prussian War in 1870, the application of science to firearms was surprisingly slow. Advances came incrementally from individual experimentation rather than from army or corporate research. In the 1830s, the British army still used muzzle-loading muskets with flintlocks – versions of the Brown Bess that had served for over a century. Muzzle-loaders gradually became more reliable and by the 1850s Enfield rifles, made at the Royal Small Arms Factory in London, were adopted by the British army. Their rifled bores spun the bullets and so gave greater range and accuracy. Improved muzzle-loading rifles were used in the Crimean War, the American Civil War as well as in many colonial conflicts in the 1850s and 1860s.

A central problem of muzzle-loaders was that gunpowder and bullet had to be loaded separately, with the projectile pushed down a long barrel. This slowed the process of reloading. Breech-loaders were a multinational innovation that decisively shifted the performance of rifles in the 1860s. A British army officer experimented with manufactured brass cartridges that contained both powder and bullet; a Swiss inventor, Friedrich von Martini, developed the breech-loading device; and a gunsmith in Edinburgh, Alexander Henry, improved rifling in the barrel. In turn, these weapons required industrial advances: mass-produced, strengthened steel for the breech so that it did not explode or backfire; a new type of gunpowder; and standardised, manufactured bullets of the correct calibre.[9] They could be loaded more quickly with a lever next to the butt that also extracted the spent cartridge. Smaller bullets with a flatter flight enabled greater accuracy and a range of about 800 yards (730 metres). With smokeless powder, the shooter

[9] Storey, *Guns, Race and Power*, 113.

could remain concealed while the lever action allowed easier reloading when prone or on horseback. The British army upgraded quickly to adopt these more expensive Martini-Henry rifles in 1866 and fought the major and decisive southern African campaigns of the 1870s with improved versions. The bullet tip was composed of hardened lead that flattened on impact 'causing massive tissue damage and splintering bones lengthways, stopping its victim in his tracks'.[10]

Around 1860, many of the major African kingdoms in southern Africa – the Zulu, Mpondo, Thembu, Sotho, Tswana, Pedi and Swazi – remained relatively independent even though constrained and hemmed in by ambitious neighbours and expanding European settlement. Xhosa defensive strength had largely been broken after the frontier war of 1850–3 and the Cattle-Killing of 1856–7.[11] By the 1860s all the chiefdoms had to varying degrees acquired horses and firearms – part of a global spread of such technologies.[12] The Sotho Kingdom in particular developed mounted militias on a large scale.[13] In 1843, missionaries thought that Moshoeshoe's army had more guns and horses than any other African force in South Africa. This helped to keep British and Boer forces in the Free State at bay, but only till the late 1860s. By then Sotho muskets were outdated and the Boers were rapidly improving their weapons.[14] By 1868 the Sotho had to forfeit valuable land in the Caledon River valley and accept British, and then Cape colonial, rule.

As the threat to the African chiefdoms from both colonial and Boer states increased, the arms trade escalated. Storey estimates that between 1857 and 1881, over 300,000 firearms were imported through the Cape and Natal and many others through Delagoa Bay. By the 1870s half of all Cape imports, perhaps 150,000 weapons, went to Kimberley, largely for sale to Africans working on the diamond

[10] J. Laband, *Zulu Warriors: The Battle for the South African Frontier* (New Haven, CT: Yale University Press, 2014), 125.

[11] Peires, *The Dead Will Arise.*

[12] S. Swart, *Riding High: Horses, Humans and History in South Africa* (Johannesburg: Wits University Press, 2010); P. Mitchell, *Horse Nations: The Worldwide Impact of the Horse on Indigenous Societies Post-1492* (Oxford: Oxford University Press, 2015).

[13] A. Atmore and P. Sanders, 'Sotho Arms and Ammunition in the Nineteenth Century', *Journal of African History*, 12, 4 (1971), 535–44.

[14] Atmore and Sanders, 'Sotho Arms', 540,

fields.[15] Over the four years from 1873 to 1876, Pedi migrants constituted an average of 58 per cent of black mine workers at Kimberley, not least because of the attraction of guns as they sought to defend their lands against the Transvaal Boers. At this time the Zulu Kingdom was the most populous unconquered African state in the subcontinent. Cetshwayo, Zulu king from 1873, ruled over roughly 300,000 people and encouraged the purchase of arms through Delagoa Bay via middlemen.[16] A British consul in Delagoa Bay estimated that between 1875 and 1877, 20,000 muskets and 500 breech-loaders passed through the port annually and that three-quarters of these were destined for Zululand; another estimate suggested 10,000 were imported to Zululand in 1877.[17]

One reason that so many firearms poured into the African kingdoms is that European armies were selling off obsolete muzzle-loaders. These were available in South Africa at £2 to £4 each and in Zululand, weapons that had cost four head of cattle a decade before could now be traded for a cow. Breech-loaders cost about £25 and required the added expense of cartridges, which were available in only a few major centres such as Kimberley. Muzzle-loaders used the old black gunpowder that could be widely obtained in rural trading stores and they could fire a range of projectiles. Unlike breech-loaders, they were rough and ready guns which could be repaired by local craftsmen. In addition to warfare, firearms were used for hunting and for protection from local threats, human as well as animal. Many African purchasers did not have warfare against Europeans as their only concern. At the end of the 1870s, Storey calculates that a majority of adult men in South Africa, black as well as white, possessed firearms.

The Zulu victory at Isandlwana against an invading British force in January 1879, and subsequent engagements during the Anglo-Zulu War, have attracted enormous attention from popular historians. About 900 British troops and 400 African auxiliaries died at Isandlwana – heralded as the most significant British defeat in a

[15] W. Worger, *South Africa's City of Diamonds* (New Haven, CT: Yale University Press, 1987), 74; R. V. Turrell, *Capital and Labour on the Kimberley Diamond Fields, 1871–1890* (Cambridge: Cambridge University Press, 1987), 249.

[16] J. Guy, 'A Note on Firearms in the Zulu Kingdom with Special Reference to the Anglo-Zulu War, 1879', *Journal of African History*, 12, 4 (1971), 560.

[17] Knight, *The Anatomy of the Zulu Army*, 168. Guy believes that these numbers are exaggerated.

colonial battle in the nineteenth century. By comparison, in three years of the bloody Crimean war, Britain lost 4,500 men in combat – although perhaps 17,500 died of disease. It is useful to assess the balance of military technology in the light of such figures. Although Cetshwayo did not wish to engage British forces, he made some preparations during the tensions leading up to the outbreak of war.[18] Sotho advisors and gunsmiths were employed to assist with training and weaponry. Some Zulu combatants went into battle with firearms – muzzle-loaders with limited range – as well as spears and shields.[19] But 'in all the major clashes with the British', Guy argues, 'the Zulu based their attacks on conventional tactics'.[20] They used the well-entrenched cattle horn formation, in which smaller contingents would spread out and try to encircle the enemy, while the central core of warriors staged a mass attack at a vulnerable point.

Zulu soldiers did fire muskets at Isandlwana, but only once, at a distance, as they attacked. This had little effect and it was too time-consuming to reload.[21] Some threw their assegais when they came closer, and this made an impact as the British troops were without shields. But they depended very largely on reaching and breaching British defences in close-quarter combat where they could use their spears. Even at Isandlwana, where they won the battle, the Zulu army probably lost as many men as the British on the battlefield; others, who were injured, died on their way home. The victory had depended on surprise, on the inexperience of the British contingent in dealing with this frontal attack, and on the extraordinary persistence of Zulu soldiers in the face of concentrated rifle fire. Their firearms were of limited value.

Firearms did prove of some effect when they were deployed by the Zulu in a different way. Around 1,000 British breech-loaders, as well as ammunition, were captured at Isandlwana, some of which were used by the Zulu in the subsequent engagement at Rorke's Drift. Most of the British casualties at Rorke's Drift were the result of sniper fire, which was also responsible for the death of the French 'Prince Imperial', the only son of Napoleon III, whose newsworthy death sent a further frisson throughout Europe. Devolved skirmishes along British

[18] Guy, 'A Note on Firearms'; J. Guy, *The Destruction of the Zulu Kingdom* (Johannesburg: Ravan Press, 1979).
[19] Knight, *Anatomy*. [20] Guy, 'A Note on Firearms'.
[21] Knight, *Anatomy*, 213–14.

supply routes caused some problems and commentators noted the skill with which small Zulu groups used rocks and long grass as cover – only to be revealed by the puffs of gunpowder smoke from their muzzle-loaders. In this respect, Zulu contingents developed tactics which resembled those deployed by the Xhosa in earlier decades on the eastern Cape frontier. Zulu snipers with breech-loaders also seem to have made some impact at Khambula in March 1879, when seventy British soldiers died, and at Ulundi in July, where the British army lost seventy-nine men.[22] But the Zulu 'were merely an armed people, not a standing army'.[23] The regiments did not stay in the field, nor was there much central control. They lost all these battles.

Zulu commanders only partly transformed their approach in this six-month war. At Rorke's Drift, Zulu impis again charged in a mass attack. British troops were better prepared and opened fire with devastating results when the advancing Zulu were 400–500 yards away.[24] At Khambula the Zulu army staged a further mass attack, with more warriors than at Isandlwana. Here too British troops maintained a more concentrated formation and the breech-loaders enabled them to reload quickly. Altogether 785 Zulu bodies were counted on the battlefield and it is possible that more than 2,000 died. This loss effectively ended the war.[25] Subsequently, scorched earth policies, burning Zulu homesteads, as later in the South African War, ensured that there was little sustained guerrilla resistance. There were perhaps 8–10,000 Zulu casualties, maybe more, in the invasion. Unlike the South African War, British forces did not provide medical facilities for Zulu soldiers.

In the history of colonialism Maxim guns have been emblematic of overwhelming military power and technology. In the last phases of the war against the Zulu, the British introduced its forerunner, the Gatling gun, a hand-cranked repeater rifle which could fire 150 bullets per minute, but it was not used to significant effect in that conflict. South Africa had largely been conquered by the time that Hiram Maxim, a Scottish American, fully mechanised loading so that this machine gun could fire about 600 rounds per minute. Patented in the United States in 1884, his gun was adopted and improved by the British army from

22 Guy, 'A Note on Firearms', 562–3. 23 Guy, 'A Note on Firearms', 564.

24 I. Knight, *Zulu Rising: The Epic Story of Isandlwana and Rorke's Drift* (London: Pan Macmillan, 2011).

25 Guy, 'A Note on Firearms', 561

1889 and used in battle by Rhodes's British South Africa Company against the Ndebele in 1893. Even though they possessed some breech-loaders, the Ndebele generals also deployed a mass charge and were outgunned: 'the relatively small force fielded by the Chartered Company owed its success to the steady pounding fire from its Maxim guns, which produced a profound moral, as well as a devastating physical, effect'.[26] Perhaps 2,000 Ndebele were killed in battle and overall losses were probably around 4,000. Rhodes is reputed to have ordered a demonstration of the Maxim gun, mowing down a field of maize, in front of chief Sigcau at the time of the annexation of Mpondoland to the Cape in 1894. Whether this incident in fact took place is unclear, but the story is symbolic of the underlying character of colonial power.[27] Sigcau, weakened by three years of civil war, and resigned to the inevitable, did not try to fight the Cape government with weapons.

During the 1870s, key British and Cape political actors advocated control of the proliferating arms trade in southern Africa. They believed that regional political hegemony, at a time when British politicians were seeking confederation in the subcontinent, would be facilitated by a monopoly, rather than merely a superiority, of arms. Gordon Sprigg succeeded in implementing legislation when he became Cape prime minister in 1878. When the Cape attempted to disarm the Sotho, who had already been annexed and brought under Cape control, a war was triggered, leading to the takeover of Basutoland by Britain rather than the Colony. British and Cape imperial officials did succeed in reducing the flow of arms to the annexed African kingdoms, but they were less successful with the Boer republics, which were able to accumulate formidable arsenals prior to the South African War,

26 D. L. Golfsmith, *The Devil's Paintbrush: Sir Hiram Maxim's Gun* (Toronto: Collector Grade Publications, 1989), 101–2.

27 It is mentioned, perhaps for the first time, in J. G. McDonald, *Rhodes: A Life* (London: Philip Allan, 1927), 167. McDonald says that Rhodes told him this. Some of McDonald's detail is wrong in that he claims that the episode preceded the act of submission. In fact, Walter Stanford and Henry Elliot, the two key local officials, had already negotiated Sigcau's agreement before Rhodes's visit. The episode is not mentioned in Stanford's reminiscences although he does give details of Rhodes's visit. It was picked up in later biographies, as well as in Monica Hunter's anthropology of Mpondoland. Rhodes did send a telegram to his officials at the time suggesting he would deploy force if there was any resistance and perhaps the story grew out of his memory of this missive.

including some artillery and Maxim guns. By this time, the Boers favoured Mausers, the latest German bolt-action breech-loaders.

Superior firearms were only one element in a complex configuration of wealth, supplies, organisation and technological power that enabled political decisions about colonial expansion to be put into practice. Multiple agents were involved, not only the British but also settler states, settler militias, and Griqua and African forces. Yet at critical moments weaponry that was dependent on scientific and technical innovation did play a significant part in military encounters. Even so, weapons did not entirely determine the political outcomes of conquest. The fact that Tswana, Sotho and Swazi chiefdoms managed to achieve protectorate status directly under British authority, rather than absorption into South Africa, had more to do with political than military factors.

Farming: Sheep, Ostriches and Scientific Innovation

General histories of South Africa tend to see the mining revolution as the main driver of modernity and of agrarian change. Mining was geographically concentrated, attracted massive overseas investment, and was geared to technological innovation. But farming in the nineteenth-century Cape was also deeply influenced by new technology and knowledge. While this may be more difficult to discern, because the processes were dispersed and uneven, agrarian interests and the merchants associated with them sustained networks of experimentation and innovation. As in the case of weaponry and mining strategies, many of the technologies were pioneered elsewhere. Yet they found their way to South Africa relatively quickly along diverse and sometimes unpredictable channels of communication.

In 1872 the Cape exported wool worth £3,275,150, and the average from 1872 to 1875 was nearly £3 million annually. This was a relatively sudden new stream of income: thirty years earlier, in 1841, wool exports totalled little over £50,000. Wool exports dwarfed any other at that period. Even when diamonds had caught up by the early 1880s, values were barely higher – hovering at just over £4 million annually for the rest of the century with occasional dips. Angora goat numbers surged from 121,000 in 1865 to 3 million in 1891, their mohair fetching over £500,000 annually; in the 1890s, the average exports exceeded £500,000 annually.

Wool yields increased from about 1.7 lb (0.7 kg) per sheep in 1855 to 4.25 lb (nearly 2 kg) in 1891. Over the period after the discovery of diamonds in 1867, the value of pastoral product exports was much the same as that of diamonds, and after 1905, with the rapid rise of ostrich feathers, the value of pastoral products outpaced diamonds. For half a century from 1915 onwards, wool exports alone from South Africa exceeded diamonds in every year. This sustained intensification was possible not so much through specific scientific discoveries, but through the application and development of a host of agrarian technologies, especially in connection with breeding, water provision, fencing and disease control. Merino sheep, well known for their fine wool, were introduced to the Cape during the late eighteenth century. Originating from Spain, they had already been successfully farmed in France, Mexico and Saxony. Attempts to rear them in Britain failed, probably because they were better adapted to drier climates with less grass. They were gradually established at the Cape, which was also the source for the first Australian flock.

Charles Rubidge (b. 1818) was one of the most successful merino breeders in the nineteenth-century Cape.[28] His detailed diaries have survived, and his strategies provide a useful prism through which to view the cutting edge of agrarian innovation. Rubidge's family came from Britain in 1821. His father was sufficiently wealthy to buy a farm near Grahamstown and another in the Karoo. In 1840 Charles moved onto about 12,000 morgen (10,000 ha) of land in Graaff-Reinet and renamed it Wellwood. This area had been colonised by Boers since the 1760s, but a number left on the Great Trek or to seek better pastures to the north. The Karoo was already attracting English-speaking farmers because merinos seemed to thrive there.

The Mosenthals, Anglo-German Jewish merchants based in Port Elizabeth, imported pedigree French merinos and the first angora goats.[29] Rubidge purchased from them Rambouillet rams that had

[28] Beinart, *Rise of Conservation*; S. Dubow, *Land, Labour and Merchant Capital in the Pre-Industrial Rural Economy of the Cape: The Experience of Graaff-Reinet District (1852–1872)* (Cape Town: Centre for African Studies, University of Cape Town, 1982); F. Lilja, *The Golden Fleece of the Cape: Capitalist Expansion and Labour Relations in the Periphery of Transnational Wool Production, c. 1860–1950* (Uppsala: Uppsala University, Studia Historica Upsaliensia, 247, 2013).

[29] D. Fleischer and A. Caccia, *Merchant Pioneers: The House of Mosenthal* (Johannesburg: Jonathan Ball, 1983).

been developed from the original Spanish breed into bigger sheep, with longer wool and also good mutton. They were cross-bred with local merinos, and by the 1850s Rubidge's sheep were yielding an average of 2.5 kg each, perhaps three times the colonial average.[30] He visited France to purchase rams himself in 1861 and 1873, and his success in breeding enabled Rubidge to sell stud animals at a much higher price than those sold for meat. On the eve of his first trip to Europe he calculated his assets in land, house, improvements and livestock at £23,000, equivalent to about £2.6 million (R48 million) in 2018. In 1875 he recorded his operations for the colonial census: 54 people lived on the farm, including 24 farm workers; they managed 23 horses, 2,500 woolled sheep, 60 angoras, 700 goats, 48 oxen and over 100 other cattle as well as irrigated crops and an orange orchard.

Rubidge's investment in the best livestock clearly followed a pattern familiar on estates in Britain, but he faced entirely different environmental conditions with an annual rainfall of about 13 inches (300 mm). Sheep farming on this scale required some control of space, water and wild animals. In the era before fencing, stone kraals were built by specialist migrant Sotho-speaking Mantatee workers. Named after the Tlokwa chieftainess Manthathisi, who established a polity in the Caledon River valley in the 1820s, migrants and refugees from this area moved to the Cape, bringing skills in building with dry-packed stone. These were used in some parts of the Highveld for homesteads and walls; they were also a familiar part of the British landscape.[31] Rubidge commissioned a wall enclosing 5 acres (2 ha) around his new homestead in 1851, as well as kraals to protect and house sheep. By 1875 he had invested in at least four large earth dams fed by channels that captured water from surrounding hills when it did rain. His son Walter, established on a separate farm, famously said that 'to save South Africa you must dam it'. Such operations required specialist dam-building contractors.

Wealthier farmers who were investing in improved livestock increasingly recognised that fencing of farms as a whole would be beneficial. It promised more effective control of predators – a scourge for Cape

[30] Information drawn from Rubidge Diaries, Wellwood farm, Graaff-Reinet; Beinart, *Rise of Conservation*; Lilja, *The Golden Fleece.*

[31] P. Delius, T. Maggs and A. Schoeman, *Forgotten World: The Stone Walled Settlements of the Mpumalanga Escarpment* (Johannesburg: Wits University Press, 2015).

farmers. And if sheep could be left out at night, they would be saved from lengthy walks every day from kraals to pastures and back, which also damaged the veld. Progressive farmers, as they began to call themselves, as well as the first colonial veterinary surgeon, appointed in 1876, associated kraals with the spread of disease. Fenced pastures in which the sheep could run free would lead to healthier, better-fed sheep, more wool and richer veld. As stone walls were too expensive to enclose farms of this size, Rubidge experimented early with wire fences in 1861. Steel wire on the scale required was a relative new product of the Industrial Revolution. Various barbed wires were patented in the United States in the 1860s and 1870s. By the mid-1870s they were being imported into the Cape and Rubidge used barbed wire as his main fencing material. Imports of wire fencing to the Cape increased from £14,000 in 1886 to £114,000 in 1896.[32]

Rubidge also established a citrus orchard, one of the few in the Colony, which required irrigation and provided a valuable annual return of oranges. He maintained a wide network of information through newspapers, merchants and farmers' groups. His diaries show frequent visitors to his farm, and he provided credit to other farmers. He married locally, a granddaughter of Dr Liesching, the leading physician in Cape Town in the early nineteenth century, whose son had settled in the newly buoyant Graaff-Reinet district. His brother Richard Rubidge was a medical doctor, a member of Atherstone's circle, who pursued interests in geology and botany as well as medicine. His grandson Sidney Rubidge was responsible for building a renowned fossil museum at Wellwood, which attracted luminaries like Robert Broom and James Kitching; his great-grandson Bruce Rubidge is today a palaeontologist specialising in Karoo therapsids. As previously shown in the case of the Bowkers, the scientific and landowning elites of the eastern Cape were closely interconnected.

Whereas sheep farming was rooted in popular scientific knowledge and rapid practical innovation, ostrich farming required more direct scientific understanding and experimentation. This connection was explicitly made in 1877 when Julius de Mosenthal and James Harting published a pioneering book on *Ostriches and Ostrich Farming*.[33] Mosenthal came from the Port Elizabeth merchant family

[32] Lilja, *The Golden Fleece*, 86

[33] J. de Mosenthal and J. E. Harting, *Ostriches and Ostrich Farming* (London: Trübner and Co., 1877).

and was elected as a member of the Cape Legislative Assembly. He developed particular expertise in the ostrich feather market and became the commissioner for the South African colonies in Europe. Harting was a British ornithologist at the very heart of growing scientific research on birds. He was on the staff of *Field* magazine, edited *The Zoologist* from 1877 to 1896, and was a life member of the Zoological Society of London, as well as fellow, secretary and librarian at the Linnean Society of London. Most of the book on ostriches was written by Harting as a lengthy comparative scientific treatise on the classification and behaviour of large flightless birds. He drew heavily on earlier South African travellers such as Lichtenstein and Burchell as well as Layard's compendium. Harting discussed how the behaviour of ostriches facilitated the development of farming. He had an eye not only on feathers but also the potential for meat, which, he reported, tasted like young beef and might therefore be popular among British consumers.

In the book his co-author, Mosenthal, focused on the budding Cape feather industry. He noted that Sparrman had recorded tame ostriches on Cape farms a century beforehand, and that African people used the feathers. These had long commanded a good price in Europe. Attempts at domestication by hatching eggs in captivity began in North Africa and Italy in the 1850s, and accounts of these were published by the Acclimatisation Society of Paris, an offshoot of the Museum of Natural History. Jules Verreaux and others experimented at the Cape. In the 1865 Cape census, eighty tame birds were recorded; by 1875, the number had risen to 32,000 birds widely distributed through the Colony, with the highest numbers in Riversdale and Oudtshoorn. Most farmers kept a limited number of birds in large, fenced paddocks where they grazed mainly on veld.

Mosenthal's account relied heavily on Atherstone, who wrote a report on ostrich farming at the request of Governor Henry Barkly in 1874 which was republished in the *Cape Monthly Magazine*.[34] Atherstone concluded that ostriches thrived on sweetveld and alkaline soils rich in natural minerals and salts. In the sourveld areas near Grahamstown and on the coast, ostrich diets had to be supplemented with crushed bone to supply minerals. In Algeria, they were fed with chopped prickly pear leaf – a practice also adopted at the Cape.

[34] De Mosenthal and Harting, *Ostriches*, 198.

Although Cape feathers were not the most valuable, they dominated the London market. Mosenthal suggested crossing North African 'Barbary' birds, whose feathers were favoured in Paris, to improve quality in South Africa and he shipped two to the Cape. As he had earlier noted, 'we live on acclimatised products'.[35] There was, nevertheless, some concern to ensure that domestication did not displace wild ostriches, whose feathers were also highly valued, and an Act for the Better Preservation of Wild Ostriches was passed by the Cape Legislative Assembly in 1870. This attempted to control hunting, legal only with an expensive permit, as well as the taking of eggs.

Arthur Douglass, a friend of Atherstone, published a more extended treatise on ostrich farming in 1881.[36] Douglass tried to develop a detailed knowledge of ostrich behaviour for the purposes of breeding, farming and feather production. But he also realised the potential of controlling reproduction by artificial hatching and read about ancient strategies of incubating hen's eggs in Egypt and duck's eggs in China. Experiments with incubation in Europe followed a major scientific innovation, the development of reliable mercury thermometers and scales of measuring temperatures by Fahrenheit (1714) in the Netherlands and Celsius (1742) in Sweden. Even then, one of the key difficulties experienced in artificially hatching hen's eggs was maintaining the right, constant temperature. Artificial incubators were thus only successful from the 1840s when thermostats were improved by the Scottish chemist Andrew Ure.

Encouraged by Atherstone, Douglass imported a machine in 1869, rebuilt it, and worked by trial and error on ostrich eggs to find the ideal, steady temperature. He kept ostriches on his farm and took fertilised eggs from their nests. It was well known, since precolonial times, that if an egg was removed from a nest, the female bird generally laid another to replace it, so he had a steady supply. A critical discovery was his recognition that the natural process should be as far as possible emulated. Eggs required heat concentrated at the upper surface, mimicking a brooding bird; all-round heat prevented germination. He discovered that the germ always gravitated to the top of the egg. Successful breeding also required reduction in heat

35 J. Mosenthal, *Cape Monthly Magazine*, 7 (1860), 299.

36 A. Douglass, *Ostrich Farming in South Africa* (London: Cassell, Petter, Galpin and Co., 1881).

immediately before hatching, and – as in nature – some chicks needed help to break out of their shells. By incubating the eggs, as opposed to letting the ostriches brood on them, Douglass was convinced that a higher rate of hatching was achieved. In nature ostrich pairs shared the brooding, but cocks, especially on farms with lots of birds concentrated together, were unpredictable and might easily get distracted. Lower rates of hatching resulted. Some ostrich farmers continued to hatch in this more natural way, but the process was vulnerable to environmental hazards.

Figure 3.1 Ostrich Farming in South Africa, 1878.
Credit: *Illustrated London News*, 30 March 1878

Douglass patented his own incubator, which he then also marketed. Chicks required careful feeding and continual warmth. After initial success in the late 1860s and early 1870s, yellow liver disease spread in flocks. Birds were also assailed by tapeworms and a smaller worm, 'totally new to science', described by Douglass and named by a British scientist *Strongylus douglassii*.[37] He conducted post-mortems to unravel the causes of yellow liver and came to the conclusion that the major problem was dietary. Birds were not getting the range of minerals and nutrients that they would find in the veld. His diagnosis chimed with that of farmers trying to understand sheep diseases; they were also assailed by worms. The remedy was to ensure fastidious cleanliness in incubators and to provide large camps in which ostriches could roam freely and eat natural vegetation. As it was difficult to herd ostriches, expansion of ostrich farming hastened the erection of wire fences in the Cape – but not barbed wire, which damaged the birds. Douglass himself supervised construction of 40 miles (64 km) of fencing on his farm. In 1915, a Department of Agriculture scientist found that the disease was caused by a mould, *Aspergillus fumigatus*.[38] Douglass's instincts about hygiene were right, and his solution effective, even though he did not identify the cause of yellow liver. Some ostrich owners reverted to controlled natural breeding in outdoor nests as a solution to diseases that seemed to assail incubated birds.

Douglass was primarily concerned to develop a financially profitable enterprise and paid close attention to shape of birds and the demands of the market for feathers – 'good breadth, softness, closeness and droop'.[39] A breeding pair could fetch over £100 – a high price even in relation to pedigree rams and horses. Douglass was justifiably optimistic about his successes. He exhibited at the 1876 International Exhibition in Philadelphia and was awarded a prize for his incubator. The citation generously noted that he had 'added a most important industry to the world' as well as 'averting the threatened extermination of this species'. At the Royal Agricultural Show in Derby in 1881, he presented the Prince of Wales with a copy of his book and two plumes. In 1882, the year after his book was published, Cape feather exports first topped £1 million. This was the nineteenth-century peak; lower prices and disease hampered growth until the Oudtshoorn-based boom in the early twentieth century, when export values climbed briefly to £2 million a year. Over the long term, however, the fashion for feathers never rivalled the value of wool.

[37] Douglass, *Ostrich Farming*, 155; A. Martin, *Home Life on an Ostrich Farm* (London: George Philip, 1890).

[38] J. Walker, 'Aspergillosis in the Ostrich Chick', Union of South Africa, *Department Agriculture Annual Report*, 3–4 (1915), 535–74.

[39] Douglass, *Ostrich Farming*, 97.

Mining: Kimberley and the Rand, Cyanide and Dynamite

As Storey suggests, the historiography of mining in South Africa has been rich in the analysis of socio-economic issues, but less so in the discussion of the evolving technology that made mining on this scale possible. In Kimberley, pumping machines proved to be critical. Within a decade of the start of mining, in the mid-1870s, water hampered excavation of diamondiferous earth. Rob Turrell argues that 'the pumping contractors were in a position of great power in the mines'.[40] Charles Rudd, entrepreneur and speculator, bought machines and employed twenty-one-year-old Cecil Rhodes to tender for pumping contracts; in 1874 they were successful at Dutoitspan, initially using a second-hand pump bought from a farmer. Their interests spread to De Beers mine and, when he went to Oxford to study in 1876, Rhodes shipped a 12-horsepower machine to Rudd which brought them a considerable profit. The water removed could also be used to improve diamond recovery because from 1875 gemstones were increasingly 'extracted from the blue [ground] in large rotary washing machines'.[41] Other sources of water were in short supply until a pipe was laid from the Vaal River in 1883.

Hauling machinery was initially horse-drawn, giving rise to the intricate spider web of ropes and pulleys captured in early photographs. But increasing concentration of ownership removed many of the smaller diggers, both white and black. Steam-powered earth-hauling machines spread in the late 1870s, unlocking a technological logjam, and this favoured bigger operations. Rhodes argued in 1882 that working costs could be reduced by 27 per cent and yield increased fourfold by deeper mining, with better engines, as well as by providing barracks for workers.[42] In his costings, more than half the savings would come from further mechanisation as opposed to reducing white labour and black wages. The early attempts at amalgamation by Rhodes, the Rothschilds, and J. B. Robinson, failed. But as landslips increasingly hampered production, a centralised approach to pumping and ore extraction had clear advantages. As the diamond mines became deeper and reached the harder blue ground, dynamite became essential, and drilling machines were needed for making holes

[40] Turrell, *Capital and Labour*, 83. [41] Turrell, *Capital and Labour*, 9.
[42] Worger, *South Africa's City of Diamonds*.

in which to fix it. Technological innovation thus facilitated amalgamation and strengthened the arguments for concentration. By 1886 Rhodes's De Beers mine was the most profitable, and by 1891 his company achieved a monopoly of diamond production. Investment in machinery was an element in his growing wealth and pre-eminence. A monopoly also facilitated control of diamond sales, which kept prices relatively high.

In the case of gold-mining, scientific innovation proved critical in the shape of 'best-practice technology, brought to the Witwatersrand from all over the world by highly paid mining professionals', many of whom had 'academic as well as working skills'.[43] Technological expertise, as well as capital, was highly internationalised. On the Witwatersrand, gold was deposited in reefs sloping at an incline underground. Initially, mining focused on the outcrops of gold where the reefs met the surface. Here the exposed rock was softer, and the best locations contained high grades of ore. As the reefs sank downwards, the rock became harder, the proportion of gold smaller, and the ore more difficult to extract. Fred Struben, a prospector and amateur geologist, whose brother owned land near Johannesburg, understood by April 1885 that the gold-bearing outcrop sloped down into the earth.[44] Within five years, by about 1891, the surface and shallow mines had largely been exhausted and many operators went out of business.

Geologists and engineers had, by this time, shown that there was a regularity in the incline of gold-bearing reefs (although the specific slope varied in different areas). The major discoveries were made by American geologists and mining engineers such as J. S. Curtis and Hennen Jennings, working with the mine owner Julius Wernher. Jennings had developed his skills in difficult conditions in Venezuela before being recruited, through London investors, to South Africa. He was probably responsible for engineering the first deep shaft.[45] The

[43] E. N. Katz, 'Outcrop and Deep Level Mining in South Africa before the Anglo-Boer War: Re-examining the Blainey Thesis', *Economic History Review*, 48, 2 (1995), 304–28.

[44] E. Mendelsohn (ed.), 'A Geological Section of the Witwatersrand Drawn by Fred Struben in 1885', *Transactions and Proceedings of the Geological Society of South Africa*, vol. L, 177–8.

[45] M. Nkosi, 'American Mining Engineers in the Kimberley and Witwatersrand Mines *circa* the Turn of the 20th Century', presented at the International Mining History Congress, Johannesburg, 17–20 April 2012, unpublished paper;

results of these boring operations were kept secret for a time, and this gave Werhner and his partners, Beit and Eckstein, an advantage in purchasing land under which the reef sloped. Deep-level shafts and diamond rock drills enabled the company to exploit the buried ore. Owners could make reasonable assessments of the potential of deep-level mines: 'the pragmatics of the borehole assays convinced the overseas mining engineers and scientists [of the] presence of payable gold at depth.'[46]

Mining knowledge was soon exchanged through a plethora of new professional networks: the South African Association of Engineers and Architects (1891), the Association of Mine Managers (1892), the Chemical and Metallurgical Society of South Africa (1894), and the South African School of Mines and Technology, launched in Kimberley in 1896 and translocated to Johannesburg in 1904. The Chamber of Mines acquired a collection of rocks and minerals, which it housed from 1890 in its headquarters on Market Street in collaboration with the Geological Society of South Africa (est. 1895) – the nucleus of the country's first Geological Museum.[47]

The key figure in establishing the Geological Society was David Draper, brought up in Colesberg, who prospected for diamonds, coal, and gold, and whose expertise was recognised by his election in the early 1890s as the first South African born fellow of the Geological Society of London. Draper collaborated with Gustaaf Molengraaff, a Dutch geologist and botanist who published on Timor and Borneo before taking, in 1897, the position as state geologist in Kruger's South African Republic. Molengraaff and Draper were both fervent supporters of the Boer cause during the war, providing them with expert hydrological advice with a view to diverting the Molopo River so as to deprive the besieged British army of water at the siege of Mafikeng.[48] After imprisonment by the British for his republican sympathies, Draper sought work in Brazil where he identified kimberlite diamond pipes similar to those in South Africa. (This subsequently provided geologist Alex du Toit with key evidence for his theory of continental drift.)

M. Nkosi, *Mining Deep: The Origins of the Labour Structure in South Africa* (Cape Town: David Philip, 2013).

46 Katz, 'Outcrop', 312.

47 H. S. Harger, 'Historical Notes on the Geological Museum of Johannesburg', *South African Museums Association Bulletin*, March 1942, 1–2.

48 S. Chetty, 'David Draper: The Making of a South African Geologist', *Historia*, 63, 2 (2018), 17–39.

Molengraaff mapped the geology of the Transvaal and published on its rock formations and stratigraphy. In 1901 he travelled to Sulawesi to report on gold discoveries, returning to South Africa where he expanded his academic interests by working on the Bushveld Complex and the Vredefort Dome, formed by an enormous asteroid crash 2 billion years ago. Draper was given an honorary doctorate by the University of the Witwatersrand (Wits) in 1928 and Molengraaff in 1929.

South Africa's mineral revolution benefitted from and gave an incalculable stimulus to geology, drawing initially both on professionals and self- trained experts. The expanding capital-intensive world of deep-level mining was highly cosmopolitan in its reliance on international networks of specialists with diverse experience and talents. By 1895, at least half of the new gold mines were managed by Americans, including the chief engineers of the Wernher-Beit group and Rhodes's Consolidated Gold Fields.[49] A visiting American mining engineer wrote in the *National Geographic Magazine* of November 1896 that 'the mining industry is chiefly under the guidance of American mining engineers'.[50] They also supported urgent development and a transformation of the Transvaal state.[51]

During previous rushes, gold had been panned in a relatively simple process that separated the earth and rock by running water over the mix. This was not very efficient, and it only worked on the surface outcrops where the rock and soil had been weathered. Underground gold could only be freed from the surrounding rock by mechanical stamping or crushing into relatively small particles, but even then it was not easy to extract gold effectively with water into a 'free metallic form'.[52] A process using mercury enabled some gold to be captured but was inefficient and expensive.

Australia's gold rush in the 1850s predated South Africa's and attracted large numbers of settlers so that by 1860 the white population of Australia far outstripped that of South Africa. When the surface

[49] Marks and Trapido, 'Lord Milner and the South African State' quoting T. J. Noer, *Briton, Boer and Yankee: The United States and South Africa, 1870–1914* (Ohio: Kent State University Press, 1979).

[50] G. F. Becker, 'The Witwatersrand and the Revolt of the Uitlanders', *National Geographic Magazine*, VII, 11 (1896), 349–67.

[51] C. van Onselen, *The Cowboy Capitalist: John Hays Hammond, the American West and the Jameson Raid* (Johannesburg: Jonathan Ball, 2017).

[52] J. Todd, *Colonial Technology: Science and the Transfer of Innovation to Australia* (Cambridge: Cambridge University Press, 1995), 112.

deposits in Australia were exhausted, the difficulty of gold extraction from hard underground rock undermined further growth. By the 1870s, this problem was focusing the minds of entrepreneurial chemists. Gold was the basis for world currencies and demand for it remained high at the same time as the supply was threatened. Glasgow, the second-largest British city, with extensive imperial trading networks, was the heartland of industrial innovation in spheres such as steam engines; it was also the site of major chemical industries.[53] Scottish entrepreneurs, working initially with an American metallurgist, formed the Cassel company to finance research that might improve methods of gold extraction.[54]

In 1886 John MacArthur, working in Glasgow with Robert and William Forrest, developed a process that could be used for the extraction of gold particles from crushed rock on an industrial scale. Their major innovation was to use a solution of potassium cyanide that could leach out gold and dissolve it, leaving behind the tailings; they were also able to neutralise and separate other minerals present.[55] They patented the cyanide process in both Britain (1887–8) and South Africa (1889).[56] Although some elements of this technique had been registered a few years earlier in the United States, the MacArthur-Forrest process proved more adaptable for industrial usage than other existing methods; silver could be extracted by the same means. MacArthur visited South Africa in 1890 at the very moment that the mining industry needed to improve gold extraction. While the country's unique mix of abundant minerals and exploitable labour formed the basis of its rapid industrial expansion, it is unlikely that gold could have been mined on so large a scale at this time without the adoption of the MacArthur-Forrest cyanide process.

[53] J. M. Mackenzie, 'The Second City of the Empire: Glasgow – Imperial Municipality' in F. Driver and D. Gilbert (eds.), *Imperial Cities: Landscape, Display and Identity* (Manchester: Manchester University Press, 1999), 215–37.

[54] Balliol College Archives and Manuscripts, Oxford, Papers of John Stewart MacArthur 1856–1920. MacArthur's son trained as a clergyman at Balliol College and deposited three boxes of his father's papers. There are draft notes for an article 'The Elementary Principles of the Cyanide Process' and for a 'Cyanide book' that was not published.

[55] J. MacArthur, 'Gold Extraction by Cyanide: A Retrospective', *Journal of the Society of Chemical Industry* XXIV 7 (1905), 311–15.

[56] MacArthur papers, II, B, 1–3, patent: 'Improvement in treating ores to obtain gold, silver, lead, zinc copper and other metals or useful compounds'.

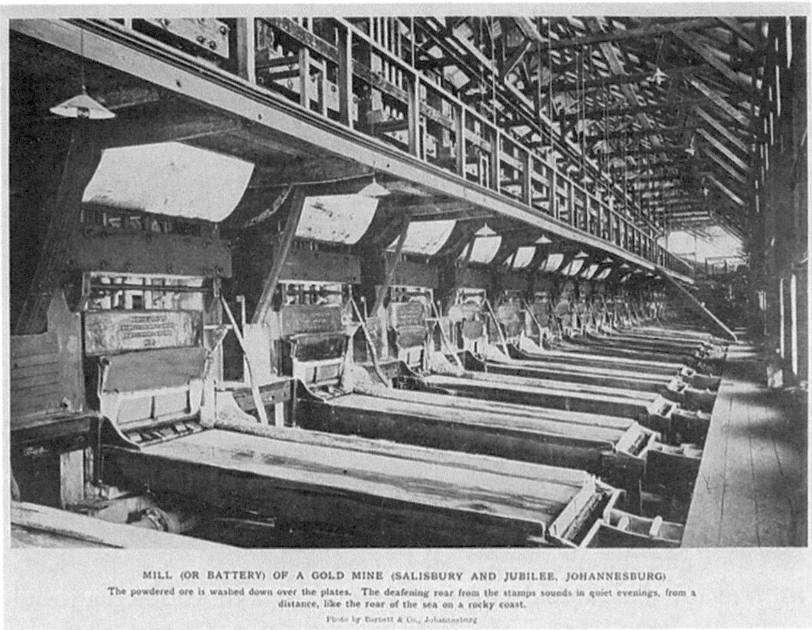

Figure 3.2 A Mill or Battery of a Gold Mine, Johannesburg, 1903.
Credit: Barnett and Co., Johannesburg, WikiCommons

American engineers had already developed and installed machinery to crush rock, and in the early 1890s they experimented with a variety of techniques to find the most effective and profitable method of using the cyanide process.[57] Within a couple of years the percentage of gold extracted from ore doubled, and by 1895 the Transvaal mines had become the largest producers in the world. Over 2,000 stamps were being used for crushing ore and the cyanide process was developed for large-scale extraction, giving the mounting mine-dumps their distinctive brownish-yellow colour.

Cassel initially charged royalties for use of the MacArthur-Forrest technology, but the mining companies were reluctant to pay the rate requested. By 1893, the Chamber of Mines set up a committee of investigation into the MacArthur-Forrest patents. When negotiations failed, the mine owners blamed this on the inflated demands of the

[57] F. H. Hatch and J. A. Chalmers, *Gold Mines of the Rand: Being a Description of the Mining Industry of the Witwatersrand, South African Republic* (London: Macmillan, 1895), 217–18; A. Scheidel, *The Cyanide Process: Its Practical Application and Economic Results* (Sacramento: State Office, 1894).

agents of the patent holders. The latter 'chose to ignore', it was later claimed, 'the sound principle which has always been upheld by South African lawyers, that a patent must not be allowed to overburden the industry it is intended to help forward'.[58] They were also accused of seeking to get a monopoly concession from the Transvaal government for the manufacture of cyanide solutions. Two major court cases resulted – one in London (1895) and the 'Great Cyanide Case' in Pretoria (1896).

In both cases, the key issue was not the rate of royalty but the validity of the patents. Those representing the mining industry sought to show that patents taken out in the United States, and especially one by Simpson in 1885, revealed a 'general chemical knowledge of the fact that cyanide of potassium would dissolve and thus extract gold from its ore'.[59] The Court of Appeal judges in London argued that this might have been general knowledge, but that the previous patents did not set out the same detailed, multi-stage methods followed in the MacArthur-Forrest process.

A long and complex court case ensued in the Transvaal High Court for which remarkably detailed evidence was gathered internationally and the history of the relevant chemistry rehearsed. There was exhaustive discussion on complex issues such as the significance of the strength of the solutions required and the originality of the process of separating unwanted metal residues. The mine owners' lawyers drew much of their evidence from the United States, using their engineers' knowledge. Although the judges were split, two out of three, including Chief Justice Kotze, decided in favour of the mining companies in November 1896. Kotze said that 'the fact that the process had been successful on the Witwatersrand was not enough'; it had to be shown that MacArthur's 'was the first or original invention'.

It is notable that the South African Republic judiciary, part of a state that was much maligned by the mine owners for corruption and restrictive practices, delivered a judgment that was highly favourable to the industry. Pressure may have been placed on the Transvaal judges, some of whom had personal debts to mining capitalists.[60] Either way, the innovative MacArthur-Forrest process resulted in

[58] MacArthur papers, III, B, 1–7, *South African Mining Journal*, 7.11.1896, 103.
[59] Judgement, Court of Appeal, London, 9 April 1895, case between Cassel and Cyanide Gold Recovery Syndicate.
[60] van Onselen, *Cowboy Capitalist*, chapter 1.

greatly increased rates of gold extraction with little financial reward for its inventors.

The mining of rock in huge volumes also required explosives.[61] In 1896, the Nobel company began to produce dynamite at Modderfontein, 20 km north of Johannesburg. Operated as a government monopoly by Kruger's South African Republic and staffed by a cosmopolitan collection of European experts under the direction of Dr Franz Hoenig, it immediately became the largest factory manufacturing dynamite in the world and remained so for the best part of a century. Dynamite accounted for around 9 per cent of the working costs of the mines in 1896, and Kruger's control of its production was greatly resented by Rand mining magnates. Milner considered the dynamite monopoly to be 'a capitalists' question pure and simple'; it was one of several greed and grievance factors which led to the outbreak of war in 1899.[62]

Soon after the start of dynamite production at Modderfontein, two other competitor factories were established: in the Cape, at the behest of Rhodes (1898), and in Natal (1909). One of the by-products of explosives production was sulphuric acid, used in the manufacture of superphosphate fertiliser. In 1931–2 a novel synthetic ammonia process was adopted in South African explosives plants which had the advantage of dispensing with imports of nitrates from Chile. The ammonia process, pioneered by Nobel Prize-winners Fritz Haber and Carl Bosch, was seen as a great German 'scientific triumph'; in South Africa it was developed at scale and became a critical factor in agribusiness.[63]

Railways, Food and Refrigeration

Railways constituted the largest state investment by the South African colonies in the late nineteenth century. They borrowed well-developed technology from British and, in the Transvaal, from Dutch companies. As in the case of mining, Glasgow was a key centre of innovation,

61 A. P. Cartwright, *The Dynamite Company* (Cape Town: Purnell, 1964) 191–2.

62 A. Milner to Selborne, 29 December 1897 in C. Headlam (ed.), *The Milner Papers. South Africa 1897–1899* (London: Cassell, 1931), 138.

63 B. Fine and Z. Rustomjee, 'The Political Economy of South Africa in the Interwar Period', *Social Dynamics*, 18, 2 (1992), 41–2; Freund, *Twentieth-Century South Africa*, 58.

especially in the construction of steam engines. The Cape decided on a narrower gauge of rail than Britain, largely as a result of the cost and the difficulty of constructing lines through the mountains surrounding Cape Town, where the railways started. Natal made an early beginning with standard British gauge but later came into line. In the longer term, this made it difficult to increase the speed of trains. The railway reached Kimberley in 1885 and Johannesburg in 1892; a line from Delagoa Bay to the Transvaal was completed in 1894. By 1894 the Cape Government Railways were purchasing substantial quantities of coal from the Lewis and Marks coal mines in the Transvaal. In 1895, the rail systems of the four colonies and states plus Mozambique were linked to provide interconnected routes from the coastal cities to Johannesburg and Pretoria. It is difficult to imagine an industry on the scale of the Witwatersrand gold mines without such rail connections and associated industrial enterprises.

Johannesburg, whose population increased from about 40,000 in 1892 to 100,000 in 1904 and over 200,000 in 1911, soon developed its own food market. By 1892 an extraordinarily high number of over 12,384 wagons (carrying perhaps 3 tons each) were registered as depositing goods there.[64] They transported fresh food and meat produced largely by white and black Highveld farmers. But the sudden concentration of so many people, in the largest African conurbation south of Cairo, created a demand for food and supplies that could not be fully supplied locally. New techniques of preservation and more rapid shipping connections had recently become available. Britain itself shifted dramatically towards the industrial production and preservation of food during the nineteenth century. This process also shaped tastes and influenced diets in the anglophone world at this time. Shipping and railways in particular enabled such products from the food industries of Britain to reach this new concentration of people in southern Africa.

Between 1850 and 1880, Britain started to import meat, which soon made up around 10 per cent of its consumption. At the same time, livestock numbers in Australia, New Zealand, Argentina, Uruguay and South Africa expanded rapidly. There was thus a strong incentive on both sides to preserve and transport meat. Different strategies were

64 E. A. Cripps, 'Provisioning Johannesburg, 1886–1906', MA thesis, University of South Africa (2010).

developed for drying, powdering or canning meat and producing bottled extracts. Tinning began in Australia in the 1840s and tinned mutton was displayed at the Great Exhibition of 1851. In the 1860s Liebig carried out experiments in Argentina to improve tinned meats and extracts. Once his techniques were transposed to Australia, the amount of tinned meat exported to Britain increased by 1880 to 16 million tons.[65] South Africa did not develop a similar capacity for meat export.

While global markets for meat increased, the costs of transporting live animals remained high. Refrigeration offered the possibility of keeping meat relatively fresh. The general scientific principles of refrigeration had been known for some decades, but the technical application at scale was lacking. As in the case of the cyanide process, many recognised that market conditions were right if the technical solution could be found. In the United States, ice had been harvested in great quantities in the colder northern states and transported by rail for commercial use to cool farm produce such as dairy and meat. Experiments in using compressed gas, especially ammonia, for cooling had been pursued throughout the nineteenth century and by the 1870s they were sufficiently effective to be applied commercially in transporting meat by rail in the United States.[66]

In the late 1870s American exporters, relying on Glasgow-built ships, sent chilled rather than frozen meat to Britain; Argentinian consignments came soon after. Australia was also a site of experimentation and the first successful Australian shipment, fully frozen, arrived in London in February 1880 amid considerable publicity. New Zealand farmers geared up for this trade from 1882 and exported 78 million sheep carcasses over the next thirty years.[67] New Zealand was less suitable for merinos and thus a variety of English mutton breeds were introduced and a cross was developed known as the Corriedale, which became the mainstay of frozen lamb exports.[68] As

[65] J. T. Critchell and J. Raymond, *A History of the Frozen Meat Trade: An Account of the Development and Present Day Methods of Preparation, Transportation, and Marketing of Frozen and Chilled Meats* (London: Constable, 1912), 11.
[66] Critchell and Raymond, *Frozen Meat Trade*, 1.
[67] Critchell and Raymond. *Frozen Meat Trade*, 41
[68] R. Woods, *The Herds Shot Round the World: Native Breeds and the British Empire, 1800–1900* (Chapel Hill: University of North Carolina Press, 2017).

earlier with merinos, improving the breeds of empire proved important in maximising production.

Although the Suez Canal offered an alternative route after 1869, much of the Australian and New Zealand shipping still called at Cape Town. When the first refrigerated ships stopped on their way to England, David Graaff, brought up on a Villiersdorp farm and working for the local butcher Combrinck, became interested in the potential of the frozen meat industry.[69] In 1881, aged twenty-two, he travelled to Britain and the United States gathering information on techniques and markets. Over the next couple of decades, Combrincks developed refrigeration plant and distribution networks in Cape Town and other leading cities. Most of the butchers in these cities were contracted to them for frozen meat, and Combrincks also controlled the Johannesburg Cold Storage Company. When rinderpest decimated the South African cattle herds in 1896–7, Graaff had the facilities to import and distribute frozen meat. He then formed the South African (later, Imperial) Cold Storage and Supply Company in 1899, which, despite his political sympathies with the Afrikaner Bond, won a bid to supply the British army with meat in the South African War.

During this conflict, over 450,000 troops were recruited from Britain and the British colonies. Though they were not all enlisted at one time, the Cold Storage Company had to supply an estimated peak of 250,000 men, while the concentration camps housed at their height more than 120,000 people removed from the farms by the British army.[70] Many of the military installations and camps were sited along the railways to facilitate the movement of armaments and supplies. Local animals, wild and domestic, could not provide an adequate or hygienic supply of meat. Graaff's company transported large quantities of frozen supplies from Cape Town and Durban and had to distribute it quickly because the railway carriages were not refrigerated. During the two years in which they held the contract, the company made profits of about £1.5 million.[71] On the basis of this success, Imperial Cold Storage became a major South African company. Just as electricity proved valuable for mining operations, it was also important for

[69] E. Dommisse, *Sir David de Villiers Graaff: First Baronet of De Grendel* (Cape Town: Tafelberg, 2011); P. B. Simons, *Ice Cold in Africa: The History of the Imperial Cold Storage & Supply Company Limited* (Cape Town: Fernwood Press, 1999).

[70] Simons, *Ice Cold in Africa*, 58.

[71] Simons, *Ice Cold in Africa*, 67.

cold storage plants. Graaff, as a city councillor and mayor of Cape Town in 1891–2, played a major role in establishing the first power station in the city. He also invested in the first cable car up Table Mountain.

Veterinary Medicine, Insects and Conflicts over Science

Much of the literature on the agrarian revolution in South Africa links it to the transformation of Highveld farms in the Free State and Transvaal in the wake of the mineral revolution. But, as we have indicated, forms of commercial and capitalist agriculture were in many respects rooted elsewhere. Agricultural intensification had a different momentum in the Cape and Natal around vineyards, wheat, wool, ostrich feathers and sugar production. Similarly, the development of a modern interventionist state with distinct departments and research capacity has been linked to Lord Milner's mining-oriented reconstruction government after the South African War. Yet the Cape in particular had already appointed a significant number of scientific officers. From the final quarter of the nineteenth century, governments became principal employers of a variety of experts, including veterinarians, hydraulic engineers, agronomists and entomologists, whose job was to serve the interests of the farming community, enhance the fiscal resources of the state, and prevent diseases that undermined agriculture.

Many parasites and diseases affecting humans, animals and plants spread along new routes opened up by the European maritime empires; such biological transfers have become a major focus of environmental history.[72] While these historiographical strands are receiving more attention with respect to southern Africa, there is as yet no overarching attempt to assess 'ecological imperialism' or its limits in the region. The significance of smallpox in decimating Khoekhoe people has been debated in respect of Dutch colonisation in the eighteenth century. Ross and Adhikari argue that disease was not the major factor in the decline of the Khoekhoen, and the latter proposes that the term 'geno-

[72] A. Crosby, *Ecological Imperialism: The Biological Expansion of Europe, 900–1900* (Cambridge: Cambridge University Press, 1986); J. R. McNeill, *Mosquito Empires: Ecology and War in the Greater Caribbean, 1620–1914* (Cambridge: Cambridge University Press, 2010).

cide' by colonial agency is more appropriate.[73] Nevertheless, in the second half of the nineteenth century, diseases of people, animals and plants, such as smallpox, plague, venereal disease, scab, lungsickness, rinderpest and phylloxera, all made an impact on southern African society both in the losses they caused and in measures taken to combat them.

Scab had long been endemic in Cape flocks and lungsickness (bovine pleuropneumonia) decimated cattle herds in the 1850s. With these and other scourges in mind, the Cape government, under the sheep-farming prime minister John Molteno, created the post of colonial veterinary surgeon in 1876. The first incumbent, William Branford, strongly influenced a landmark government commission into the diseases of cattle and sheep (1877), which emphasised environmental causes of disease. His successor, Scots-trained Duncan Hutcheon, appointed in 1880, was a protagonist of the new germ theory. He gradually gained a growing understanding of the Cape's complex map of animal disease and began to investigate the causes of a number of locally named diseases, such as heartwater, which killed sheep, goats and cattle.[74]

In 1880 Peter MacOwan, a science teacher and enthusiastic botanist, was appointed as professor of botany at the South African College and curator of the neighbouring government botanical garden. Hutcheon, MacOwan and the other early scientific officers initially worked alongside amateur experts. For example, the first experiments on heartwater were conducted on a Fort Beaufort farm. Eastern Cape farmers had already deduced some connection between the disease and ticks although Hutcheon did not accept this theory at the time. MacOwan collected plants with Atherstone's circle in Grahamstown and with Harry Bolus in Graaff-Reinet. Bolus moved to Cape Town in 1874 to pursue a successful business career at the same time as continuing botanical work. His collection of plant specimens formed the basis for the University of Cape Town's Bolus Herbarium. In other regions, also, amateur experts, missionaries and farmers read and

[73] R. Ross, 'Smallpox at the Cape of Good Hope in the Eighteenth Century' in *African Historical Demography* (Edinburgh: Centre for African Studies, University of Edinburgh, 1977; M. Adhikari, 'A Total Extinction Confidently Hoped for: The Destruction of Cape San Society under Dutch Colonial Rule, 1700–1795', *Journal of Genocide Research*, 12, 1–2 (2010), 19–44.

[74] Beinart, *Rise of Conservation*; D. Gilfoyle, 'Veterinary Science and Public Policy at the Cape Colony, 1877–1910', DPhil thesis, University of Oxford (2002).

collected widely, wrote about their observations, and provided a network of applied and scientific knowledge (see Chapter 4).

In 1886 phylloxera swept through the vineyards of the western Cape and devastated one of the Colony's most deeply rooted agricultural enterprises. This microscopic insect, similar to an aphid, fed on grapevines, disfigured the roots, and rapidly created conditions for fungal infection. The parasite was introduced to Europe on American vines. Mediterranean and European vines, the core of an ancient industry, had no resistance to phylloxera and the associated fungal infection. Most of Europe's vines were wiped out. It was already known that European cultivars could be successfully grafted onto American rootstocks, and this, together with a process of hybridisation, provided the basis for a solution. The Cape parliament appointed a scientist to investigate and supervise the process, and within twenty years, production at the Cape had recovered. Simultaneously a Department of Agriculture was established in 1887 following the Australian example. An *Agricultural Journal* was launched in order to communicate improved methods and to publicise the department's activities. It also provided a vehicle for reporting developments in other countries and for farmers to write about their own experiences. In 1892 the Cape created a separate Ministry for Land, Mines and Agriculture, headed initially by John X. Merriman, who was also colonial treasurer.

After Hutcheon was incorporated in the department, he was able to expand the veterinary service. Jotello Soga, appointed as one of two additional government vets in 1889, worked with white farmers and black shepherds in his quest to find the causes of the goat disease called nenta. Son of the first black Presbyterian minister in South Africa, Tiyo Soga, and his Scottish wife, Janet Burnside, Jotello trained in Edinburgh in the 1880s – he was probably the first South African-born vet to qualify. Soga's appointment at a professional level in the colonial civil service was unusual and, along with court interpreters, he was among the few senior black civil servants. Opportunities for training and recruitment at this level were closed down as segregation became entrenched in the early twentieth century.

Nenta was noted by Thunberg in the 1770s as a Khoekhoe name for an animal disease – later called *krimpsiekte* in Afrikaans.[75] Thunberg's

[75] Veterinary Surgeon [J. F.] Soga, 'Disease "Nenta" in Goats', *Agricultural Journal of the Cape of Good Hope*, 29 January (1891), 140–2; C. J. Botha,

informants associated it with a particular plant, also called nenta, which MacOwan identified as *Lessertia annularis*. Hutcheon was not convinced about its association with the symptoms of paralysis that he identified as nenta/*krimpsiekte*. In August 1890 Soga expanded research in the eastern Cape where angoras had become valuable. The black shepherds to whom he talked seemed to apply the term to a range of poisonous plants. White farmers were also uncertain about the specific cause, but one had experimented with a plant and Soga decided to test it. He first conducted post-mortems on animals said to have died from nenta. This indicated that it affected all goats, local Boer and African types as well as angoras. He collected leaves from the plant indicated, cut and bruised them, and fed them to eight goats selected from an area free of the plant. All became ill and four died.

Soga's may have been the first attempt in South Africa at a scientific experiment to test for a toxic plant. He was able to ascertain the period of greatest toxicity – towards the end of the year, from the period of flowering to the setting of seed. From January to May, when there were no leaves on the plant, disease incidence abated. MacOwan identified the plant that Soga tested as *Cotyledon ventricosa* (now *Tylecodon ventricosus*), a type of crassula; this was a different species from the one that MacOwan originally associated with nenta. Soga recorded in detail both the external symptoms of afflicted goats and the post-mortem indications. He thought it worth trying to eradicate the plant on farms and also experimented with cures: doses of Epsom salts (a laxative) and chloride of ammonia; bleeding; and a mix of chloral hydrate, creosote and Boer brandy. Veterinary pharmaceutical products were then in their infancy.

Hutcheon was a key figure in the early Department of Agriculture and used his increasing authority to drive through legislation to control scab in sheep. Such measures had already been implemented in Australia. The disease not only damaged the health of sheep, but severely diminished the quality of wool. Parasitological research by microscope showed the disease to be caused by an acari mite barely visible to the naked eye. It burrowed into the skin of sheep, causing a surface scab and loss of wool. The acari could be killed by the

'Krimpsiekte in South Africa: Historical Perspectives', *Journal of the South African Veterinary Association*, 84, 1 (2013); Beinart, *Rise of Conservation*, 145–6.

application of water-soluble chemical compounds, either by spraying or by dipping the animal in a trough.

Various attempts at voluntary, partial legislation were proposed from the 1870s and a Scab Act was passed in 1886. Proponents of state control argued that the disease could only be eradicated through systematic and compulsory intervention because any remaining reservoirs would rapidly re-infect flocks. John Frost, Queenstown farmer and member of parliament, who had also served on the 1877 inquiry, chaired a commission on scab from 1892.[76] An exhaustive investigation over two years provided Rhodes's government, in alliance with leaders of the Afrikaner Bond, with the evidence to legislate the first major, scientifically based veterinary public health measure in 1894.[77]

The Scab Act was a far-reaching measure, in certain respects more intrusive than smallpox vaccination and controls over venereal diseases in human medicine. There were about 12–15 million sheep in the Cape in the 1890s.[78] Every animal had to be dipped or sprayed on a regular basis until clean of scab, and the movement of small stock was constrained through a system of permits. Inspectors were appointed and fines imposed for contravention. The Act applied across the Colony as a whole, to white and black sheep owners alike; scab had no respect for racial boundaries and any flock could be a reservoir. But the principle of universality in eradication also meant that all potentially benefited from this public veterinary health measure.

The Scab Act precipitated deep political divisions, especially among Afrikaner farmers. Moderchai Tamarkin analyses the battle over science, government intervention, interpretations of disease and ethnic identity.[79] Most of the Bond leaders supported the Act, but many rural Afrikaners, including landowners, were suspicious of Rhodes's motivations and British domination. Attitudes to scientific knowledge lay at the heart of the debate. Even Afrikaner members of parliament declared they knew nothing about science. They did not believe that

[76] Cape of Good Hope, *Report of the Scab Disease Commission, 1892–94* (Cape Town: G.1 – 1894, Government Printers).

[77] M. Tamarkin, *Cecil Rhodes and the Cape Afrikaners: The Imperial Colussus and the Colonial Parish Pump* (London: Frank Cass, 1996).

[78] Beinart, *Rise of Conservation.*

[79] M. Tamarkin, *Volk and Flock: Ecology, Identity and Politics among Cape Afrikaners in the Late Nineteenth Century* (Pretoria: University of South Africa Press, 2009).

specialists such as the veterinary scientists, or authors published in the *Agricultural Journal*, had any authority: 'It was useless teaching farming from books.'[80] Hutcheon took to the road with his microscope in order to demonstrate the acari at farmers' gatherings. In the western and north-western districts, sheep owners either avoided these gatherings or declared themselves uninterested in the revelations of the microscope.

Opponents of scab legislation argued that experience and practical learning were the most reliable source of knowledge. Australia could provide no model because Cape farmers' experience was specific; D. P. van den Heever, a leader of the anti-dipping movement, contended that God gave different regions and locales their characteristics. Even in the Cape itself, knowledge from one district was not necessarily applicable elsewhere: it had to be tested by practice locally.[81] Some disputed causation by acari, and the capacity of these mites, which they could not see, to reproduce and transfer to other sheep. This was important because dipping aimed to break the cycle of reproduction that took fourteen days from the laying of eggs to hatching; it had to be done regularly. Even those who agreed to dip, in the hope it might improve he condition of sheep, dipped when they saw the disease rather than as a prophylaxis to break the life cycle of acari. Frequent dipping, they feared, could result in more losses than scab itself.

Interpretations of disease prior to major developments in parasitology and germ theory in the late nineteenth century often included environmental explanations. These were widespread among both white and black livestock owners.[82] Many Afrikaner sheep owners believed that scab was prevalent during droughts and dry seasons, when water and natural veld were in short supply. It seemed to be a seasonal disease that could best be avoided by moving animals to better pastures. This had long been their practice – especially in the drier and poorer northern and western districts of the Cape. Such observations were to a degree justified in understanding the context, if not the cause, of the disease; sheep that were malnourished and in poor condition were usually more susceptible.

Some saw the diseases as internally generated or 'spontaneous', 'in the blood' or in the 'nature of the animal'. Interpretations drew on

[80] Tamarkin, *Volk and Flock*, 16–17. [81] Tamarkin, *Volk and Flock*, 20–1.
[82] Beinart, *Rise of Conservation*; Beinart and Brown, *African Local Knowledge*.

environmental knowledge and experience of pastures, poisonous plants, seasons and types of animal. Religion shaped the mindset of many rural Afrikaners. Their reading of the Bible saw scourges as an expression of God's will or as a punishment for sins. God was the most likely genesis of healing: prayer was better than legislation. In many respects rural Afrikaners functioned in a secular and market-linked world, selling sheep and wool annually. But when the issue of intervention became politicised, they linked science and progress with secularism, and therefore as inimical to the religious order.[83]

Most English-speakers drew on different post-Enlightenment, hybrid ideas, that allowed their religious beliefs to accommodate elements of science, such as parasitology. They linked these interpretations with the powerful Victorian metaphors of light and darkness, progress and backwardness. Those identifying as 'progressives' saw their Afrikaner opponents not only as narrow-minded 'back velders' but as undermining the economic progress of the Colony as a whole. In some respects, they echoed Barrow's early nineteenth-century views.

Afrikaners also had material concerns. Controls over livestock movement would threaten their established patterns of transhumance, which they saw as essential in maintaining the condition of their livestock as well as their way of life. Scab, they argued, was 'caused by poverty' of the veld which could be ameliorated.[84] Yet, transhumance, long central to South African pastoralism, was becoming increasingly constrained as private property spread and fences were erected.[85] Veterinary officers discouraged transhumance because they believed that movement spread disease rather than cured it. When it came to preventing scab, Hutcheon and his colleagues, as well as many self-consciously 'progressive' farmers, were much more confident about dipping. They saw transhumance as socially retrogressive – an enemy of investment, education and a settled rural society. English-speakers were more content with state intervention at this stage, although Afrikaners became dependent on the state in the twentieth century. In a country where the majority of people still had daily interactions with livestock, veterinary regulation was a key point of contestation between the state, its citizens and subjects.

[83] Tamarkin, *Volk and Flock*, 40–1. [84] Tamarkin, *Volk and Flock*, 30.
[85] W. Beinart, 'Transhumance, Animal Diseases and Environment in the Cape, South Africa', *South African Historical Journal*, 58 (2007), 17–41.

Resistance to scab control was not violent – opponents 'shed many words but no blood'.[86] But it did polarise politics in the Colony and deepened the gap between Afrikaners and anglophone South Africans for whom Rhodes was a popular, if not always trusted, leader. The 1895–6 Jameson Raid exacerbated these divisions and became one of the triggers of the South African War. New lines of division, occasionally violent, opened up over the response to rinderpest – a highly infectious cattle virus that caused sudden and more serious losses.[87] Rinderpest moved down the African continent from 1890, killing over 80 per cent of cattle in some areas, to arrive in Botswana and Zimbabwe in 1896. Hutcheon drew on earlier British experience in order to control the disease by shooting infected animals, applying quarantines to infected areas, and restricting movement. These ideas acknowledged the latest scientific understanding of infection and the epizootic character of the disease.

Rinderpest entered the colony in May 1896 in the Tswana reserves of the Northern Cape and gradually spread to neighbouring areas.[88] As in the case of scab, some Afrikaner farmers were strongly opposed to culling; when threatened with mass killing of their cattle, the African people of the Taungs reserve in the northern Cape rebelled. 'Stamping out' was abandoned by October. A fence was tried but again the disease jumped boundaries. The Cape turned instead to inoculation. In 1896, Alexander Edington, director of the Bacteriological Institute, was working to develop an effective injection with methods that had been developed in Europe over the previous couple of decades by Robert Koch, Louis Pasteur and their associates. Theiler and Watkins-Pitchford, leading vets in the Transvaal and Natal, combined forces to work together on a serum in the Marico district of the Transvaal, adjacent to Bechuanaland. Rhodes thought it worthwhile to import Koch himself, a global leader in the field, and set up a state-of-the-art laboratory in December 1896.

86 Tamarkin, *Volk and Flock*, 9.

87 C. van Onselen, 'Reactions to Rinderpest in Southern Africa, 1896–1897', *Journal of African History*, 13, 3 (1972), 473–88; P. Phoofolo, 'Epidemics and Revolutions: The Rinderpest Epidemic in Late Nineteenth-Century Southern Africa', *Past and Present*, 138 (1992), 112–43.

88 D. Gilfoyle, 'Veterinary Research and the African Rinderpest Epizootic: the Cape Colony, 1896–1898', *JSAS*, 29, 1, (2003), 133–54.

Koch, picking up from Edington's experiments, focussed initially on the blood of animals that had been infected, and recovered, as a possible immunising agent. 'Salted' animals, as they were called at the Cape, were highly valued. He found that it had a noticeable, if weak effect, and tried to strengthen this by combining it with blood from infected cattle – following methods he had used for diptheria. He then switched to experiment with techniques, known in South Africa since the lungsickness epidemic of the 1850s, of injecting a mixture of blood and bile from animals which had died from rinderpest. Theiler was working along similar lines.

Koch refined this method, using bile or gall, and by March 1897 thought that he had found an effective prophylaxis. Field tests in infected areas seemed to spread as much as contain the disease, and it was not easy to produce bile on a large scale. When Koch left the Cape in March 1897, Edington developed an alternative strategy returning to bile and blood. Hutcheon initially backed him, but evidence accumulated that this too had limited effect, at most giving some protection for ten days. Although there was competitive confusion between Edington and other scientists experimenting with this strategy, and Hutcheon found it difficult to regulate or recommend the best method, collectively they achieved some success.

Daniel Gilfoyle argues from the statistical evidence that in mainly Anglophone districts where inoculation was largely accepted and widely practiced, survival rates were 70–80 per cent. In districts affected earlier, or where inoculation was opposed or less available, such as in the Transkeian Territories, survival rates were 30–40 per cent and probably less in the African districts of Natal. The rinderpest epidemic petered out in South Africa in 1898 and methods of containment, further improved by Theiler, proved sufficient to control further outbreaks. A more successful attenuated vaccine was developed in the 1940s. The rinderpest epizootic expanded South Africa's veterinary research capacity and important advances were made in the control of a global disease.[89]

Medicine, War and Segregation

Medical provision expanded rapidly in the late nineteenth-century Cape with the number of qualified, registered practitioners

[89] Gilfoyle, 'Veterinary Research'.

quadrupling to 617 between 1864 and 1904.[90] The great majority were British men, attracted by the opportunities in a Colony where population and wealth were growing rapidly. In response to a smallpox epidemic, a limited Public Health Act was passed in 1883; the state also enhanced the system of district surgeons and, by the end of the century, the government Department of Health.[91] Provision was largely private, however, and highly uneven by race, class and region.

Many Afrikaners self-medicated, using a wide range of patent medicines and plants, and African people were largely immersed in their own systems of healing. Missions provided a bridge, and served as the vehicle for both the first women and black physicians. Jane Elizabeth Waterston taught at Lovedale before qualifying in London in 1880, worked briefly as mission doctor in Malawi and the Eastern Cape, and then established a very diverse private practice in Cape Town. William Anderson Soga, brother of Jotello, was educated at Lovedale, qualified in Glasgow, and worked as a medical missionary in the Transkei during the 1880s and 1890s. Anne Digby has rediscovered his Doctor of Medicine thesis which includes an early analysis of diseases in a rural African community, as well as discussion of African ideas of healing.[92]

Medical provision was overwhelmed when South Africa was drawn to the centre of British imperial and world events in the final years of the nineteenth century: tensions between the Witwatersrand mine owners, British imperial politicians and the Boer republics triggered the South African War in October 1899. Whatever the precise causes of the war, it was rooted in Britain's long-standing imperialist ambitions in the region and the scale of the gold-mining industry. After initial Boer military successes, Britain poured major new resources into the conflict and its army successfully occupied Bloemfontein, the Free State capital, in March 1900, Johannesburg in May, and Pretoria, capital of the South African Republic, in June 1900. Faced with likely

90 A. Digby, 'Making a Medical Living: The Economics of Medical Practice in the Cape c. 1860–1910' in Deacon, Phillips and van Heyningen, *The Cape Doctor*, 249–79.

91 E. B. van Heyningen, 'Agents of Empire: The Medical Profession in the Cape Colony, 1880–1910', *Medical History*, 33 (1989), 450–71.

92 A. Digby, 'On the Notable MD thesis of William Anderson Soga, First Black Doctor in South Africa', *South African Medical Journal*, 97, 5 (2007), 345–6; Digby, *Diversity and Division in Medicine*.

defeat in set-piece battles, the Boers changed their strategy and developed guerrilla warfare. Although their military capacity gradually declined, they were able to extend their struggle for close on two further years until the Treaty of Vereeniging was signed in May 1902.

As in most nineteenth-century wars, disease killed more people than did weapons. But to a greater extent than in earlier conflicts, the British army had medical professionals at its disposal. In the sphere of surgery, they probably made a significant difference to the scale of casualties. The war was fought with improved breech-loading rifles – largely Mausers on the Boer side and Lee-Metfords on the British. These had smaller bores and, in contrast to Martini-Henry rifles, used nickel rather than lead bullet heads. Surgeon-historian J. C. de Villiers argues that British surgeons treating war wounds were often struck by the cleanness of incisions with relatively few bullets lodging in bodies. If wounds were kept dry, many healed quickly within a couple of weeks.[93] In this case, ironically, improved weaponry as well as surgery seems to have reduced the number of deaths. This was one reason why British forces initially distributed some lead 'dum-dum' bullets, named after an imperial arsenal near Calcutta, and used by British forces in India. In retaliation, the Boers filed bullet heads to increase their explosive effect. An intense war of words ensued, probably leading to a reduction in the deployment of such bullets.

There was far less success in treating disease. Over 450,000 troops were mobilised on the British side, some from South Africa itself but most from Britain, in the largest overseas mobilisation since Crimea in the 1850s. Of an estimated 22,000 British deaths, about 7,600 resulted directly from conflict and over 14,000 from disease.[94] This ratio was an improvement on Crimea, where an even greater percentage of the 20,000 British deaths are attributable to disease. (The overall scale of casualties in this multinational war was far higher.) On the Boer side, about 87,000 were mobilised and they probably lost fewer in war, around 6,000 – not least because they increasingly avoided set-piece battles. However, the shift of Boer military tactics led the British to adopt scorched earth policies, burning farmhouses and removing Boer families as well as African servants and tenants to concentration

[93] J. C. de Villiers, *Healers, Helpers and Hospitals* (Pretoria: Protea Boekhuis, 2008).
[94] de Villiers, *Healers, Helpers.*

camps. In this way, they hoped to undermine the Boer fighters' capacity to live off the land. Over 120,000 people were pushed into the camps, and of these an estimated 28,000 Boers and 20,000 Africans died of disease and neglect – a staggering death rate of perhaps one-third – far higher than that experienced by combatants. The casualties of this war, perhaps 75–80,000 in all, included many children and women.

Elizabeth van Heyningen concludes that 'camp policy was not deliberately genocidal nor was it unremittingly atrocious'.[95] Nevertheless this was a mass killing. There were in fact two medical scandals during the war, one in connection with typhoid among British troops, and the other concerned with deaths in the concentration camps. The British had the capacity and, to a significant degree, the knowledge, to provide adequate care for their own soldiers and for the Boers and Africans in the camps, but they failed – at least initially – to do so. Kitchener and Roberts were focused on trying to win the war quickly without counting the costs. Emily Hobhouse, a liberal British pacifist linked to Quaker networks, arrived in Bloemfontein early in 1901 to alleviate the parlous situation in the camps. She proved highly effective in exposing what Liberal politician Henry Campbell-Bannerman referred to as 'methods of barbarism'. Hobhouse's searing report on the camps in June 1901 prompted reforms, though these were too late to stem the calamitous epidemics of typhoid and measles.

The cause of typhoid had been identified by 1880 as a bacteria or bacillus *Salmonella typhi*. Scientists knew by the 1890s that it was transmitted through water and food contaminated by infected human faeces and that, like cholera, it was only passed between humans. There was not, as in the case of plague, an animal reservoir. Nevertheless, typhoid was difficult to distinguish quickly from other gastric diseases that caused fever and it could be carried by humans who did not have obvious symptoms – even by those who had recovered from the disease. This made it particularly difficult to control. Although the precise epidemiology of typhoid was not fully understood, the disease was associated with war, and it was known that clean water and food were key prophylactic measures. In 1896 British and German scientists developed a vaccine.

[95] E. van Heyningen, *The Concentration Camps of the Anglo-Boer War: A Social History* (Auckland Park: Jacana, 2013), ix.

Typhoid was already endemic in South Africa, and British doctors knew that there were likely to be carriers among the hundreds of thousands of recruits. It broke out among troops in December 1899 after the British defeat at Magersfontein, on the Modder River, about 35 km south of Kimberley. Infected soldiers then took it into Kimberley when they relieved the siege there in February 1900. The disease travelled with the British army and struck with a vengeance when it reached Bloemfontein in March. This small town was not prepared for so large an army contingent and supply lines were stretched. Despite 2,000 sick and wounded, 'there were no sanitary officers in the army that entered Bloemfontein'.[96] The first field hospital arrived at the end of March equipped to deal with 100 patients. The pressure to deliver food and military supplies by rail from Cape Town was huge, but sections of the line from Port Elizabeth to Bloemfontein were in Boer hands. Railways were at the heart of this war and could have provided the logistical back-up to diminish typhoid.

Roberts wanted to march on Pretoria in order to maintain the momentum of conquest and continued to assemble troops in Bloemfontein. But in April, the Boers succeeded in destroying some of the water supply, which exacerbated the incidence of typhoid and dysentery. Figures suggest 961 deaths among 4,959 cases, a death rate of 19 per cent.[97] Van Heyningen reckons that in total 10 per cent of British soldiers contracted typhoid and 8,255 died of the disease – around 18 per cent of those infected and 38 per cent of British casualties.[98] It was the last of the typhoid wars.

About 400,000 doses of vaccine were sent to South Africa, enough for most of the British soldiers, but relatively few were administered. There were fierce debates as to the effectiveness of the vaccine, with medical personnel taking different sides. Some soldiers were suspicious as they did not understand the mode of transmission, others were careless, despite warnings, about drinking contaminated water. William Burdett-Coutts, a British MP, was sent by *The Times* to South Africa. Arriving in Bloemfontein in early April 1900, he wrote a scathing indictment of the army's treatment of sick troops, which was censored at the time. But he and others persisted in trying to embarrass

[96] De Villiers, *Healers, Helpers*, 108. [97] de Villiers, *Healers, Helpers*, 111ff.
[98] van Heyningen, *Concentration Camps*.

Prime Minister Salisbury and the army in public; his book *The Sick and Wounded in South Africa*, prompted a commission of inquiry.[99] Though the report was reluctant to assign blame, this episode contributed to significant improvements in medical provision during the war and, for the first time, women nurses were recruited on a large scale – about 1,000 served in this role.[100] Effective treatment of typhoid required skilled nursing to enforce isolation and rest, cleaning of patients, and high standards of hygiene. As a direct result, a British military nursing service was established in 1902.

While a similar number of Boers, and probably Africans, succumbed to typhoid in the concentration camps, measles proved an even bigger killer. Although there were some precursors, the camp system was started on a large scale in April 1901, in order to resolve the problem created by scorched earth policies and to neutralise grass-roots material support of the Boer guerrillas. There were forty-two created for whites and twenty-five for blacks, mostly along the railway lines. The measles epidemic swept through soon after their establishment, killing especially the young and old who had not gained immunity through earlier infection. Conditions in the camps were poor, and most inhabitants lived in tents with limited water supplies and medical care. In the cold winter months of 1901, diseases were rife.

Camps were not compulsory, but the majority of relatively poor and uneducated Afrikaner women could not sustain themselves away from their farms. 'In the camps', van Heyningen writes, 'much of the resistance of Boer women to the cultural onslaught of British imperialism took place in respect of matters of health' – a striking replication of the lines of conflict over the Scab Act.[101] Boer women came from an agrarian, largely pre-industrial context, in which the locus of health was the home and the carers were women. Boer women wanted to keep tents closed when someone was ill, to keep the patient warm and swaddle those with fever in clothing, rugs and skins. They thought that a fever should be sweated out and treated with home remedies such as herbs, vinegar, aloe (usually used as a laxative), patent medicines and even animal dung. Boer women preferred their own practitioners, especially when it came to childbirth.

99 W. Burdett-Coutts, *The Sick and Wounded in South Africa* (London: Cassell, 1900).
100 de Villiers, *Healers, Helpers*, 177.
101 van Heyningen, *The Concentration Camps*, 208–9.

British doctors, by contrast, wanted fresh air, sanitation and cleanliness. They saw medicine as part of an imperial and, to some degree, a class mission. They were critical of Boer practices as dirty and inimical to health, just as they despaired of careless soldiers who drunk unboiled water from rivers. They paid inmates, particularly the young men who had ended up in these institutions, to clean camps. Toilet wars or latrine fights were triggered in the camps because physicians feared typhoid-infected stools and were convinced that it was best to treat any enteric diseases in hospital where faeces could be effectively disposed of. Boer mothers did not like this: visiting was restricted, food was seen as poor, and the hospitals were associated with death. They suspected that their children were being starved or poisoned.[102]

War also brought plague to the South African ports. The global plague pandemic that started in China in 1894 spread along the shipping routes of the British Empire.[103] New scientific work, originating in Hong Kong, identified the plague bacillus and connected it with rats. The disease was also strongly associated with poverty and slums. When it arrived in Cape Town in 1900, the medical officer of health believed that African settlements close to the city, perceived as unsanitary, should be removed. Between 6,000 and 7,000 Africans were pushed further out of town to Uitvlugt, near present-day Langa. As in the Boer concentration camps, slum dwellers, as well as those in Muslim areas, attempted to resist the removal and isolation of patients. Maynard Swanson has linked the official response to plague with a 'sanitation syndrome' that fed both white racial prejudice and more rigorous urban segregation in South Africa.[104] Elsewhere, such as in Alexandria, Egypt, sanitary reforms and control of plague were achieved without such drastic removals.[105] While disease was by no means the only factor in the long road to spatial segregation, scientific ideas about sanitation and the isolation of infected people fed into emerging patterns of racial domination. Medical and biological

[102] van Heyningen, *The Concentration Camps.*
[103] M. Echenberg, *Plague Ports: The Global Urban Impact of Bubonic Plague, 1894–1901* (New York: New York University Press, 2007).
[104] M. W. Swanson, 'The Sanitation Syndrome: Bubonic Plague and Urban Native Policy in the Cape Colony, 1900–1909', *Journal of African History*, 18, 3 (1977), 387–410.
[105] Echenberg, *Plague Ports.*

thinking – including metaphors like 'segregation' – were absorbed into eugenicist social engineering.

In the last few decades of the nineteenth century, science and technology were an intrinsic element in colonial power. They underpinned the subjugation of the remaining independent African kingdoms as well as British conquest of the Boers. At the same time, their application in agricultural innovation and the mineral revolution triggered a phase of rapid economic growth and urbanisation that would lead both to war and to the creation of a unified South African state. Different understandings of medicine and healing triggered less violent conflicts over the treatment of animal diseases such as scab, as well as human diseases such as typhoid, measles and plague. In the post-war era, science and technology remained a matter of considerable contestation, though there were now conspicuous efforts to portray it as a neutral and progressive force with the capacity to assuage the bitterness of the past – at least as far as whites were concerned.

4 *Science, Reconstruction and the Imagining of the First 'New' South Africa, 1902–29*

The aftermath and resolution of the South African War brought the Union of South Africa into being as a self-governing, unified state. In the era of reconstruction (1902–10) new political, administrative and ideological arrangements were imagined and negotiated. Proponents of white unity and power, who promoted an ethos of South Africanist nation-building, also articulated ideas that would lead to the systematic exclusion of black South Africans on grounds of race. The rhetoric of science and progress played an important role in enabling and legitimising the emergence of new state structures and the political alliances underpinning Union. Science was also a beneficiary of these processes through the growth of state bureaucracy and development of a university system.

Much of the historical literature about this period focuses on the ideological, political, economic and constitutional aspects of reconstruction and segregation. Science, in the guise of efficiency and progress, was a key component of these debates. As discussed in the previous chapter, ideas about sanitation and public health fed into white racialised concerns about impoverished black areas as reservoirs of disease. Attempts to control plague fuelled urban segregation in Cape Town and Port Elizabeth.[1] Hygienic cleanliness implied racial cleansing. In the rural areas, intensified interventions to control livestock diseases impinged on the lives of the great majority of people, black as well as white. Scientifically conceived ethnographies contributed to policy formation around African reserves and the restoration of chieftaincy.

While scientific funding and research were partly concentrated in government departments, South African scientific initiatives were increasingly multi-sited within companies, universities, hospitals, and museums. Most scientists were to some extent constrained by the priorities of their organisations, but there were also opportunities for free-ranging research. The diversity of approaches in some fields is

[1] Swanson, 'Sanitation Syndrome'.

striking. In entomology, for example, Henri Junod continued a tradition of systematic collection and classification of species; Eugene Marais, a major Afrikaner literary figure, began careful observation of termites, developing theories about their organisation and the natural world more generally; Charles Lounsbury used his training as an applied entomologist to unravel the life cycle of ticks, thus revealing their capacity to transmit dangerous animal diseases. In geology, Hans Merensky was drawn into practical prospecting and commercial mining, while Alex du Toit theorised continental drift while working as a geological consultant to the mining industry. Clarence 'Peter' van Riet Lowe, a government civil engineer and bridge-builder, became a professional archaeologist, stimulated by his finds of flints and artefacts in the field. Gifted individuals could encompass different spheres in aspects of their work; an expanding national economy and unified state bureaucracy helped to broaden intellectual horizons.

The growing institutional base in which scientists operated, as well as the requirements for funding, produced an increasingly complex politics of science that was also imbricated with nation-building. We focus on this at the beginning and end of the chapter. We also illustrate the activities and passions of a minority with less defined roles – missionaries, mineral prospectors, enthusiasts, and self-appointed experts – who were able to pursue relatively independent explorations but nevertheless made a major impact both on the store of scientific knowledge and on the popular scientific imagination. In this chapter, our field of vision moves away from British, Cape, and anglophone networks – the centre of gravity in the nineteenth century – towards the growing dynamism of the mineral-rich Transvaal.

New Institutions: The South African Association for the Advancement of Science, and the Royal Society

The South African Association for the Advancement of Science (which has adopted the abbreviation S2A3) was established in 1902–3 with high-level political support as part of the process of creating a new South Africa. Modelled on the British Association, the impetus behind the S2A3 was provided by a British-born mining engineer, Theodore Reunert, while its founding president was the Scottish-born astronomer at the Cape, David Gill. In his inaugural address, Gill argued that science could play a vital role in healing the bitterness of the South

African War while cementing the imperial connection in ways that transcended politics. 'Science', he maintained, 'knows no nationality, and forms a meeting-ground on which men of every race are brethren, working together for a common end – and that end is truth.'[2]

Gill was an astronomer of international repute, known for measuring the distance of the sun from the earth in 1877 and photographing the Great Comet of 1882. Knighted in 1900, he was appointed as scientific adviser to the colonial governments of the Transvaal and Orange Free State in a clear indication of the importance that Britain now accorded to its expanding southern African empire. Gill also aimed to promote science in the public and political domain, associating himself closely with Lord Milner's post-war objectives for anglicising South Africa. In this regard he was more overtly a supporter of imperialism than his predecessor as Her Majesty's Astronomer at the Cape, Thomas Maclear.

Like Maclear, Gill's approach to astronomy was applied as well as theoretically informed. Gill's desire to map the arc of the meridian from the Cape to Cairo, which he proposed to Rhodes and the British high commissioner, Milner, appealed to their celestial imperial imaginations and also their appreciation of the need for accurate survey and mapping. As elsewhere in the world, mapping, warfare and imperialism were closely connected. When Gill died in 1914, the 30th arc of the meridian had been surveyed as far as Tanzania. It was finally completed in 1954.[3]

It suited Gill to project science as a non-partisan, apolitical pursuit. But the S2A3 was premised more broadly on British authority, especially the notion that the universality of science was coterminous with the pursuit of 'civilisation' and progress through technical achievement. In the nineteenth century, such ideas were underpinned by the Christian gospel and a view of moral progress. In the twentieth, they were increasingly secularised in the language of science. Colonial

2 Address by Sir David Gill to the SAAAS, in *Report of the South African Association for the Advancement of Science. First Meeting Cape Town 1903*, 20, 36.

3 'Obituary: Sir David Gill', *The Observatory*, 37, 472 (1914), 115–17; D. Gill, 'On the Origin and Progress of Geodetic Survey in South Africa, and of the African Arc of Meridian', *Report of the Seventy-Fifth Meeting of the British Association for the Advancement of Science. South Africa 1905* (London: John Murray, 1906), 228–48; J. Smith, 'From LaCaille to Gill and the Start of the Arc of the 30th Meridian', 'Shaping the Change' XXIII FIG Congress, Munich, 8–13 October 2006, 1–17, www.fig.net/resources.

surveying and mapping were intimately connected with power and the drawing of new boundaries, and these often contested.[4]

The S2A3 started with a foundation membership of 268 (including 14 women), growing rapidly to a peak of 1,322 in 1906. Amateur as well as professional scientists were welcomed into the Association. The disciplines were represented in six discrete sections ranging from astronomy and mathematics to engineering, education and anthropology. Applied as well as pure science was encouraged. Thus, membership of the S2A3 frequently overlapped with the specialist professional and technical organisations that had been constituted from the 1890s, including the Geological Society of South Africa, the South African Institute of Electrical Engineers, the Chemical and Metallurgical Society of South Africa, the Transvaal Medical Society, and the South African Association of Engineers and Architects.[5] Reunert had in mind an annual congress of engineers when he first proposed the organisation that became the S2A3.

The broad geographical reach of the S2A3 represented a conscious attempt to create a positive spirit of national unity and reconciliation. The naming of the Association as 'South African' was itself aspirational in a divided country that had yet to come into being.[6] Members of the Association were drawn from schools, university colleges, missions and museums. Although predominantly anglophone at its inception, the S2A3 aimed to include Afrikaners as well. An article in the Grahamstown-based *African Monthly* in 1907 welcomed the S2A3 as 'evidence of true progress towards nationality' and cast the institution as 'an important factor in the scheme of Federation', but in likening the S2A3 to a 'parliament', it also observed that the future South Africa would likely be 'British in character and therefore Imperial in tendency'.[7] Sentiments like these aroused strong Afrikaner resentment and

[4] L. F. Braun, *Colonial Survey and Native Landscapes in Rural South Africa, 1850–1913: The Politics of Divided Space in the Cape and Transvaal* (Leiden: Brill, 2015).

[5] Brown (ed.), *A History of Scientific Endeavour*, 70; B. F. J. Schonland, 'The South African Association for the Advancement of Science, Its Past and Present', *SAJS*, 49 (October–November 1952), 61–8.

[6] C. Plug, 'The Origin and Early Development of S2A3' in *S2A3. Rudolf Marlow Centenary Brochure-Brosjure 1902–2002* (Pretoria: South African Association for the Advancement of Science, 2002); S. Dubow, 'A Commonwealth of Science: The British Association in South Africa, 1905 and 1929', in S. Dubow (ed.), *Science and Society in Southern Africa* (Manchester: Manchester University Press, 2000), 66–99.

[7] *African Monthly*, 2, 8 (1907), 227.

encouraged the formation of rival cultural and scientific societies, notably the Suid-Afrikaanse Akademie vir Wetenskap en Kuns, which constituted itself as an academy dedicated to science and the arts in 1909, a key year in the Unification of South Africa. There were no black members of either the Association or the Akademie.

The S2A3's strategy of organising annual meetings in different cities, coupled with its network of local chapters, did much to engender a sense of wider belonging. A moderate subscription fee of £1 per annum helped to make the Association accessible to the educated white citizenry. Its house journal, the *South African Journal of Science* (SAJS), served as a vehicle for ideas and as an accessible outlet for local research. The composition of the S2A3 was thus notably broader than the more exclusive Royal Society of South Africa, established by Royal Charter in 1908, whose statutes were modelled on the Royal Society of London. The inaugural Council of the Royal Society was composed of seven leading men of science: astronomer S. S. Hough (president), botanist Harry Bolus, marine biologist J. D. F. Gilchrist, geologist George Corstorphine, palaeontologist Robert Broom, mathematician 'Jock' Beattie (later first principal of the University of Cape Town) and Louis Péringuey of the South African Museum. All had been born abroad and four were Scots.[8]

The Royal Society owed its immediate origins to the South African Philosophical Society, established in 1877 under the presidency of the British high commissioner Bartle Frere. Its fellowship was the preserve of the country's scientific elite and its house journal, *Transactions of the Royal Society of South Africa*, reflected this seniority. The working arrangement implied that the S2A3 should not attempt to emulate the research intensity of the Royal Society; instead, the two organisations strove to work in tandem. As president of the Philosophical Society in 1901–2 as well as inaugural president of the S2A3, Gill straddled both scientific organisations. Many fellows of the Royal Society were also active participants in the S2A3, including Gilchrist, who served in senior positions for both organisations. Gilchrist bore the title 'government biologist', perhaps the first such designation in the world.[9]

[8] J. Carruthers, 'Scientists in Society: The Royal Society of South Africa', 3; J. M. MacKenzie with N. R. Dalziel, *The Scots in South Africa: Ethnicity, Identity, Gender and Race, 1772–1914* (Manchester: Manchester University Press, 2012).

[9] *Nature*, 118, 2978 (1926), 776–7.

As national academies, the Royal Society and S2A3 symbolised the melding of colonies into a self-governing state as well as optimism about the future. Nevertheless, they also reflected the divisions of the present. The Royal Society was predominantly Cape-oriented and unsure of how to negotiate the growing industrial and technical power of the Transvaal. The S2A3 was more regionally balanced, but its largely anglophone composition became clear in 1905 when it hosted an official visit from the British Association for the Advancement of Science, involving nearly 400 overseas delegates. The project was led by Gill, who secured the support of the British colonial secretary, Joseph Chamberlain, in the hope that the British Association's visit would assist the 'work of conciliation in South Africa'.[10] Sir Donald Currie offered reduced fares on his Union-Castle line on the grounds that the enterprise was 'thoroughly in accord with the spirit of Imperialism'. Travel to the various cities in which the joint meeting was held, as well as tourism (the event was referred to as a glorified 'picnic'), required the provision of special trains and close coordination of the colonial railway systems, the integration of whose infrastructures was one of the major appeals, and payoffs, of political unification.

Figure 4.1 Members of the British Association for the Advancement of Science en route to South Africa aboard SS Saxon as photographed by Mrs C. Henderson.
Credit: Wellcome Collection, London

[10] Dubow, 'A Commonwealth of Science', 66–7, 69.

In countless official speeches, the visit of the British Association was celebrated for conferring stature on the S2A3 and as a testament to the benefits of the imperial connection. Although gestures of inclusiveness towards Afrikaners were routinely made, there was also plenty of evidence of imperial triumphalism. This was plainly evident in the opening address of Sir George Darwin (the eugenicist son of Charles), who observed with satisfaction 'the dominance of the European race' in America, Australasia and South Africa before proceeding to expound on the links between natural selection and forms of political governance. Later in the trip, Darwin opened the new railway bridge for passenger travel at the Victoria Falls and took the opportunity to laud the achievements of Cecil Rhodes.[11] Theodore Reunert cabled his 'warmest congratulations', observing that 'more links had been formed in the chain of civilization'.[12]

The published proceedings and a commemorative volume, *Science in South Africa* (1905), offered an overview of the state of South African research and also a statement of the organisation's ambitions. The format served as a template for future publications and as a benchmark of the 'state of many fields of formal study at this time'.[13] One of the purposes of the 1905 meeting was to attract interest and funding for ambitious new projects. Cambridge anthropologist A. C. Haddon used his presidential address to Section H (Anthropology) to survey existing knowledge about the different 'races' of South Africa and spoke loftily about 'doomed' 'primitive' people who merited thorough investigation before they disappeared. In subsequent years, Haddon's manifesto became a touchstone for South African anthropologists seeking government and institutional support for research.[14] Henri Junod won a prize in 1907 for an essay calling for the comprehensive study of every African 'tribe'.

David Gill's proposal to expand the geodetic survey to the whole African continent was carefully crafted to appeal to utilitarian as well as pure scientific interests. He assisted in fixing the position of the 20th

[11] Dubow, 'A Commonwealth of Science', 70, 72.
[12] *The Times*, 14 September 1905; *Nature*, 72, 1868 (1905), 68–72.
[13] J. Carruthers, 'Reviewing Historical Books: "Science in South Africa, 1905"' (2014), www.royalsocietysa.org.za.
[14] A. C. Haddon, presidential address to Section H Anthropology in *Report of the Seventy-Fifth Meeting of the British Association for the Advancement of Science*, 525.

meridian from the Orange River towards Rietfontein on the border of Namibia, arising out of a boundary dispute between Britain and Germany in South West Africa.[15] During Gill's tenure 'Observatory mean time' was established and disseminated by telegraph to urban communities in the Cape. In 1903, all clocks in South Africa were adjusted to correspond to two hours east of Greenwich: uniform time therefore became a meaningful reality in the period leading up to the country's unification.[16] Funding for Gill's African meridian project was one of the principal outcomes of the 1905 joint meeting. Another result was the decision to transform the Transvaal Observatory in Johannesburg into the Union Observatory, with R. T. A. Innes, a stalwart of the S2A3, designated Union astronomer. This was a double win for Gill whose tenure at the Cape was an especially productive one for astronomy.

The relationship between the Cape and the Union Observatory reflected the emerging importance of the Transvaal and the reorientation of scientific as well as the political centres of gravity to the north. In respect of astronomy, Johannesburg's industrial and technical prowess was evident in Innes's collaboration with the newly established Campos Rodrigues Observatory in Lourenço Marques (Maputo), Mozambique, in matters of meteorology and timekeeping. The ambitions of the Campos Rodrigues Observatory testified to Portugal's reform-minded republican regime and its desire to solidify economic ties to Johannesburg in respect of labour supply and rail transport. The S2A3 gathered twice in Lourenço Marques, in 1913 and 1922, thereby confirming the port city's regional importance for Johannesburg, its principal source of traffic and trade.[17]

The confluence of scientific and industrial expertise in Johannesburg was also signalled by moves to establish a South African Institute for Medical Research (SAIMR) in 1912. At a well-attended ceremony to mark the laying of its foundation stone in 1913, Irish-born surgeon Sir Kendal Franks invoked 'the good of mankind' and looked forward to

[15] D. Gill, *A History and Description of the Royal Observatory, Cape of Good Hope* (HMSO, London, HMSO, 1913), civ–cv.
[16] Gill, *A History and Description of the Royal Observatory*, cxliv.
[17] P. M. P. Raposo 'Time, Weather and Empires: The Campos Rodrigues Observatory in Lourenço Marques, Mozambique (1905–1930)', *Annals of Science*, 72, 3 (2015), 279–305.

the day when 'our land' would be freed of the scourge of disease.[18] Housed in a splendid building designed by Herbert Baker next to the Johannesburg General Hospital, it opened two years later. The SAIMR was established as a collaborative enterprise by the government and the Witwatersrand Native Labour Association, the principal recruiting agency for the gold mines. In addition to providing laboratory and diagnostic services, the SAIMR pursued research into diseases endemic to the mines such as tuberculosis, silicosis, hookworm and typhoid.

The SAIMR's research remit was broad. It made its mark at a time when the first university medical chairs were only just beginning to be established and when veterinary research was arguably more advanced than human medicine. As a joint creation of the mines and the state, the SAIMR was the outcome of a growing realisation – in the context of appalling living conditions endured by black migrant workers in particular – that sustained national prosperity required investment in 'the prevention and treatment of human disease', a mantra that served as the SAIMR's founding charter.[19] Yet this does not mean that patient-centred welfare or health was its foremost priority.

In 1912, South Africa became the first country in the world to compensate financially for silicosis and tuberculosis as occupational diseases, albeit on a racially differentiated scale of entitlement. But Randall Packard has argued that for all the concern about TB amongst black miners, including a Commission of Enquiry in 1914, medical research focussed largely on the requirements of the mines. In this respect, some physicians favoured migrant labour because it was seen to give underground workers annual respite from the conditions that facilitated spread of the disease.[20] Jock McCulloch's history of silicosis similarly argues that there was a tight circle of medical expertise with unhealthily close interconnections between state, industry, and the SAIMR. In South Africa, perhaps more than Australia, Britain, or the United States, 'it was science which defined risk and therefore the legal

[18] *Rand Daily Mail*, 24 April 1913.

[19] J. F. Murray, 'History of the South African Institute for Medical Research', *South African Medical Journal*, April 20 (1963), 389.

[20] R. M. Packard, *White Plague, Black Labor: Tuberculosis and the Political Economy of Health and Disease in South Africa* (Berkeley: University of California Press, 1989).

obligations of employers and the state.'[21] This network nevertheless developed particular interests and skills in connection with lung disease. William Steuart, who qualified as an engineer as well as a radiologist in London, constructed a mechanical ventilator to assist the breathing of miners on the Witwatersrand suffering from lung disease. He completed it in 1918 during a polio epidemic, but it became available too late to save the last victim of the outbreak.[22] This wooden 'iron lung' predated the well-known 'Drinker respirator' by a decade.

The SAIMR's first director, Wilfred Watkins-Pitchford, was a bacteriologist who qualified as a doctor in 1891 at St Thomas' Hospital, London. He came to South Africa with the military during the South African War via Mumbai where he had worked as a government plague officer, with particular interests in the efficacy of disinfectants, from 1898. This circuit was not unprecedented: Robert Koch, discussed in the previous chapter in connection with rinderpest, identified the cholera bacterium *Vibrio cholerae* in India in 1884; F. E. Kanthack cut his teeth as a civil engineer in India before becoming director of Irrigation in the Cape and then Union.

In 1902 Watkins-Pitchford moved to work with his brother, Herbert, a veterinarian bacteriologist at the Allerton Laboratory in Natal, and in 1911 he took up a role as the Transvaal government pathologist and bacteriologist. At the SAIMR, Watkins-Pitchford became expert on silicosis, a consequence of breathing in mine dust, and he published several important papers on the topic. In 1916 he became chair of the Miners's Pthisis Medical Bureau, the first such organisation in the world. When Watkins-Pitchford retired in 1926, the SAIMR's medical staff had expanded to more than seventy. It served as an important node for the promotion of research in public health, publishing research on bacterial mutations by G. D. Maynard, on pneumococcal vaccines by F. S. Lister, as well as pioneering work on yellow fever by A. J. Orenstein.[23]

[21] J. McCulloch, *South Africa's Gold Mines and the Politics of Silicosis* (Suffolk: James Currey, 2012), 59.
[22] M. Wade, 'Straws in the Wind: Early Epidemics of Poliomyelitis in Johannesburg, 1918-1945', MA thesis, University of South Africa (2006), 92.
[23] 'Obituary. Wilfred Watkins-Pitchford', *The Lancet*, 1 November 1952, 890–1; J. F. Brock, 'Some Highlights of Medical Research in South Africa', in Brown (ed.), *A History of Scientific Endeavour*, 192–3.

Born in Odessa (or to Russian parents in the US) in 1879, Orenstein was educated in Philadelphia and went on to develop a remarkable international career as a medical scientist and occupational health reformer.[24] From 1905–12 he worked with US army physician William C. Gorgas on the eradication of malaria and yellow fever during the construction of the Panama Canal. The collaboration with Gorgas brought Orenstein to German East Africa (Tanzania) in 1913 and then to South Africa where Gorgas was asked to advise on tropical diseases among gold miners. The following year he was appointed as a sanitation expert by Rand Mines in Johannesburg, rising to become the company's chief medical officer. As a strong proponent of health reform on the gold mines, Orenstein sought to alleviate diseases such as pneumonia, tuberculosis, and hookworm. Improving sanitation and relieving overcrowded accommodation was a major concern – though McCulloch argues that Orenstein's closeness to the mining industry disposed him to play down the deleterious effects of the migrant labour system on black workers' health.[25] Twice a delegate to International Labour Conferences in Geneva in the mid-1920s he enjoyed strong links to the International Labour Office. Orenstein also served with high rank as director general of medical services in the South African defence force during both world wars.

Swiss Missionaries and Science in the Transvaal and Mozambique

By the late nineteenth century, governments and private corporations in South Africa employed professional scientists to address areas that they believed needed urgent attention. Yet some of those engaging in scientific activity in the country were not formally trained, and the gap between experts and amateurs in certain fields could be narrow. Those who bridged the divide included missionaries educated at universities

24 'Obituary Notices. A. J. Orenstein', *British Medical Journal,* 19 August 1972, 478; 'Orenstein, Dr Alexander Jeremiah', S2A3 Biographical Database of Southern African Science, www.s2a3.org.za/bio.

25 J. McCulloch, 'Medicine, Politics and Disease on South Africa's Gold Mines, *JSAS*, 39, 3 (2013),543–56 and 'Mining Evidence: South Africa's Gold Mines and the Career of A. J. Orenstein', *Social History of Medicine*, 31, 1 (2018), 61–78.

in Europe who evinced an intense curiosity about the world around them. Some were medical men while a number were directly involved in education as well as Christian conversion and the management of mission stations. To a greater extent than most government scientists, they were in daily interaction with African people.

Although based at some distance from South Africa's major towns and institutions, missionaries were not necessarily isolated. Many had frequent contacts with their home societies in Europe or with southern African missionary networks and with African intellectuals. Major mission stations were hubs of activity in the social geography of southern Africa in the early twentieth century, with their expanding churches, educational institutions, hospitals and agricultural initiatives. A limited number of mission societies bought extensive areas of land and ran communities that were economic as well as religious focal points. The expansion of Christianity and education among African people at this time was a critical element of social change.

Patrick Harries has written the most detailed study of missionary scientists in southern Africa, focusing on Swiss mission societies based at Neuchâtel. In an edited collection with David Maxwell, they talk of the 'spiritual in the secular' – though the converse also applies.[26] Paul Berthoud and Henri Alexandre Junod from Neuchâtel were among the most talented university students of their generation. They found opportunities in southern African to develop as intellectuals, bridging fields as diverse as cartography, entomology and anthropology. Not only did they connect with international networks, they also operated across colonial boundaries in areas removed from most anglophone South African scientific work.

Paul Berthoud initially joined the well-established Paris Evangelical Mission Society in Lesotho in 1873. Two years later he started a new mission at Valdezia in the Soutpansberg, and from 1883 the Neuchâtel church combined with two others to form the Swiss Romande Mission. In addition to keeping roots in Lesotho and the Transvaal, they established a major centre at Rikatla, in coastal Mozambique, about 30 km north of Maputo. When Berthoud made his first trip from Lesotho to the Soutpansberg in 1873, he carried an aneroid barometer and found

[26] Harries, *Butterflies and Barbarians*; P. Harries and D. Maxwell (eds.), *The Spiritual in the Secular: Missionaries and Knowledge about Africa* (Grand Rapids, MI: Eerdmans, 2012).

that some of the landmarks on earlier maps had been misplaced. These were relatively new devices, developed in France in 1844, which could measure air pressure, and hence also height, without the use of liquids. Berthoud's observations were used for the first detailed map of the Soutpansberg and southern Mozambique. His younger brother, Henri, who arrived in 1881, travelled widely with map-making instruments, both while on mission work and on cartographic expeditions. He first mapped the course of the Levubu River to its confluence with the Limpopo and on later trips surveyed the coastal route from the mouth of the Limpopo to Maputo.

This missionary cartography was incorporated into Fred Jeppe's maps of the Transvaal. The German-born Jeppe was appointed postmaster-general of the South African Republic in 1868 and he issued his first map in that year.[27] Assisted by Alexander Merensky of the Berlin Mission Society, who was attempting to gain a foothold in the Pedi kingdom, Jeppe incorporated information from such missionary sources to improve his coverage. As there was still no systematic survey of the Transvaal, he was dependent, even up to his final map in 1899, on 'a hodgepodge of sketch inspections, block surveys, and a few astronomically fixed positions, which made it as much art as science'.[28] When the Swiss missionaries published their maps, their information was incorporated into his updated versions, the last and best of which was printed in Switzerland. In 1895 the Portuguese army used a Swiss missionary map when invading the Gaza kingdom and capturing King Gungunyane; British armies relied on Jeppe's map to assist with their invasion of the Transvaal in 1900.[29]

Map-making used well-established scientific skills to measure, record and help understand environments that had not been fully incorporated into the colonial world. Africans and early settlers were deeply familiar with the landscape and routes of this terrain, but mapping offered greater capacity to define boundaries, to plan journeys or campaigns, and to modify environments. Maps also validated

[27] J. Carruthers, 'Cartographic Rivalries: Friedrich Jeppe and the Transvaal', in N. Etherington (ed.), *Mapping Colonial Conquest: Australia and Southern Africa* (Crawley: University of Western Australia Press, 2006), 102–22.

[28] L. Braun, 'An Agent in Pretoria? Fred Jeppe, the Cartography of the Transvaal and Imperial Knowledge before 1900', *Cartographic Journal*, 55, 2 (2018), 111–20.

[29] Harries, *Butterflies and Barbarians*, 104–5.

new place names. Valdezia was the Latin name for the Vaud area of Switzerland; other stations such as Elim and Antioka were derived from the Bible. Mountains were named in Dutch/Afrikaans such as Soutpansberg and Spelonken. Harries writes of the way in which 'missionaries appropriated and Christianized the land' through scientific mapping and naming.[30] This, he argues, helped them signify increasing familiarity with the terrain. Yet they also used African names for rural places and settlements, such as Chief Magude's site near the Nkomati River, and Berthoud recommended African clan names for geographical districts. The record of African settlement largely dropped off South African maps in the twentieth century so that densely populated areas were sometimes depicted as devoid of habitation, except for small magisterial towns and mission stations.

Junod arrived from Neuchâtel in 1889 to start work at Rikatla station. He plunged immediately into botanical and entomological fieldwork, at the same time throwing himself into the mission field: 'nature is the work of God', he wrote, 'and merits attention and calls for study'.[31] By the time he left for a sojourn in Switzerland in 1896, he had collected over 500 species both to the south and north of the station. He continued his botanical researches in various parts of southern Africa and by 1906 had amassed specimens of 2,800 different plants; at least thirty species and a genus were named after him. Initially most of his specimens went to William Barbey in Geneva, who played a role as informal coordinator of global records, similar to the Hookers at Kew. Plant scientists were then moving their attention from collection and classification, using their increasingly powerful microscopes to focus on morphology and physiology (the structure and functioning) of plants. Yet there was still ample space for amateur collectors in areas such as Mozambique and the eastern Transvaal, which had been less systematically studied than the Cape.

Junod was based for a period after 1899 at the school for evangelists in Shiluvane, eastern Transvaal. In a diverse ecological area, with lowveld, escarpment forests, and high plateau all within reach, he concentrated increasingly on beetles, caterpillars, moths and butterflies. He found many species new to science, published widely, and thereby 'left an indelible mark on the entomological map of south-

30 Harries, *Butterflies and Barbarians*, 115.
31 Harries, *Butterflies and Barbarians*, 123.

east Africa'.[32] Most of his insect species went to Neuchâtel where the Natural History Museum assisted him with equipment and identification. Junod increasingly understood, like Mary Barber some decades earlier, the complex interactions between species. He employed African collectors, particularly Elias Libombo, who was skilled in trapping butterflies and, as a trained diviner, was able to explain African knowledge and practices. In the process, Junod 'developed a deep respect for indigenous powers of observation and an admiration for the indigenous system of classification. He also had a high regard for the way in which local people exploited plants for qualities that were at once medicinal, magical, sartorial and nutritional in edible or liquid form'.[33] Junod found value in African ways of thinking about nature and the connections made between different plants on the basis of their appearance, utility and properties, both material and supernatural. This knowledge, he wrote, was 'a true and, in a certain sense, scientific observation on their part'.[34]

In southern Mozambique, Junod learnt the Thonga language and worked with African intermediaries to publish Christian texts, a book on elementary science (1904), and eventually a translation of the Bible. He was largely responsible for recording the Ronga version of the language, based on the dialect in southern Mozambique. His extensive contacts with African people, and his interests in local knowledge and language, intensified his focus on the growing field of ethnography or anthropology. After an initial publication in French on 'Les Ba-Ronga' (1898), he wrote increasingly in English and published a two-volume work on *The Life of a South African Tribe* (1912–13, updated 1926–7) – at that time the most detailed and systematic description of a southern African society. Junod developed links with anglophone academics and government officials in the Transvaal and through his diverse connections provided a flow of ideas and material from the southern African periphery to Europe. In Junod's eyes, he was using a scientific method, akin to entomology and botany, for his work on people. In 1905, he presented a paper to the newly founded S2A3 and in 1920 he served as president of its Anthropology section.

[32] Harries, *Butterflies and Barbarians*, 136.
[33] Harries, *Butterflies and Barbarians*, 220–1.
[34] Harries, *Butterflies and Barbarians*, 138.

Though he and other Swiss missionaries were not unique in their range of interests, Junod was, in his time and place, a formidable intellect who was determined to understand the world around him. He was driven by religion and the wish to Christianise but became increasingly sympathetic to African society and knowledge. This led him to attempt, with the assistance of Libombo and others, a reconstruction of Thonga society as they imagined it had been before colonisation. In the tradition of Kolb and other empathetic observers, he presented an ordered society that worked for its members and to a significant degree carried effective moral as well as functional components. His was a representation of a 'good' society as opposed to a 'savage' society.

Junod was, however, reluctant to engage with issues of social change, despite his familiarity with mission activity, colonisation and extensive labour migration from Mozambique, as well as the alienation of land in the Transvaal. Harries calls his work 'salvage anthropology'. By the 1920s, his political views, informed by these ideas, were tending in the direction of liberal segregationism: he thought that African people needed protection from the new forces of social change, which he saw, in the main, as destructive.[35] Junod was sensitised to the negative impact of the migrant labour system, which fifty years later was to influence social historians such as Patrick Harries in South Africa .[36]

Mavericks: Merensky and Marais

In South Africa, Hans Merensky was the first to understand the country's platinum deposits. He was the son of Alexander, a central figure in the Berlin Missionary Society, which sought to minister to the Pedi kingdom in the nineteenth-century Transvaal.[37] Alexander Merensky had acted as a go-between with the Boers and the British in the 1870s when the Pedi Kingdom faced overwhelming colonial forces. He purchased extensive tracts of land for the Botshabelo mission station near Middelburg in the south-eastern Transvaal and constructed a Christian

[35] Dubow, *Racial Segregation.*

[36] P. Harries, *Work, Culture, and Identity: Migrant Laborers in Mozambique and South Africa* (London: James Currey, 1994).

[37] P. Delius, *The Land Belongs to Us: The Pedi Polity, the Boers and the British in the Nineteenth Century Transvaal* (London: Heinemann, 1984).

society that served as a cultural and economic as well as religious hub. He was also active as a map-maker.

Hans Merensky was born on Botshabelo in 1871 and educated in Germany, acquiring the highest qualification in mining studies, as well as extensive practical experience in geology. He settled in Johannesburg in 1905, advertising himself as a 'Consulting Geologist and Mining Engineer (previously of the Prussian Mines Department)'.[38] As Transvaal mining law gave rights over minerals to landowners, many farmers tried their hand at prospecting, but they needed experts who could identify minerals. Merensky quickly found work and travelled widely in the subcontinent. When the first diamonds were identified on the Namibian coast in 1908, he explored the geological terrain and developed a distinctive theory about their origins. Merensky did sufficiently well, earning about £2,000 a year, to stay at the Rand Club in Johannesburg – a focus of information and political gossip about mining. He also acted as an investor, but in this he failed; his financial troubles were compounded when he was interned as a German alien during the First World War. In 1918, at forty-seven, he had to start again, initially prospecting for coal. Merensky gained a first-hand overview of deposits in South Africa and Zimbabwe and became involved in early research with German companies on the liquification of coal for petrol. Neither Germany nor South Africa had known oil reserves, and demand was increasing sharply with the growth of motorised transport.

Platinum deposits were known to exist near the Waterberg, and in the early 1920s prices were high. In 1875 the international standard metric measures for a metre and a kilogram were made of an alloy of platinum and iridium because of their inert qualities. As smelting techniques improved, it was used as a setting for diamonds and in Fabergé's jewelled eggs.[39] Demand increased greatly in the early twentieth century for use in engines, ignition systems and electrodes, where metals had to withstand high temperatures. The main global supply was from newly communist Russia and this, together with inflated prices, triggered a particular urgency to find alternative deposits. It

[38] E. W. Machens, *Platinum, Gold and Diamonds: The Adventure of Hans Merensky's Discoveries* (Pretoria: Protea Book House, 2009), 42.

[39] Machens, *Platinum*, 93.

was, however, not easy to locate and exploit in payable quantities since surface deposits were scarce and not generally visible.

In 1924 a farmer near Burgersfort in the eastern Transvaal found a greyish-white metallic substance in an erosion channel and sent a sample to Merensky who had 'an astounding sense for a find, phenomenal powers of observation and considerable experience'.[40] Confirming the metal as platinum, he joined the farmer to track the origins of the small alluvial trace. He detected a platinum reef upstream from the lighter colouring in the rock and, with assistants, traced this in a broken line of about 160 km.[41] Without ready cash, he persuaded German contacts to invest in a mining enterprise named L. P. (Lydenburg Platinum).

Merensky could see that the reef, a few metres wide, ran diagonally underground in a way similar to the gold-bearing rock. But within a few months, his geological experience and knowledge of the Transvaal led him to surmise a different pattern of deposits. He understood that the Bushveld geology was distinctive, marked by its vegetation. He theorised that the platinum-bearing reef was in a thin saucer-shaped layer between other rock layers or pancakes. They had found the eastern protruding section but there were likely to be other outcrops. Late in 1924 he travelled to Mokopane (then Potgietersrus) where he and others traced a new reef; he then moved on to Rustenburg to find what he predicted was the western rim of the saucer. This proved to be the richest, stretching over nearly 300 km. Although the rock formations holding platinum did not always replicate the Burgersfort deposit, he found ways to detect them, and by the end of 1925 the surface outcrops of the reef, named after him, had largely been mapped. Merensky took Prime Minister Hertzog to inspect the Mokopane deposit that year.

The original Lydenburg Platinum investors were able to start a financially successful mine. Anglo American, formed in 1917, which monopolised diamond production together with its affiliate De Beers, focused with partners on Mokopane, where over £1 million was invested. Merensky wrote at least thirteen reports on various platinum

[40] Machens, *Platinum*, 68; *South African Journal of Geology*, 102, 3 (1999), Special Issue Commemorating the 75th Anniversary of the Discovery of the Merensky reef.

[41] L. B. H., 'Dr Hans Merensky: Centenary of the Discoverer of the Platinum Reef', *Platinum Metals Review*, 15, 3 (1971), 102.

deposits and had the imagination to predict that South Africa would soon become the largest producer in the world.[42] Percy Wagner, senior geologist in the Geological Survey of South Africa, popularised the coinage 'Merensky reef' and published a summary of the finds in 1929.[43]

The earlier Namibian diamond finds were assumed to be a result of stones washed down the Gariep river and its tributaries. Some diamonds did seem to lie in the coastal sand. But Merensky had speculated in 1908 that the primary source for the deposits was not in the interior but in what had been part of the Atlantic coastal shelf. While others rushed to the coast, he instructed his prospectors to dig trenches along the 'oyster line' buried below the sand. Determined to start his own operation, he was able to buy out other claim holders in the Alexander Bay dunes. Backed by the mining and property magnate Julius Jeppe, and the platinum miner Gustav Becker, his enterprise found around 7,000 stones of over 12,000 carats worth £150,000 within a couple of months. In 1929 Ernest Oppenheimer of De Beers bought out Jeppe and Becker for £500,000 each and Merensky for £1.1 million.

Merensky was later involved in consultancy work on the Free State goldfields and in chrome prospecting where he made a large amount of money by starting his own mining company and then selling out to Union Carbide of the US in 1952 for £500,000. Obsessively interested in geology and prospecting, Merensky spent an extended period at Phalaborwa in the 1940s. At this important site of precolonial African mining, the phosphate content of the rocks was known, but had not attracted investment. In the early 1940s, South Africa's fertiliser supply was threatened by the war and prices were rising. Merensky invested over £60,000, employed 200 workers, and mapped the phosphate deposits.[44] Anglo American attempted to purchase Merensky's claims, but the new Afrikaner Nationalist government, reluctant to see this potentially vital resource for agriculture in private hands, bought out Merensky for £370,000. This purchase became the basis for the parastatal FOSKOR (see Chapter 6),

42 O. Lehmann, *Look Beyond the Wind* (Cape Town: Howard Timmins, 1959).
43 P. A. Wagner, *The Platinum Deposits and Mines of South Africa* (Edinburgh: Oliver and Boyd, 1929).
44 Machens, *Platinum*, 246.

Merensky found fame in the first half of the twentieth century as 'the Wizard Geologist'.[45] While all of these deposits would most likely have been found and exploited in the longer term, he had the geological imagination to understand their potential. Like mining magnates Cecil Rhodes, Percy FitzPatrick and Lionel Phillips before him, Merensky invested his riches into land and specifically into fruit production. In 1929 he bought Westfalia Estate, near Tzaneen, from Phillips and developed it into a model farm for avocados, forestry and soil conservation projects. This investment in land was in part a statement of his South African identity.

The scientific imagination of Eugene Marais, born in the same year as Merensky, was also stimulated by wandering around the rural Transvaal.[46] A law student in London in the 1890s, and then a Pretoria-based Afrikaans newspaper editor and poet, he moved to the Waterberg in 1907. Dependent on morphine and seeking refuge in a simple, rural life, Marais settled on the farm Doornhoek, which had a large troop of baboons. He was not a trained scientist but loved animals and soon decided to habituate the baboons with food and record his observations of them. In the resulting book, *My Friends the Baboons*, he wrote: 'I do not think that I ever passed a happier time in my life than the three years we spent in the mountains of Doornhoek.'[47] He purchased a baby baboon, named it Piet, and kept it as a pet.

Marais may have been among the first to observe primates in their natural environment with clear scientific intentions. He emphasised their capacity for learning and their need to acquire knowledge in order to survive in the wild. He tried to test the limits of their educability with captive baboons who were trained to use tools, put on clothes and show emotion towards humans.[48] Marais explored animal memory and pain in other species, as well as the rumoured resistance of honey badgers to snake venom. His biographer, Leon Rousseau, wrote that he was 'a lover of animals, ... he could become deeply attached to individual animals, but he was by no means sentimental' and was prepared to kill them in experiments.[49]

45 Machens, *Platinum*, 7.
46 L. Rousseau, *The Dark Stream: The Story of Eugene Marais* (Johannesburg: Jonathan Ball, 1982).
47 Rousseau, *The Dark Stream*, 202.
48 Rousseau, *The Dark Stream*, 228.
49 Rousseau, *The Dark Stream*, 230.

Initially planning to train as a doctor, Marais worked as a healer in the Waterberg, learning local remedies.[50] He conducted rainfall measurements and interviewed farmers to write about environmental change, espousing the widely held theory that South Africa was drying out. His environmental curiosity took shape especially in observations of termites.[51] Marais recorded their life cycle and nest-building, and distinguished between the different roles, and appearance, of queens, workers and 'hornsmen' or soldiers. Watching termites cope with the stress of drought, he investigated their fungus gardens and the deep tunnels they made to find water.

Marais began to think more broadly in the early 1910s and the 'theory about the unity of the termitary took shape in his mind'.[52] Driven by curiosity and an increasingly strong overarching analysis of the natural world, he was widely read in entomology but felt that much European specialist writing was insufficiently informed by observation. This fed off his strong sense that rural Afrikaners had a 'special knowledge' about the South African environment. He linked Afrikaners to a love of nature and landscape, and associated this more generally with Afrikaner nationalist claims to the country, especially through his supporter and interlocutor, Gustav Preller, an important cultural figure in his own right.

Despite his strong familiarity with English, Marais chose to write most of his poetry and scientific work in Afrikaans and he aimed to appeal to a popular audience. His ideas were tried out in Afrikaans-language newspapers and magazines such as *Die Burger* and *Huisgenoot.* An early article on termites in 1923 developed the idea of an 'animal soul' and *Die Siel van die Meer* (Soul of the Ant), published in 1933, argued that termites acted as a unitary organism with a group soul: the queen acted as the brain; food was absorbed and processed collectively; and worker termites would sacrifice themselves for the greater good. 'The death of a thousand individuals', he wrote, 'made not the least impression on that living stream'. In developing this

50 Rousseau, *The Dark Stream*, 208.

51 S. Swart, 'The Ant of the White Soul: Popular Natural History, the Politics of Afrikaner Identity and the Entomological Writings of Eugene Marais', in W. Beinart and J. McGregor (eds.), *Social History and African Environments* (Oxford: James Currey, 2003), 219–39.

52 Rousseau, *The Dark Stream*, 276.

approach, he affirmed that the species – and its survival – was more important than the individual.

Sandra Swart discusses the incident in 1927 when the Nobel Prize-winning Belgian scientist, Maurice Maeterlinck, published a similar idea in *The Life of the White Ant.* Marais accused him of plagiarism and it is likely that he had read Marais's articles. A version of this theory was already current in William Morton Wheeler's article, 'The Ant Colony as an Organism', acknowledged by both Marais and Maeterlinck.[53] Maeterlinck had written in a similar vein about bees, and these routes of thinking can be traced in the lineage of Darwin and Ernst Haeckel. Later entomologists were sceptical of the value of using the idea of a soul to understand termites, and suggested that Marais wrote poetry rather than science. Maeterlinck attracted a similar critique.

Yet, in many senses, Marais achieved what he set out to do. He alerted a broader range of Afrikaners to the fascination of science and by 1938 his book had become a set work for Afrikaans schools. Whatever the scientific value of the overarching theory, Marais cultivated an interest in nature and observation. Some saw in Marais's work an analogy for nationalism, with groups of people as organisms. 'The termite articles', Swart argues, 'were part of the socially constructed ideology for unifying purposes, promoting a shared aesthetic of Afrikanerdom, insisting on the connection between land and people.'[54] His poetry achieved a similar purpose and demonstrated the possibilities of lyrical Afrikaans writing in descriptions of nature.

Marais reached a larger South African audience than most contemporary scientists and, in the longer term, he also achieved international recognition. Though disparaged by the British-based South African-born zoologist Solly Zuckerman, who conducted observations of baboons in the 1920s, Marais was later credited as a significant precursor of the discipline of ethology – a field that only took off when Konrad Lorenz and Niko Tinbergen published in the 1950s and 1960s. Marais was also taken up by Robert Ardrey, who dedicated his best-selling, socio-biological book *African Genesis* (1961) to the long-deceased Afrikaner naturalist.[55] Ardrey championed Lorenz's theories

[53] Swart, 'The Ant', 236. [54] Swart, 'The Ant', 236.

[55] S. Swart, 'The Construction of Eugène Marais as an Afrikaner Hero', *JSAS*, 30, 4 (2004), 847–67; R. Ardrey, *African Genesis: A Personal Investigation into the Animal Origins and Nature of Man* (London: Collins, 1961).

about the significance of aggression and territoriality, which Marais had suggested in his writing on baboons. Another key inspiration for Ardrey's view of the links between territorialism and innate violence was the Wits comparative anatomist and palaeontologist Raymond Dart.

Science, Race and Segregation

Segregationist policies were gradually implemented by law during the early decades of the twentieth century. These included: comprehensive restrictions on African voting rights; the partial separation of urban residential areas by race; preference for whites in the labour market; a migrant labour system on the mines; and an increasing stranglehold by white owners on the bulk of agricultural land. These measures ensured white political power and economic advantage. Segregation was informed and shaped by forcefully articulated ideas about race.[56] Claims that Africans were naturally 'tribal', that they were liable to undergo 'degeneration' in modern urban environments, and that their purported intellectual inferiority or irrationality meant that they should not aspire to equal political rights, all formed part of this discourse.

Theories that conflated observed human difference with biological conceptions of race were given particular relevance during South Africa's mineral and industrial revolution, because economic transformation required radical new thinking about labour control and rights of citizenship. From the beginning of the twentieth century, social Darwinist theories were taken up by South African policymakers and politicians. Senior British administrators formed the select Fortnightly Club in Johannesburg in 1906 established to debate a range of issues related to the challenges of post-war reconstruction. Papers were delivered by influential policy-minded figures like Lionel Curtis, Howard Pim, and the American mining engineer W. L. Honnold, which addressed reported differences in racial capacities in the context of segregation. Comparisons were made with other parts of the world, including the southern states of America.[57]

[56] Dubow, *Racial Segregation and the Origins of Apartheid* and *Scientific Racism in Modern South Africa*.

[57] Dubow, *Racial Segregation*, chapter 1; Paul Rich, *White Power and the Liberal Conscience: Racial Segregation in South Africa and South African Liberalism*

The Native Affairs Society of the Transvaal, formed in 1908, provided another outlet for the discussion of segregationist racial theories that were extreme even by the standards of the day.[58] Authors dealing with this subject frequently claimed the authority of 'science'. In a more popular vein, a series of sensationalist ethnological books produced by Dudley Kidd with lurid titles like *The Essential Kafir* (1904) and *Savage Childhood* (1906) made free use of social Darwinist and psychological theories of 'arrested development'. Kidd's prolific use of photographs and testimony from missionaries and administrators gave the misleading sense that his academic ideas were grounded in deep ethnographic research.

Henri Junod was also much taken by Darwinian theories of evolution in the natural world and to some degree shared the view that African mentality and character was more limited than European. Yet he explicitly distanced himself from Kidd's generalising approach.[59] Whereas Kidd sensationalised cultural differences and saw these as permanent, Junod's religious outlook, coupled with his profound interest in African knowledge systems, led him to draw different conclusions. He resisted the application of the same 'pitiless' laws that he observed in the natural world to human society. While he advocated the value of education, he was inclined to protectionist policies for 'primitive humanity' (a term he used with caution) threatened with extinction in the face of all-conquering Europeans. Missionaries, Junod argued, 'had a moral duty to intervene on the side of the weak in the unequal struggle for survival'.[60]

In late 1908 or early 1909, Junod visited Lovedale Institution in the eastern Cape. Founded in 1824, this missionary school produced generations of African intellectuals and politicians. Junod met David

1921–1960 (Johannesburg: Ravan Press, 1984); Wits Manuscripts, F. W. Bell papers (A417, 'Native Affairs' box) and J. H. Pim papers, A881, Fa 18/4; M. Legassick, 'British Hegemony and the Origins of Segregation in South Africa, 1901–14', in W. Beinart and S. Dubow (eds.), *Segregation and Apartheid in Twentieth Century South Africa* (London: Routledge, 1995); V. Thakur and P. Vale, *South Africa, Race and the Making of International Relations* (London: Rowman & Littlefield, 2020).

58 Wits Manuscripts, F. W. Bell papers (A417, 'Native Affairs' box) and J. H. Pim papers, A881, Fa 18/4; Legassick, 'British Hegemony'.

59 H.-A. Junod, *The Life of a South African Tribe* (Neuchâtel: Attinger frères, 1912), 8, 100.

60 Harries, *Butterflies and Barbarians*, 142, 144.

Hunter, father of the anthropologist Monica Wilson, and also A. W. Roberts, a Scottish-born educator and talented amateur astronomer. Another visitor to Lovedale at this time was Samuel Shaw Dornan, a Paris Evangelical missionary based in Lesotho who was interested in many aspects of African cosmology, including astronomical knowledge. In 1910, Roberts described the appearance of Halley's Comet and gained an international reputation for his assiduous observation of variable stars. A stalwart of the S2A3, Roberts came to the attention of General Smuts, who, in 1920, nominated him to the Union Senate as well as the newly created Native Affairs Commission, which contributed to the articulation of segregationist philosophy. In this context, Roberts sought, with little success, to defend a form of moderate liberalism.[61]

While Junod and Roberts were directly involved in African education, a variety of other scientists and scientific communities participated in developing eugenicist and segregationist ideas. The medical profession was actively involved in discussion of eugenics and compulsory sterilisation of the 'unfit' in the *South African Medical Journal*, where doctors like J. T. Dunston, H. Egerton Brown, P. W. Laidler and E. G. Dru Drury sought to shape public policy.

Two leading zoologists, both prominent figures in the S2A3, did much to map out the shape of eugenic debates in the interwar era. James Duerden, professor of zoology at Rhodes University, Grahamstown, from 1905, studied ostrich plumages and revealed that defects in feathers, known as bars, were the result of a reduction in overnight blood pressure.[62] After the collapse of the ostrich industry in the First World War, Duerden switched his expertise to sheep and applied his genetic knowledge to the improvement of wool. Evidently influenced by his studies of livestock, Duerden was a moderate eugenicist who served as president of the S2A3 in 1921. His eugenic theories were considerably influenced by the American zoologist, H. S. Jennings, who criticised the 'false biology' around race. Duerden was therefore interested as much in environmental factors as biological determinants. He considered that the plight of poor whites was 'remedial' and thought that Africans, though undoubtedly racially distinct

61 K. Snedegar, *Mission, Science, and Race in South Africa: A. W. Roberts of Lovedale, 1883–1938* (Lanham: Lexington Books, 2015), 70–1 and *passim*.

62 'J. E. Duerden', *Nature*, 140, 576 (1937), 576; Dubow, *Scientific Racism*.

from whites, had yet to prove their 'full cumulative capacity under a new and sympathetic environment.'[63]

By contrast, Duerden's counterpart, H. B. Fantham, who took up the chair of zoology at Wits in 1917, was a hard line Mendelian eugenicist and avid follower of Weismann's germ plasm theory. Together with his wife, Annie Porter, a parasitologist based at the SAIMR, Fantham published many studies purporting to show that hereditary differences were ineradicable. Fantham was strongly in favour of sterilising the 'feeble-minded' and advocated firm action to prevent 'race degeneracy'. In his presidential address to the 1927 meeting of the S2A3, Fantham came out in favour of racial segregation, with dire warnings of the effects of racial intermixture.[64]

Biologically based racial determinism was taken up into justifications of segregation, as were some elements of a more sympathetic Christian and liberal protectionist view. Another important constitutive element in the segregationist mix was anthropological writing which, in one form or another, emphasised cultural aspects of racial difference. In the field of physical anthropology and comparative anatomy, which flourished from the second decade of the twentieth century, research into human evolution initially seemed to provide new evidence to support the idea that humans, like plants and animals, were subject to forces of speciation and that this process was confirmed by the emergence of different racial 'types'. However, from the 1920s anthropology pursued new perspectives, and Junod was on the cusp of this reorientation.

Liberal social anthropologists, now mostly university-trained, renounced reductive biological modes of explanation and developed more relativist and tolerant understandings of human difference. Monica Wilson, who was nurtured in the eastern Cape liberal missionary environment and trained in Cambridge, epitomised such approaches in her study *Reaction to Conquest* (1936), which also included chapters explicitly analysing social change.[65] Hers was one

[63] Dubow, *Scientific Racism*, 137–40; H. S. Jennings, 'Heredity and Environment', *The Scientific Monthly*, September (1924), 237; J. E. Duerden, 'Social Anthropology in South Africa: Problems of Race and Nationality', *SAJS*, 18 (1921), 29.

[64] Dubow, *Scientific Racism*, 133–6.

[65] Hunter, *Reaction to Conquest*; A. Bank and L. J. Bank (eds.), *Inside African Anthropology: Monica Wilson and Her Interpreters* (Cambridge: Cambridge University Press, 2013).

of a number of innovative anthropological studies that provided a bedrock of information about African societies in the 1930s, just at the time when the migrant labour system was becoming thoroughly entrenched.

With the development of humanities and social sciences in universities, social anthropologists ceased to view themselves as scientists. It became a discipline of strong opinions but equally strong doubts and came to recognise that scientific methodologies and testing sat uneasily with the complexity of social life. Rather than objectifying those they studied, social anthropologists tended to identify with 'their people' in a protective, if paternalist, manner. However, the tendency to focus on small-scale units of study, to see tribes as self-contained totalities, and to underestimate social change all fed later critiques of the role of anthropologists and linguists in the formation of political ethnicity or 'tribalism'.[66]

State Science: Applied Entomological and Veterinary Research

The national Department of Agriculture was, alongside the Department of Health, a key employer of applied scientists. This grew particularly out of the urgent need to address problems of agricultural production. Entomology and veterinary science were intimately linked because advances in parasitology in the 1890s demonstrated the significance of insect vectors in a range of animal as well as human diseases: trypanosomiasis carried by tsetse flies, malaria by mosquitoes, and tick-borne maladies. In the United States, Smith and Kilborne demonstrated that redwater disease, or Texas fever, in cattle was caused by ticks. A very similar disease had killed large numbers of cattle in the Cape and Natal in the 1870s and 1880s. Their research was published in the *Cape Agricultural Journal* in 1894.

The appointment of twenty-three-year-old American Charles Lounsbury as official Cape entomologist in 1895, during Rhodes's premiership, was aimed at more systematic investigation and treatment of insect pests in agriculture.[67] By this time, applied entomology was a well-developed field in the United States and Australia, where

[66] A. Mafeje, 'The Ideology of "Tribalism"', *Journal of Modern African Studies*, 9, 2 (1971), 253–61.

[67] K. Brown, 'Political Entomology: The Insectile Challenge to Agricultural Development in the Cape Colony, 1895–1910', *JSAS*, 29, 2 (2003), 529–49.

specialists attempted to address such problems as locust swarms that devastated crops and fruit.[68] Scientists had begun to study the life cycle of insects and the effects of insecticides. Lounsbury was initially encouraged to focus on insect pests in the vineyards and fruit-growing areas of the western Cape. The pressure to appoint a dedicated specialist came from the wealthier Afrikaner farmers in these districts; Rhodes himself was on the point of investing heavily in fruit farms.

Lounsbury focused initially on *Icerya purchasi*, a scale insect introduced from Australia that thrived on fruit trees. He made one of the first attempts at biological control by introducing a ladybird that might, if bred on sufficient scale, devour the scale insect. This proved unsuccessful, as well as potentially dangerous, because the introduction was unregulated and untested. Lounsbury then turned to insecticides, drawing on American research and experimenting with various mixes in Cape conditions. He bicycled widely around the western Cape to advise farmers and insisted on fumigating imported fruit trees – a measure made compulsory by legislation in 1905 – to ensure that they brought no damaging insects with them. Introduced codling moth had already attacked deciduous fruits such as apples, pears and quinces.

Junod's and Marais's entomological studies grew out of a fascination with diversity and beauty as well as insects' complex social organisation. Lounsbury self-consciously saw his role as the reordering of nature for the purpose of material gain. He described his profession as 'the study of insects injurious to the agricultural and other interests of man with a view to discovering methods for preventing or checking their ravages'.[69] Beneficial insects were to be encouraged and the injurious ones extirpated. He saw the adoption of scientific methods as essential to ensure food security and combat commercial competition from other countries.

In a context of rising scientific commitment and rapidly expanding understanding of insect vectors, Lounsbury and his associates, including the Swiss veterinarian Arnold Theiler in the Transvaal, started to investigate ticks. By 1900 he confirmed that the bont tick carried heartwater, the disease that had baffled Hutcheon in the early 1880s.

[68] I. Tyrell, *True Gardens of the Gods: Californian-Australian Environmental Reform 1860–1930* (Berkeley: University of California Press, 1999); P. Palladino, *Entomology, Ecology and Agriculture: The Making of Scientific Careers in North America, 1885–1985* (Amsterdam: Harwood Academic, 1996).

[69] Brown, 'Political Entomology'.

In 1902 he identified the tick that carried biliary fever in dogs. His work received new impetus as the cattle disease of east coast fever headed southwards towards South Africa. Newly provided with a laboratory, he showed by 1904 that the brown tick, *Rhipicephalus appendiculatus*, was its principal propagator in the nymph stage of their life cycle. He was able to clarify the length of time it needed to adhere to animals to feed and reproduce. No effective vaccine could be found, unlike rinderpest, and attention therefore turned to methods of insect control. Dipping and controls over animal movements were already well-established strategies for scab. Lounsbury identified arsenic compounds as the most effective for tick control and established that weekly dipping was needed to break the brown tick's reproductive cycle during the seasons that it attacked livestock. Within a decade, he had become a global leader in researching ticks and tick-borne diseases.[70]

It is remarkable that the implications of Lounsbury's findings were incorporated so quickly into state policy. Attempts to keep east coast fever at bay with fences failed, and it proved uncontrollable wherever the brown tick thrived; death rates of cattle could be as high as 80 per cent in the coastal districts, including the African reserves. In Mpondoland, where livestock were central to livelihoods, a count in 1911 before the disease struck gave about 280,000 cattle, but in 1915, after the disease had raged for a few years, numbers declined to 54,000.[71] The only solution seemed to be compulsory dipping, including of cattle owned by African people, because infected areas could act as a reservoir for re-infection. Starting around 1910, the state enforced dipping of millions of cattle throughout South Africa – including weekly dipping in the wet and warm summers when the tick thrived. Following the measures against scab, it was one of the most systematic interventions into rural life, especially in African areas. Brown notes that 'in many respects East Coast fever was the turning-point for South African veterinary science, both in terms of the confidence it ultimately imparted to its scientists and in the credibility it gave them in the eyes of the public and the state'.[72]

70 Brown, Political Entomology'.

71 W. Beinart, *The Political Economy of Pondoland 1860–1930* (Cambridge: Cambridge University Press, 1982).

72 K. Brown, 'Tropical Medicine and Animal Diseases: Onderstepoort and the Development of Veterinary Science in South Africa 1908–1950', *JSAS*, 31, 3 (2005), 514.

David Bruce, another graduate of Edinburgh Medical School, made similar breakthroughs researching tsetse fly in northern Natal at the same time as parasitologists working in other parts of Africa.[73] They were able to show that some species of fly carried the trypanosome that transmitted sleeping sickness in humans and *nagana* (a Zulu word) in cattle. In 1903 Bruce was probably first to identify the protozoan parasite that used the fly as vector and actually carried the disease to cattle and horses: it was named *Trypanosoma brucei*. Working with his wife, Mary, Bruce distinguished between 'native', lay 'European' and his own bacteriological approach, which in some ways synthesized these ideas.[74] The governments of Natal and Transvaal made no rapid chemical intervention to combat *nagana*, as they did against east coast fever, because it was more restricted spatially. However, livestock farmers near the tsetse belts opposed the expansion of wildlife numbers, which they believed would further spread *nagana* among their cattle. During the early twentieth century, after small game reserves such as Umfolozi were demarcated in Zululand, farmers were determined that they should be deproclaimed.[75] Other techniques of control, such as fly-trapping, had limited effect. In this case, two new scientific ideas – one identifying an agent of livestock disease, and the other arguing for the preservation of wildlife before they were finally exterminated – came into direct conflict. Eventually conservationists were able to argue that by corralling wildlife spatially, and destroying those on the farms, they could contain the disease.

Within South Africa, which was on the southern borders of endemic tsetse terrain, sleeping sickness among humans was relatively rare, but it still posed a seasonal threat. It was only after the Second World War

[73] K. Brown, 'From Ubombo to Mkhuzi: Disease, Colonial Science, and the Control of Nagana (Livestock Trypanosomosis) in Zululand, South Africa, c. 1894–1953', *Journal of the History of Medicine and Allied Sciences*, 63, 3 (2008), 285–322.

[74] J. A. Skotnes-Brown, 'Pests, Knowledge and Boundaries in the Early Union of South Africa: Categorising, Controlling, Conserving', PhD thesis, Cambridge University (2020), chapter 2.

[75] S. Brooks, 'Changing Nature: A Critical Historical Geography of the *Umfolozi* and *Hluhluwe* Game Reserves, Zululand, 1887–1947', PhD dissertation, Queen's University, Kingston (2001).

that tsetse flies were systematically attacked by aerial spraying with the new chemical DDT. This effectively controlled both sleeping sickness and *nagana* in the 1950s, but DDT – which featured so strongly in Rachel Carson's landmark book *Silent Spring* (1962) – was soon shown to be dangerous and its deployment was curtailed in the 1970s. In the early twenty-first century, *nagana* was reported to be spreading on the northern borders of South Africa around the Kruger Park, and northern KwaZulu-Natal, but not on a significant scale.

Arnold Theiler started working for the Transvaal republican government before the South African War and survived the transition to British rule. He initially made his name by developing a vaccine for rinderpest and, working with Lounsbury, by identifying the specific tick-borne protozoan (a single-cell organism named after him as *Theileria parva*) that caused east coast fever. In 1908 he was appointed as the head of the new veterinary research centre at Onderstepoort, immediately north of Pretoria. At the heart of a vigorous new national Department of Agriculture, and with British support, Onderstepoort was designed as a centre for the whole of the African empire. It was able to operate on a scale far greater than the earlier colonial initiatives and became a key international locus of research on tropical veterinary medicine, parasitology, toxicosis and environmental causes of disease.[76] The first veterinary training facilities in the country were launched in the early 1920s. Together with the South African Institute of Medical Research, Onderstepoort put public and veterinary medicine at the forefront of South African research.

Theiler developed long-term investigations into diseases that proved both global and local, and at the same time refined experimental techniques in veterinary science. *Lamziekte* (with symptoms of lameness and paralysis) had long been considered as an environmental or deficiency disease of some kind. Theiler was able to establish that it was most prevalent in areas where the soil and vegetation were short of phosphorus and magnesium. Cattle instinctively nibbled the carcasses of dead animals in order to increase their mineral intake. These rotting carcasses carried botulism; *lamziekte*, initially understood as a local malady, was identified as a version of this global disease.

[76] Brown, 'Tropical Medicine and Animal Diseases'.

Figure 4.2 Arnold Theiler, Marguerite Henrici and Henry Green at Armoedsvlakte, Vryburg, for *Lamziekte* research, early 1920s, as photographed by Dr Ackerman. Credit: University of Pretoria Library.

Theiler prioritised research into horsesickness, a scourge in South Africa and Africa more generally. It struck seasonally in most years on a relatively small scale, but successive epizootics had been experienced since the early nineteenth century. That of 1855–6, when the Cape lost about 40 per cent of its 150,000 horses, valued at over £1 million, was probably the worst.[77] The disease was not sufficiently devastating to preclude the use of horses in the country for personal and military transport. But horsesickness had limited their use for draught and also knocked out a promising export trade of equines from the Cape to India. Horse-breeding was restricted to drier upland parts of the country, which were less susceptible.

Early in the twentieth century, vets suspected that the disease might be carried by insects, and horse owners believed that keeping animals indoors at night seemed to diminish infection.[78] Livestock owners also

[77] C. Andreas, 'The Background to, and Impact and Management of, the Epizootics of Lungsickness and African Horsesickness in the Cape Colony, c. 1853–7', DPhil thesis, University of Oxford (2017).

[78] D. Gilfoyle, 'Veterinary Immunology as Colonial Science: Method and Quantification in the Investigation of Horsesickness in South Africa, c. 1905–1945' *Journal of the History of Medicine and Allied Sciences*, 61, 1 (2006), 26–65.

generally believed that it was dangerous to take animals out too early in the morning. Even though he could not identify the cause, Theiler first tried unsuccessfully to adapt the technique of vaccine production, using an element of infective blood, that he had helped to develop against rinderpest. He soon realised that there was more than one strain of the infectious agent. This made the production of a serum particularly difficult and the experiments complex. He aimed to maximise the polyvalency of the treatment through collecting as many strains as possible in the field and integrating them into the vaccine. However, by 1909 Theiler suspended experiments because he found he was infecting many of the horses that he was trying to immunise. He was nevertheless encouraged that mules seemed to react positively to the serum.

Theiler's son Max, born in Pretoria in 1899, completed his medical degrees at the University of Cape Town and the London School of Hygiene and Tropical Medicine. He started research work in the United States on dysentery and then focused on yellow fever. By 1928, this mosquito-borne virus was identified, and researchers linked the African and Latin American versions of the disease. In developing a vaccine, Max Theiler's team focused on a strategy that had been pioneered in Britain to attenuate the virus by passing it repeatedly through another species – in this case mice. They tested the resulting vaccine on monkeys, which were also susceptible to yellow fever, and by 1937 had found an effective solution. This also proved a valuable route for developing quantitative techniques to assess the likely protective value and danger of vaccines, thereby helping to standardise the science of immunisation. Arnold Theiler followed this work and, after he had retired in 1927, a team at Onderstepoort experimented with the technique for horsesickness. After attenuating and stabilising strains of the virus through mouse brains, Raymond Alexander and others found an effective and safe vaccine, and by 1935 managed a successful field trial. Production on any scale was labour-intensive and required a large number of mice. Nevertheless, in 1943–4 British vets in North Africa used 300,000 doses, produced at Onderstepoort and Kabete in Kenya, in order to protect army horses during the Second World War.

This work was at the cutting edge of immunology globally. Max Theiler won a Nobel Prize for his work on yellow fever in 1951, the first African-born person to do so. Horsesickness vaccination was a major achievement, which had promise throughout the African colonial world. It is striking how quickly advances were transferred between key institutions, from New York to Pretoria, Cairo and beyond. In fact, the technique developed by Max Theiler was adapted for horsesickness in South Africa before he could release it for treatment of yellow fever in 1937. The horsesickness vaccine was not entirely successful and its use was restricted by cost and the problem of multiple strains. But it is remarkable that it was developed before the epidemiology of the disease was fully understood. In the 1940s, scientists confirmed that it was another insect-borne plague, transmitted by culicoides midges. (These were more active in the morning, which may explain why horse owners thought this an infectious time.) Had scientists known this earlier, they might have tried to attack the insect vector, as in the case of ticks. The new techniques were also used at Onderstepoort to develop a vaccine for anthrax.[79]

Although the vets and scientists involved in this research originated largely outside the country, in Britain, Europe and the United States, most of them developed a long-term commitment to South Africa. They played a major role in creating a critical mass of applied science around animal diseases, parasitology, entomology, toxicology and related fields. By 1905, rinderpest was largely eradicated by a combination of vaccination and control of livestock movements. East coast fever was finally eradicated from South Africa in the early 1950s, after forty years of strict dipping and related controls. However, the dipping regime had a more temporary effect on other tick-borne diseases. Some ticks developed resistance to arsenic-based chemicals and it proved very difficult to keep ahead of new waves of resistance in the country as a whole.[80] Dipping gradually ebbed in the second half of the twentieth century and some tick-borne diseases such as heartwater and gall sickness began to spread again.

Smuts, National Institutions and 'South Africanism'

The Unification of South Africa and its emergence as a self-governing country offered new opportunities for scientific institutions, as well as demanding increased budgets. Questions arose as to which provinces

[79] D. Gilfoyle, 'Anthrax in South Africa: Economics, Experiment and the Mass Vaccination of Animals, c. 1910–1945', *Medical History*, 50, 4 (2006), 465–90.
[80] Beinart and Brown, *African Local Knowledge*.

of the country should benefit. The establishment of the Kirstenbosch Botanical Garden, encouraged by the S2A3, is a case in point.[81] Kirstenbosch's foundation was the outcome of a long tradition of Cape-based botanical expertise, given added significance through direct links to London's Kew Gardens. Harold Pearson, who occupied the newly created Bolus chair of botany at the South African College, was a leading figure in the initiative.

In 1911, Pearson proposed Kirstenbosch as a national botanical garden that would be uniquely expressive of South African identity by virtue of its dedication to the study and preservation of the country's indigenous flora. Kirstenbosch's designation as the country's 'national garden' exemplified the spirit of political Union. But it was even more an expression of a distinctive Cape identity because of the uniqueness of the fynbos biome ('Cape heaths' in the language of the day). Kirstenbosch's position below the slopes of Table Mountain added sublime aesthetic charm to scientific endeavour. For Cape Town's anglophone elite, the Cape's indigenous fynbos was an inheritance worth preserving and protecting from the incursions of alien exotic vegetation, on the one hand, and the growing economic and scientific power of Transvaal-centric South Africanism, on the other.

Botanical pride in the northern interior was, by contrast, manifested in the creation of the Transvaal Colonial Herbarium in Pretoria in 1903 by the official agronomist Joseph Burtt Davy. This was developed into a National Herbarium (1918) by Illtyd Pole Evans, a government plant pathologist and expert in aloes who strongly opposed Kirstenbosch's status as a national botanical garden.[82] Pole Evans was a close associate of Jan Smuts. He was founding editor in 1921 of the journal *Bothalia* (its title honoured South Africa's first prime minister, Louis Botha) and also the *Memoirs of the Botanical Survey of South Africa,* both published by the Department of Agriculture.[83]

Contestation over control of South African plant science continued for many years between Kirstenbosch and the better-funded

[81] L. van Sittert, 'Making the Cape Floral Kingdom: The Discovery and Defence of Indigenous Flora at the Cape ca. 1890–1939', *Landscape Research*, 28, 1 (2003), 113–29; Pooley, *Burning Table Mountain*, 135–9.

[82] Carruthers, 'Trouble in the Garden'.

[83] I. B. Pole Evans, 'Preface', in *Bothalia: A Record of Contributions from the National Herbarium* (Pretoria: Government Printer, 1921).

Herbarium and Botanical Research Institute in Pretoria; the latter benefited from stronger official links and developed applied scientific work on veld types and ecology. In Natal, Durban's botanical garden and colonial herbarium, which flourished from the 1880s under the direction of John Medley Wood, was starved of funds and it declined as a site of scientific research around the time of political Union. In 1913, the government transferred control of the garden to the Durban municipality and annexed the herbarium to the agriculture department. Now a mere provincial institution, it was one of the casualties of Union.[84]

Although government institutions in such fields as veterinary science and botany advanced scientific research, this impetus was constrained by the lack of national funding and the absence of an integrated university system. The 'University Question' was actively debated during the formation of Union. An 'inter-colonial' effort in 1908 to create a national university based either on a federal or a unitary system failed largely because of institutional resistance from existing university colleges and because of concerted opposition by Dutch-speakers based in Pretoria to English as the core language.[85]

Definitive changes were set in train by legislation passed in 1916. The University of the Cape of Good Hope, established in 1873 as an examining institution and based on the University of London, became the federal University of South Africa (now Unisa, based in Pretoria) with six constituent colleges spread through the four provinces of South Africa. Victoria College and the South African College became, respectively, the universities of Stellenbosch and Cape Town. African tertiary education was catered for at Fort Hare (formerly the South African Native College) which likewise attained the status of a university college in 1916, though treated separately in keeping with the raw politics of segregation.[86] From 1922, the various leading colleges associated with the University of South Africa gained autonomous university status (Rhodes, Natal, Witwatersrand and Potchefstroom).

84 D. McCracken, 'Durban Botanic Gardens, Natal: 1851–1913', *Garden History*, 15, 1 (1987), 64–73; Dubow, *Commonwealth of Knowledge*, 195–6.

85 M. Boucher, *Spes in Arduis: A History of the University of South Africa* (Pretoria: University of South Africa Press, 1973), 107–9.

86 Z. K. Matthews, 'The University College of Fort Hare', *South African Outlook*, April–May (1957), 1–35. Huguenot University College, which catered for women, was also treated differently; Boucher, *Spes in Arduis*, 169, 189–92.

The historian Eric Walker observed the paradox: 'Politically and economically, South Africa had just moved centripetally; academically, in the very act of political union, it ... rushed confusedly in the opposite direction.'[87]

The act of creating full-fledged English- and Afrikaans-speaking universities in all the provinces of the Union considerably augmented the capacity of the tertiary education system as a whole; it also increased claims for resources from the central state. From 1910 to 1930 the number of university lecturers in South Africa expanded from fewer than 180 to 770. Total state expenditure on universities more than tripled in real terms over this period while the number of white students in residential universities rose from 1,171 to 7,118.[88]

The expansion of university education was matched by new funding arrangements for research. In 1918 a Research Grant Board was created within the Department of Mines and Industries. This was in large measure a response to war-time concerns about the need for industrial self-sufficiency and applied scientific research. It was also the outcome of representations made by the Royal Society, whose modest annual budget for university research support, supplied by the Department of Education, was severely cut during the war. From 1919 to 1929 the Research Grant Board disbursed 216 grants of around £50 each to individual researchers. The number of grants peaked in 1929–30 when more than £1,641 was spent, but budgets were slashed over the next five years.[89] The Research Grant Board's significance lies not so much in the amount of money it distributed as in the principle that it established: scientists could henceforth look directly to the state for support.

This principle was belied by the reality that a high proportion of awards made by the Research Grant Board were underwritten by the Carnegie Corporation of New York which supported awards administered by the Board from 1928. Many of these went to support individual projects in ethnology, anthropology, and rock art.[90] As the

[87] E. Walker, *The South African College and the University of Cape Town, 1829–1929* (Cape Town: University of Cape Town, 1929), 78.

[88] E. G. Malherbe, *Education in South Africa*, vol. 2 (1923–1975) (Cape Town: Juta, 1977), Appendix 26, 727 and appendix 37, 750; *Union Statistics for Fifty Years* (Pretoria: Government Printer, 1960), table E3.

[89] N. M. Luruli and J. Mouton, 'The Early History of Research Funding in South Africa: From the Research Grant Board to the FRD', *SAJS*, 112, 5/6 (2016), 3.

[90] 'Carnegie Grants in aid of Research. Financial Statement Position as at first January 1931', Statements 1 and 2, Carnegie archives, Columbia University,

economic depression hit, the South African government's proportion of funding declined, to the extent that in 1932 the chair of the Research Grant Board complained to Carnegie that the government subvention had fallen to just £100 that year. The New York secretary of the Carnegie Corporation, James Bertram, was firmly of the view that Carnegie should support research bearing directly on South African communities and culture – rather than funding engineers or scientists who might end up working for the national steel-making company, ISCOR, which he dismissed as a 'spite' organisation with 'no possible future'. Bertram considered that South Africa had sufficient railways and that its future lay in agriculture. He was also concerned to see evidence that grant recipients were 'getting out of their racial shell'.[91]

The question of what was distinctively South African, and whether this lay in culture or in science, arose again when the young physicist, Basil Schonland, was turned down in 1932 for a grant to research thunderstorms and radiation on the South African Highveld. Schonland complained to the Washington branch of Carnegie that too much money was being spent in studying the subcontinent's flora and fauna, 'native life and customs' and musical instruments, whereas 'funding for scientific research has practically ceased': the Chairman (Young) 'does not consider that there can be any South African physics and is reluctant to make any grant from the fund for atmospheric electricity, for example.' After making discreet enquiries about Schonland from American sources (who affirmed that he was indeed a serious up-and-coming scientist) Carnegie exercised a rapid about-turn. Within a matter of weeks Schonland was given £175 for research on the diurnal variation of penetrating radiation while fellow physicist Alexander Ogg was awarded £100 towards equipping a small observatory for magnetic observations.[92] Schonland later became a trusted advisor to Carnegie on South African research priorities and toured America on a Carnegie grant.

Box III A.307 RGB files, parts 1 and 2. Beneficiaries included Maria Wilman, A. J. H. Goodwin, G. P. Lestrade, A. Bryant, P. R. Kirby, Robert Broom and T. F. Dreyer, as well as natural scientists F. E. Plummer, J. L. B. Smith and E. L. Stephens.

91 Ibid., Memo from J. Bertram to F. P. Keppel, 24 December 1931, 1–2; telegram from Keppel to R. B. Young, 24 February 1928.

92 Ibid., B. F. J. Schonland to J. A. Fleming, 15 January 1932; Beatrice Eva, RGB, to Keppel, 29 March 1932.

Growing confidence in South Africa's scientific research capacity gained its greatest advocate and interpreter in Jan Smuts, who, in his period as leader of the opposition between 1924 and 1933, devoted a great deal of time and attention to the emerging British Commonwealth. For Smuts, international statesmanship, institution-building, and scientific advance were powerfully conjoined. His fertile imagination conjured up the metaphysical idea of 'holism' which ruminated on the organic and evolutionary links between the smallest elements of matter and the largest problems of the universe. New discoveries in particle physics offered Smuts a wealth of metaphors to work with.

Smuts elaborated his ideas in a treatise entitled *Holism and Evolution* (1926) and also in his prestigious presidential address to the centenary meeting of the British Association in London in 1931. The philosophical and scientific rigour of holism has been much debated. But the concept persists, not least in the field of ecology, whose emergence as an international field of study owes much to Smuts's ideas and patronage. J. W. Bews and John Phillips, both protégés of Smuts, were botanists who pioneered the ecological 'viewpoint' through their application to South Africa of Frederic Clements's theory of plant succession in respect of forestry and grassland studies. They gained international reputations in doing so.[93]

In political terms, holism revealed how the nationality of a fledgling, newly independent country like South Africa, might find more capacious expression in the larger identity of the Commonwealth. In respect of domestic white politics, the metaphor amounted to a belief in the need to transcend ethnic differences and to strive for a greater inclusive settler nationalism that laid more emphasis on place than race. The political slogan of 'South Africanism' became a defining outlook for the South African Party under Smuts's leadership. It was intended to be more inclusive and outward-looking than the insular nationalism captured by J. B. M. Hertzog, whose catchphrase was 'South Africa first'.

In 1925, Smuts served as president of the S2A3. His official address to the organisation, titled 'South Africa in Science', represented a

93 P. Anker, *Imperial Ecology: Environmental Order in the British Empire, 1895–1945* (Cambridge, MA: Harvard University Press, 2001), chapter 2; Pooley, *Burning Table Mountain*, 50–1, 64.

remarkable effort to shift the emphasis from the 'European continent', home of nineteenth-century science, to the southern hemisphere, and so to demonstrate what this might contribute 'to the sum total of scientific knowledge'. As part of this hemispheric reorientation, Smuts posited Africa and, in particular the Witwatersrand watershed of South Africa and the Rift Valley to the north, as marking a 'great "scientific divide" among the continents'. Smuts thought that it was on either side of these great ridges that scientific prospectors – using an instantly recognisable South African metaphor – might find 'the most precious and richest veins of knowledge'.[94] He spoke soon after Merensky's well-publicised find of platinum.

Smuts began to illustrate this point by reference to a geological example – the Wegener hypothesis and the theory of continental drift. In Smuts's view, this gave Africa a new global centrality because it should be viewed as the 'mother-continent' from which South America, Australasia and India had 'calved off'.[95] By placing South Africa at the centre of this 'great divide' Smuts was making a case for the country's singularity as well as its universal significance. Here, Smuts drew on the latest work of South African geologist, Alex Logie du Toit, a University of Cape Town- and Glasgow-educated expert on Karoo stratigraphy as well as mining. In 1916, du Toit hypothesised connections between the Karoo 'system' and Australian equivalents. Then, in 1923, with the support of the Carnegie Institution of Washington and Harvard, he travelled to South America in order to better understand the ancient southern hemisphere history of Gondwanaland.[96]

From 1921, du Toit began to modify Wegener's analysis of continental drift which focused on the original supercontinent of Pangaea. This in turn separated into two smaller supercontinents: Laurasia to the north and Gondwana to the south. Du Toit's radical claim was that the fracturing of Gondwanaland gave birth to the modern land masses of Africa, South America, India and Australia. In his 1924 address to

94 J. C. Smuts, 'South Africa in Science', *SAJS*, 22 (1925), 3–4; S. Dubow, 'Global Science, National Horizons: South Africa in Deep Time and Space', *The Historical Journal* 63, 5 (2020), 1079–1106.

95 Smuts, 'South Africa in Science', 4.

96 A. L. du Toit, *A Geological Comparison of South America with South Africa* (Washington: Carnegie Institution, 1927). See also S. Chetty, '*Africa Forms the Key*': *Alex du Toit and the History of Continental Drift* (Cham, Switzerland: Macmillan Palgrave, 2021).

the S2A3, du Toit supported the Taylor-Wegener hypothesis of 'continental disruption' while foregrounding the country's pivotal position in the breakup of ancient Gondwanaland.[97] Although many had noticed the similarities in outline between, say, the east coast of Latin America and the west coast of Africa, the connections as well as the mechanisms of their purported separation were mired in controversy. Many theories supposed that there had been ancient 'land bridges' connecting the continents because it was too challenging to comprehend, let alone to account for, continental masses ploughing their way through the ocean floor. Until the 1960s, when geomagnetic evidence showing that the poles had 'wandered' combined with the paradigm-shifting theory of plate tectonics, the very idea of continental drift was treated with deep scepticism, and often outright hostility by leading scientists, American geologists in particular.[98]

Du Toit's carefully constructed arguments, which he published in *Our Wandering Continents* (1937), offered key support to Wegener's iconoclastic 'mobilist' thesis.[99] But his bold effort to 'elaborate [the] architecture of the Globe' also subtly reoriented Wegener's approach. This was signalled by a motto on the title page of his book: 'Africa forms the key'. In a nod to Smuts, fellow South African geologist T. W. Gevers, who wrote an admiring biography of du Toit, noted that his 'scientific outlook was always holistic'.[100] Few if any South African scientists in the first half of the twentieth century were as well networked as du Toit. Although not a university academic, he corresponded on an equal basis with fellow geologists in the United States, India, Latin America and Australia as well as Angola and Uganda. His work on continental drift was pieced together through detailed understanding not only of stratigraphy and glaciation, but also of mineralogy, palaeobotany and palaeontology (plant, animal and hominid fossils). Du Toit's unrivalled knowledge was gained through extensive fieldwork and travel. His career involved working for the Cape

[97] A. L. du Toit, 'The Contribution of South Africa to the Principles of Geology', *SAJS*, 21 (1924), 71.

[98] N. Oreskes, *The Rejection of Continental Drift: Theory and Method in American Earth Science* (Oxford: Oxford University Press, 1999).

[99] A. L. du Toit, *Our Wandering Continents: An Hypothesis of Continental Drifting* (London: Oliver and Boyd, 1937); S. H. Haughton, *Obituary Notices of Fellows of the Royal Society*, 6, 18 (November 1949), 385–95.

[100] T. W. Gevers, *The Life and Work of Dr. Alex. L. du Toit* (Johannesburg: Geological Society of South Africa, [1949]), 82.

Geological Commission (1903–11), the Union Geological Survey (1911–18) and the Union Irrigation Department (1918–27), followed by extensive consulting work for mining corporations.

Du Toit's compendious *Geology of South Africa* (1926) represented an extraordinary work of synthesis and interpretation, which soon gained international attention. In 1927, Harvard geology professor Reginald Daly, a supporter of mobilism, wrote an enthusiastic review in the *Journal of Geology*. Daly was well-acquainted with du Toit through the Harvard-sponsored Shaler Expedition to South Africa in 1922 which proved to be a key event in South Africa geology because it put local experts in direct contact with leading overseas scientists. Daly praised du Toit for bringing 'up to date the story of a region which is unique in illustrating principles of geological science as it is in the field of economics. South Africa has few rivals among the countries which are today most advancing the science [of geology], and clearly no American geologist can afford to be ignorant of the many discoveries and stimulating problems coming out of southern Africa.' In a letter to du Toit's publishers, Daly called *The Geology of South Africa* 'one of the very finest geological summaries of regional geology that has ever been published'. It was a local study that also implied fundamental global interconnectedness. Daly endorsed du Toit's finding that 'Nearly every important break in South Africa, paleontological as well as stratigraphical, happens to fall somewhere within one of the European Systems'.[101]

In the case of Smuts, we can be certain that it was expertise in botany – he was an acknowledged authority on veld grasses – that disposed him to accept the Wegener–du Toit thesis. Smuts's understanding of plants persuaded him to dismiss what he called the 'European fallacy' according to which migration of the northern temperate flora of Europe was the source of South African plant distribution. Drawing instead on Charles Darwin's speculation, that the development of higher plant forms may have taken place in an ancient 'isolated continent or large island, perhaps near the S. Pole', Smuts suggested that the Cape floral kingdom as well as the country's tropical and subtropical flowering plants most likely derived from

[101] A. L. du Toit papers, University of Cape Town, BC 722, box M. file 1, R. A. Daly to Oliver and Boyd, Edinburgh, 21 June 1927, and copy of Daly's review in *Journal of Geology*, 37, 7 (1927), 671–2; Du Toit, *The Geology of South Africa* (Edinburgh: Oliver and Boyd, 1926 edn), 10–11.

Gondwanaland.[102] In effect, Smuts was proposing for South Africa what Australian scientists were hypothesising at much the same time: the existence of 'great south lands' in Permian times which linked Australasia to Antarctica. In both cases, this interpretation helped to challenge deep-seated feelings of colonial isolation and inferiority based on southern hemispheric marginality.[103]

This linking of geology and botany serves as a reminder that these were long-established areas of scientific interest in South Africa. Just how long was made by clear by Edwin Percy Phillips, a Cape Town-born botanist, in his 1930 presidential address to Section C of the S2A3. In a 250-year survey, Phillips divided the history of botany into three phases. The first, beginning with seventeenth-century Dutch visitors to the Cape, was characterised by the activities of foreign and local collectors who sought to map out the taxonomy and distribution of South African plants. A second phase began in the late nineteenth century when five South African-based botanists (Harry Bolus, Peter MacOwan, John Medley Wood, Rudolf Marloth and Selmar Schonland) established herbaria in different regions of the country (the exception being the republican Transvaal).

During this period, detailed systematic botany was pursued, with particular emphasis placed on plant morphology, anatomy and development. Phillips identified the start of a distinctive new phase in 1903 with the appointment of Harold Pearson, a key proponent of Kirstenbosch, to the newly created chair of botany at the South African College. This was followed, in 1905, by the accession of Schonland to a chair of botany at Rhodes University College, followed in turn by chairs of botany at the Natal (1909) and Transvaal (1917) university colleges. The new professoriate developed a distinctly academic style of botany in the twentieth century, with different emphases from the applied concerns of the Department of Agriculture. They launched new botanical journals and series which served as an outlet for research in fields such as embryology, ecology and economic botany.[104]

102 Smuts, 'South Africa in Science', 7–8; cf. C. Darwin to J. D. Hooker, 12 August 1881, Cambridge University Library (MS DAR 95: 524–7), cudl.lib.cam.ac.uk.

103 T. Griffiths, *Slicing the Silence: Voyaging to Antarctica* (Sydney: University of New South Wales Press, 2007), 69–82.

104 E. Percy Phillips, 'A Brief Historical Sketch of the Development of Botanical Science in South Africa and the Contribution of South Africa to Botany', *SAJS*, XXVII (1930), 39–80.

In Smuts's 1925 address, his interest in botany and palaeobotany segued into meteorology. The climatic conditions of ancient Africa, he observed, had so far elicited 'only a mild scientific interest'. Yet climate change was a looming threat to South African agriculture, in particular the trend towards 'progressive desiccation of the land' as highlighted by the 1923 Drought Investigation Commission (and also by observers like Eugene Marais). The dearth of scientific knowledge about the African climate, Smuts argued, following the advice of the director of the Meteorological Office in London, necessitated establishing meteorological observatories in the Antarctic. Smuts proposed a joint Australian–South African Antarctic station for the purposes of climatic research. This would have the additional merit of reinforcing Commonwealth solidarity.[105]

Having dealt with meteorology and the possible extension of South African influence in the South Atlantic, Smuts paused to note the outstanding contributions of astronomers at the Cape, from Lacaille onwards. He singled out two Scottish-born Cape astronomers for special mention, noting the importance of Thomas Henderson's determination of the distances from the earth of Alpha Centauri, and David Gill's calculation of the distance of the sun. Missionary educator A. W. Roberts's work on variable stars was also commended, though as an amateur working with limited equipment, his considerable achievements did not rank with those of his principal mentors, Gill at the Cape Observatory and Robert Innes at the Union Observatory.

Both were Scotsmen, as was Roberts, who taught and influenced hundreds of African students at Lovedale Missionary Institution from 1883 to 1920 (including R. V. Selope Thema, Benjamin Tyamzashe, Tshekedi Khama and Florence Thandiswa Makiwane) and came to Smuts's attention as an expert on 'native affairs'.[106] St Leger Plaatje, son of the leading African intellectual of his day, Sol Plaatje, was also a student of Roberts. Roberts and Sol Plaatje were both deeply impressed by the appearance of Halley's comet in 1910. This was two years before the formation of the African National Congress of

105 Smuts, 'South Africa in Science', 11.

106 Smuts, 'South Africa in Science', 15; Snedegar, *Mission, Science, and Race*, 61–71.

which Plaatje was a founder. He named his youngest son 'Halley', thinking the comet might be a harbinger of seismic political change.[107]

Finally, Smuts turned his attention to recent developments in human palaeontology, highlighting in particular Raymond Dart's discovery of the Taung *Australopithecus* fossil skull just a few months before. Smuts did not give a firm opinion on Dart's audacious claim that australopithecines represented a crucial missing link in hominin development, but he was evidently sympathetic to the claim. As in the context of botany and Gondwanaland, Smuts cited another of Darwin's speculations in support of this theory, namely that Africa might turn out to be the 'cradle of mankind'. Considered together with the discovery of *Homo rhodesiensis* in 1921, Smuts thought that this ancestral form of anatomically modern humans indicated that South Africa might 'become the Mecca of Human Palaeontology'.[108]

For Smuts, common origins did not imply commonality or equality. On the contrary, Africa might not turn out to be *the* cradle of mankind so much as 'one of the cradles' – a clear intimation of his belief in human evolutionary divergence. 'Our Bushmen', he said, 'are nothing but living fossils whose "contemporaries" disappeared from Europe many thousands of years ago.' They were analogous to 'true "living fossils" like the cycads', botanical survivors from the Jurassic era. For the purposes of scientific research, one of South Africa's distinct advantages was that, being thousands of years behind the times, evolutionary processes and biological problems could be studied from life.[109] In other words, South Africa was important to the world by virtue of the fact that ancestral and advanced forms of humanity occupied the same space yet were products of different evolutionary time.

The title of Smuts's 1925 presidential address was well chosen. He was concerned not only to detail the history of science as practised in South Africa, but, more ambitiously, to demonstrate the country's significance for universal science. By invoking deep geological time in respect of botanical and anthropological processes, Smuts was suggesting that living remnants of the deep past offered modern scientists unique advantages as well as holding 'the key to stable and permanent

107 S. T. Plaatje, *Mhudi: An Epic of South African Native Life a Hundred Years Ago* (Alice: Lovedale Press, 1930), chapter 18; B. Willan, *Sol Plaatje: A Life of Solomon Tshekisho Plaatje 1876–1932* (Auckland Park: Jacana, 2018), 257.

108 Smuts, 'South Africa in Science', 16, 17–18.

109 Smuts, 'South Africa in Science', 17.

future progress'.[110] The message was clear: South Africa was integral to modern science, just as science was essential for the country's future.

The S2A3, the British Association and the Politics of Science

In 1929 the British Association returned to South Africa as guests of the S2A3. A high-powered delegation, including forty-four fellows of the Royal Society and the atomic physicist Sir Ernest Rutherford, were among the 500 visitors. The South African government made a substantial contribution to costs. And with 'punctilious regard to the Union's racial traditions' – code for equality between English- and Afrikaans-speakers – ten 'representative' guests were specially invited from Britain and the Netherlands.[111]

In many ways the 1929 visit reprised the form of the 1905 occasion: there were elaborate civic welcomes, extensive tours of the country, and a great deal of press coverage. The tone of the visit was, however, markedly different. From the outset, conspicuous efforts were made to reinforce the theme of ethnic equivalence within white South Africa. Sir Thomas Holland, president of the British Association, quoted the Afrikaans translation of Percy FitzPatrick's iconic South African novel, *Jock of the Bushveld*, adding for good measure that the Afrikaans language was 'capable of expressing virile sentences'. This gesture was appreciated by *Die Burger*, a leading opinion-forming Afrikaans newspaper, which also took satisfaction from the fact that not all the leading scientists were British.[112]

The spirit of bipartisanship soon came under strain. A public furore ensued over Dr E. G. Malherbe's discussion of 'Education and the Poor White'. This controversy was sparked by preliminary results from the Carnegie Commission study of white poverty (1930–2). In his presentation to the British Association, Malherbe presented poor whiteism as 'the skeleton in our cupboard', a claim that caused deep offence to politicised Afrikaner nationalists. Following the historian W. M. Macmillan, Malherbe made the provocative claim that the so-called 'native question' and 'poor white question' were intimately related. Indeed, to talk about the 'native question' was a form of psychological

110 Smuts, 'South Africa in Science', 19

111 *Cape Times*, 11 March 1929.

112 *Cape Times*, 22 July 1929; *Rand Daily Mail*, 23 July 1929; *Die Burger*, 19 July 1929; 20 July 1929; 27 July 1929.

'projection'. It was the consequence of an 'inferiority complex' arising out of the fact that the predicament of the lowest 10 per cent of whites was blamed on competition from the upper 50 per cent of blacks. Measures such as the industrial colour bar, far from protecting whites, would rob the dominant race of 'the virility bred from the buffetings of stern competition'. Behind such arguments lurked a fear that poor whites (the vast majority of whom were Afrikaans-speaking) had undergone some form of eugenic deterioration. English-speaking educational experts like C. W. Kimmins and Fred Clarke congratulated Malherbe for his bravery and 'genuine patriotism'. Conversely, Afrikaans newspapers like *Die Burger* and *Ons Vaderland* were outraged and responded by attacking the veracity of Malherbe's research.[113]

Public controversy also ensued in the context of investigations into anthropology and palaeontology, disciplines that were represented at the conference by such overseas luminaries as Henri Breuil, Leo Frobenius, Henry Balfour and Gertrude Caton-Thompson. Caton-Thompson had been specially tasked by the British Association to adjudicate a long-running debate about the origins of the archaeological ruins of Great Zimbabwe. Ever since the colonisation of Rhodesia by Cecil Rhodes, these archaeological mysteries had been attributed to foreign influence, most likely from the Near East. Speaking in Johannesburg, Caton-Thompson announced that her archaeological investigations supported Randall-MacIver's claim that Great Zimbabwe was the product of indigenous African endeavour, thus dismissing a central claim of white settler colonialists. Her talk was interrupted by a splenetic Raymond Dart, whose whole-hearted commitment to cultural diffusionism led him to protest that the discussion had been dominated by those who mistakenly supported the 'native origin' view.[114]

Afrikaner nationalist discomfort was provoked by the very idea of human evolution. Christian-nationalist thinkers took a lead from the Potchefstroom philosopher H. G. Stoker, and his theistic account of evolution. In the *Tydskrif vir Wetenskap en Kuns*, academic journal of

113 *Cape Times*, 24 July 1929; 25 July 1929; 29 July 1929; *Ons Vaderland*, 26 July 1929 (editorial); *Die Burger*, 26 July 1929 (editorial); 30 July 1929 (editorial); 'Die Arme Blanke', *Die Kerkbode*, 31 July 1929, 171.

114 *The Star*, 2 August 1929; S. Dubow, 'Human Origins, Race Typology and the Other Raymond Dart', *African Studies*, 55, 1 (1996), 1–30.

the Akademie vir Wetenskap en Kuns, intellectuals cautiously debated the idea of evolution at this time, mindful of its implications for religion. *Die Kerkbode*, voice of the Dutch Reformed Church, gave a qualified welcome to the scientific visitors, while registering disappointment that there was no place for discussions of theology in the overwhelmingly secular programme. According to this view, evolutionary theory, being man-made, was in conflict with God's word and in conflict with the story of Genesis. *Die Kerkbode* was, however, more enthused about the potential of science to advance the country's agriculture and industry, and in this magnanimous spirit it announced that it was happy to welcome the 500 learned visitors.[115]

Afrikaner nationalist political sensitivities were aroused by a cartoon that appeared in the *Cape Times* depicting two government ministers (one of whom was the future apartheid prime minister, D. F. Malan) as prehistoric cave-dwellers. One ate while the other gazed at a chunk of meat labelled 'hate'. The caption of Wyndham Robinson's cartoon read: 'Those members of the British Association especially interested in relics of the Stone Age will not have far to look.'[116] *Die Burger* pronounced this provocation a scandalous affront, but it attempted to take the high moral ground in the spirit of courtesy to its British visitors. The *Cape Times*, it loftily announced, was jingoistic and did not accurately represent British cultural values.[117]

In effect, what was being played out was drama about the politics of white South Africa. Afrikaner nationalists did not wish to be seen as anti-scientific, but they were intent on disavowing the idea that scientific advance and civilisation were identical with the British Empire or, in the Smutsian sense, coterminous with the values of the Commonwealth. It was this assumption which led the Stellenbosch physicist, A. C. Cilliers to warn in a 1938 pamphlet that Smutsian holism implied subsuming South Africa within the British Empire – and thus subverting authentic nationalism.[118]

Smuts's deputy, the precociously talented politician and administrator, Jan Hofmeyr, fastened onto the idea of 'South Africanism' which, he also hoped, could transcend the Anglo-Afrikaner divide. Hofmeyr was president of the S2A3 in 1929. The title of his much-referenced

115 'Die "Britse Association"', *Die Kerkbode*, 31 July 1929, 170–1.
116 *Cape Times*, 17 July 1929. 117 *Die Burger*, 18 July 1929 (editorial).
118 J. J. Broodryk, 'Stellenbosse Academici en die Politieke Problematiek in Suid Afrika, 1934–1948', MA thesis, Stellenbosch (1991), 15–17.

official welcoming address, 'Africa and Science', nodded to Smuts's 1925 speech.[119] Whereas Smuts foregrounded new areas of intellectual discovery, Hofmeyr emphasised the extent to which science was now institutionally rooted in South Africa. The impressive new buildings of the University of Cape Town on the slopes of Table Mountain, where the British and South African Associations convened, symbolised this achievement, as did the proliferation of specialist scientific societies and the role played by the state in encouraging and sponsoring research. Veterinary science at Onderstepoort was a shining example. So, too, were the South African Institute for Medical Research and the Miners' Phthisis Medical Bureau, both supported by industry.

Hofmeyr observed that in 1905 science was 'in large measure exotic' by virtue of the country's reliance on overseas expatriates who may not have acquired a 'South African outlook', but it was now 'essentially South African'.[120] In the quarter of a century that had elapsed between the first and second visits of the British Association, 49 university scientific researchers multiplied tenfold to a total of 467 scientists, of whom 144 were professors. This crystallised the 'South Africanisation of Science'. Yet, in a deliberate effort to avoid reading ethnic agendas into such achievements, Hofmeyr insisted that there was 'nothing narrow' about scientific South Africanism: indeed, to be so would be 'false to the spirit of Science.'[121]

For Hofmeyr, as for Smuts, science was a means to articulate an outward-looking form of patriotism and inclusive South Africanism. In line with Smuts's broader vision of the projection of European power into Africa, Hofmeyr went on to articulate an expansionist programme. Having South Africanised science, the next objective was to 'Africanise' it. Hofmeyr duly presented South Africa to his British visitors as the 'Southern gateway' by which 'Science itself can most effectively be made to permeate Africa'. This ought to be mutually beneficial to science and to Africa. In particular, he highlighted the potential of the human and social sciences: medicine, economics, anthropology, palaeontology and linguistics. These offered 'the richest opportunities to those who would investigate racial problems in the

[119] J. H. Hofmeyr, 'Africa and Science', *SAJS*, 26 (1929), 1–18. Hofmeyr's speech was given extensive press coverage in South Africa and also in Britain. See *The Times*, 27 March 1929.

[120] Hofmeyr, 'Africa and Science', 6.

[121] Hofmeyr, 'Africa and Science', 4, 6–7.

true Spirit of Science' and, in discovering solutions, avert the 'clash of colour'. Africa could be seen as a 'great laboratory' where scientists might test 'diverse social and political theories as to the relations between white and coloured races'.[122]

For all that Hofmeyr, a trained classicist and liberal paternalist, talked in terms of 'harmony' and 'common advantage', he drew a sharp line between European civilisation and African barbarism. Rome had failed to impose civilisation on Africa but 'we meet them with the advantage of having resources at our disposal which our Roman predecessors lacked': above all, a scientific understanding of the 'character and mentality of primitive peoples'.[123]

Hofmeyr's offer of South Africa as a gateway for African science was given substance by the holding, in Pretoria, of two major international meetings in 1929: the Pan-African Veterinary Congress, chaired by P. J. du Toit of Onderstepoort, and the 15th International Geological Congress, under the presidency of A. W. Rogers. This was the first occasion that this convention of geologists was held in the southern hemisphere. One of its enduring achievements was to bring new, international attention and credence to the Wegener theory of continental drift, largely as a result of fresh South African-based evidence presented by du Toit.[124]

The 1929 scientific visits to South Africa considerably raised the country's national profile as a site of scientific and social research, and the government took pride in acting as host. Prime Minister Hertzog, though much less of a scientific drawcard than Smuts, signed invitations to foreign governments to elicit their participation in the 1929 Geological Congress. He also invited the renowned French palaeontologist Henri Breuil to South Africa in 1929 and a year later gave the German ethnologist Leo Frobenius, exponent of *Kulturkreis* (cultural field) theory, a large grant of £5,000 to conduct work on Great Zimbabwe. The patronage extended by the ruling National Party offended local prehistorians, who saw the gesture as

122 Hofmeyr, 'Africa and Science', 15.
123 Hofmeyr, 'Africa and Science', 16–17.
124 *Science*, 70, 1806 (1929), 142; L. P. Mazadiego-Martínez and O. Puche-Riart, 'The 15th International Geological Congress, South Africa (1929): The Resurgence of Wegener's Continental Drift Theory', *Episodes*, 36, 2 (2013), 125–37.

an effort to outplay Smuts, the great patron of South African science, at his own game.[125]

The politics of knowledge production were also affected by the growing presence of non-state actors. Conspicuous here was the Carnegie Corporation of New York whose support of research activity in South Africa from 1929 was at once an act of international philanthropy as well as a shrewd intervention by American social scientists into Africa, hitherto almost the sole domain of European colonial policymaking and research. Mindful of the parallels between South Africa's racial problems and those of the American South, the Carnegie Corporation and its astute president, Frederick Keppel, took care to build support within moderate Afrikaner circles. Carnegie money also helped to fund the South African Institute of Race Relations (established in 1929) and in British colonial Africa it supported Lord Hailey's *African Survey*. The American, British and South African networks did much to establish the field of 'race relations' as a comparative international subject of research and policy formulation. But only within limits: a request by W. E. B. Du Bois to conduct research in South Africa was refused in 1935.[126]

Interwar South African social science was also boosted by a proliferation of international conferences and organisations. Many of these were associated with non-state actors identified with the League of Nations and the 'Geneva spirit' of international cooperation, peaceful coexistence and humanitarianism.[127] The International Institute of African Languages and Cultures, established in 1926, was designed to serve as a 'co-ordinating agency, a central bureau and a clearing-house for information'.[128] Led by

[125] N. Schlanger, 'The Burkitt Affair Revisited: Colonial Implications and Identity Politics in Early South African Prehistoric Research', *Archaeological Dialogues*, 10, 1 (2003), 21; A. H. Brodrick, *The Abbé Breuil: Prehistorian* (London: Hutchinson, 1963), 151.

[126] P. Rich, 'The South African Institute of Race Relations and the Debate on "Race Relations"', 1929–1958', *African Studies*, 40, 1 (1981), 13–22; Tilley, *Africa as a Living Laboratory*; W. A. Jackson, *Gunnar Myrdal and America's Conscience: Social Engineering and Racial Liberalism, 1938–1997* (Chapel Hill: University of North Carolina Press, 1990), 25.

[127] B. L. Estoile, 'Internationalization and "Scientific Nationalism": The International Institute of African Languages and Cultures between the Wars', in H. Tilley and R. J. Gordon (eds.), *Ordering Africa: Anthropology, European Imperialism, and the Politics of Knowledge* (Manchester: University of Manchester Press, 2007), 98.

[128] F. D. Lugard, 'The International Institute of African Languages and Cultures', *Africa*, 1, 1 (1928), 1.

British colonial administrator Lord Lugard, with French and German co-directors (Maurice Delafosse and Dietrich Westermann), the International Institute sought to make the practice of colonial administration more rigorous and scientific by utilising the insights of anthropologists and ethnologists. This helped to inspire local initiatives, such as the formation of the South African Inter-University Committee for African Studies (1932) which was affiliated to the International Institute.[129]

Smuts's vision, specifically, his capacity to imagine South African scientists taking a leading role both within the country and beyond, seemed to be finding traction. Hofmeyr echoed these ideas of inclusive South Africanism, which were then projected outwards by a muscular and confident new society taking its place in Africa. Central to these conceptions were economic growth, institutional development, and an expansion of South Africanised science. Its content could also be 'Africanised' not only through personnel but also by providing the international scientific community with a southern entry point to Africa. In 1929, South African leaders perceived that the country, and Africa, needed a diverse and buoyant science: indeed, science needed Africa as much as Africa needed science. Hertzog, with his trademark slogan of 'South Africa first', took a narrower view of science. Yet, as we shall see, he was not so much anti-science as determined to prevent science being used to reinforce the imperial or Commonwealth connection.

[129] 'Summary of Proceedings of the First Meeting of the South African Inter-University Committee for African Studies', *Bantu Studies*, 6, 1(1932), 89–92.

5 *The Commonwealth of Knowledge, 1930–48*

The joint meeting of the British and South African Associations of Science in 1929 marked a high point for those who hoped that South African science could flourish under the umbrella of the Commonwealth. Ambitions on this front were briefly shelved by the global economic Depression (1929–32), which led to substantial political turmoil and the formation of a white national government of unity in 1933–4. Hertzog, the nationalist prime minister, and Smuts, the leader of the opposition South African Party, attempted to work together as joint leaders of a Fusion (United Party) government. One of the consequences of the condordat was the passage of a major package of segregationist legislation in 1936.

The economic downturn was sharp in South Africa, but it did not last long – certainly by comparison with most European and colonial economies. Gold made the difference. When Hertzog belatedly agreed to leave the gold standard at the end of 1932, the price of gold soared, as did mining income and profits. Between 1931 and 1941, the total annual income from gold production increased 2.5 times to £120 million annually. The African workforce, at about 200,000 in 1930, peaked at over 383,000 in 1941, and the number of white miners grew from 22,000 to 41,500.[1] The government took the opportunity to impose an excess profits tax; state revenue from the industry grew from about £3 million in 1930 to £17 million in 1940 and peaked at £27 million in 1943.[2] Resources on this scale enabled ambitious and imaginative thinking on a number of fronts.

The United Party rapidly expanded state expenditure from the mid-1930s. In part, it was experimenting with new approaches, inspired by Keynesian ideas and the example of Roosevelt in the United States, to achieve economic recovery through government spending. It was also

[1] Beinart, *Twentieth-Century South Africa,* 117–18.
[2] *Union Statistics for Fifty Years*, Q-6.

somewhat committed to economic nationalism and redirected a portion of the mining taxation windfall to improve the fortunes of its political supporters. The major beneficiaries were white recipients of welfare and protected employment.[3] By the end of the 1930s, about £10 million, or 20 per cent of the total budget, was spent in this area, particularly old age pensions and support for white employment in industries such as the railways. This expenditure formed part of a far-reaching attempt at resolving the 'poor white' problem, which was becoming a source of great political concern to Afrikaner nationalists in particular. The growing predisposition in favour of state intervention was registered by the establishment of a Department of Social Welfare (1937) as well as ongoing support for nationalised steel production and 'civilised labour' policies aimed at protecting the white working class from black workers who could be paid less.

White farmers had been particular short-term casualties of the economic depression and were also a key political constituency for Hertzog. Agriculture, fencing, water resources and conservation initiatives became major elements of the new state expenditure. The Department of Agriculture budget, at around £1 million in 1930, increased to £3.7 million in 1940, and £2.5 million was spent under the Soil Erosion Act (1932), much of it on water provision.[4] Some of this bounty spilled over to African areas. Expenditure on agriculture in the Transkeian Territories increased from £25,000 in 1925 to £80,000 by 1939.[5]

State money also went into health and education as well as key industries such as the Electricity Supply Commission (ESCOM, founded in 1922–3) and the Iron and Steel Corporation (ISCOR, 1928), both with a view to fostering domestic economic expansion. The state's Industrial Development Corporation, which provided funding for infrastructure and industrial capacity, was set up in 1940 with £5 million in capital.[6] Expenditure on national roads

[3] J. Seekings, 'The Carnegie Commission and the Backlash against Welfare State-Building in South Africa, 1931–1937', *JSAS*, 34, 3 (2008), 515–37.

[4] *Farming in South Africa*, vol. 15 (1940), 478.

[5] Beinart, *Rise of Conservation*, 353.

[6] B. F. J. Schonland, 'Hendrik Johannes van der Bijl 1887–1948', *Obituary Notices of Fellows of the Royal Society*, 7, 19 (1950), 26–34; R. Christie, *Electricity, Industry and Class in South Africa* (Basingstoke: Palgrave Macmillan, 1984), 76–7.

increased more than fourfold from 1936 to 1948.[7] In all of these areas, opportunities were created for scientific and social scientific research within government service and in industry.

When Hertzog left the United Party government in 1939 over parliament's narrow and bitterly contested decision to go to war on the side of Britain, Smuts took over as prime minister (1939–48). He accorded science and technology a greatly expanded role as part of the war effort, placed accomplished scientists at the heart of the state's initiatives, and offered some hope to progressives who began to envisage science as an instrument of social transformation. These initiatives took shape in a health commission and in the wartime Social and Economic Planning Council.

The Social Sciences

As shown in the previous chapter, the Research Grant Board disbursed limited grants through the 1920s. In 1929 the social sciences were included in government funding models for the first time. This innovation was the result of the creation in 1934 of a separate National Bureau of Educational Research and its successor body, the National Bureau of Educational and Social Research.

The driving force behind this new arrangement was Ernest G. Malherbe, an Afrikaner educational sociologist who trained at Columbia University in the 1920s. Malherbe was deeply influenced by the American comparative educationist, Isaac L. Kandel, an authority on the relationship between government, culture and education, as well as the leading liberal philosopher of pragmatism, John Dewey. Malherbe was a consummate networker and committed internationalist who cultivated strong personal links with the Carnegie Corporation of New York, which provided an annual grant to the National Bureau. One of the innovative features of the Bureau was its attempt to develop research capacity in the state itself in addition to providing a bridge to academic institutions.

As we have seen, the Carnegie Commission, under Frederick Keppel, put considerable resources into a national study of poor whiteism. South Africa's problems served as a mirror for some of the United States' own dilemmas. Sponsorship of the Poor White Commission of

[7] *Union Statistics for 50 Years*, Q-18.

1930–2 proved to be one of Carnegie's great successes. From an academic perspective this innovative multi-authored interdisciplinary academic study served as a showcase for the social sciences.[8] It provided a platform for rising Afrikaner academics, such as R. W. Wilcocks, professor of psychology, who went on to become rector of Stellenbosch University and chaired the Commission of Inquiry Regarding the Cape Coloured Population (1937).

Jeremy Seekings argues that Hertzog's government was already committed to state expenditure to assist poor whites and that the Carnegie Commission signalled a clear commitment to social welfare delivered by professionals through government departments.[9] Politically, Carnegie gave fresh visibility to the problem of white poverty. Keppel considered that the Commission was one of the Carnegie Corporation's 'indisputable "strikes"' as a funding agency; subsequent efforts by the Corporation to sponsor large South African projects were nowhere nearly so successful. Keppel was fully aware of South Africa's international significance as a 'research laboratory' into race relations. The participation of American sociologists Kenyon L. Butterfield and C. W. Coulter helped to boost the study of sociology in South Africa. Expertise did not flow in one direction only. When Carnegie contributed funds towards establishing a new Department of Race Relations at Yale University in 1931, it was headed up by the leading South African educationist and liberal segregationist, Charles T. Loram, who was already a trusted adviser to the Carnegie operation.[10]

Investment by government and external funders in the social sciences was framed by a commitment to utilise social knowledge for the purposes of shaping public policy. This gave recognition to the view expounded by Malherbe (who was also responsible for one of the Poor White Commission volumes) that 'the potency of facts' could cut through the distortions of social and political prejudices.[11] Hendrik Verwoerd, who worked in the departments of psychology and

[8] J. F. W. Grosskopf et al., *The Poor White Problem in South Africa* (South Africa Report of the Carnegie Commission, Stellenbosch, 1932), 5 volumes.

[9] Seekings, 'The Carnegie Commission'.

[10] P. L. Rosenfield, *A World of Giving: Carnegie Corporation of New York – A Century of International Philanthropy* (New York: Public Affairs, 2014).

[11] S. Dubow, 'Scientism, Social Research and the Limits of South Africanism: The Case of E. G. Malherbe', *South African Historical Journal*, 44 (2001), 126–7.

sociology at Stellenbosch (1928–36), agreed that policy should be informed by social theory and research. However, Verwoerd was committed to radical Afrikaner nationalism and so diverged sharply from Malherbe's cautious progressivism, with its emphasis on education and bilingualism as a means to achieve gradual societal integration and democratisation.

In 1938 a National Research Council and Board was launched to replace the old Research Grants Board. Confident pronouncements were made about the enhanced capacity of South African science and its growing national role. R. B. Young, the Wits geologist who headed the new organisation, stated that preference would henceforth be given to research with a 'South African bearing'. Young acknowledged the generous support given by the Carnegie Corporation to the Research Grant Board, amounting to £22,000 over the period 1928–37. He stressed the need for the South African government to 'make good this loss', warning that without such support research would be seriously disadvantaged. J. H. Hofmeyr, the responsible minister, emphasised the 'national honour' that flowed from internationally significant scientific discoveries, calling the National Research Council and Board 'our South African Parliament of Research'.[12]

The amalgamated Council and Board enhanced research capacity, yet as a 'parliament' of competing interests in a politically divided government it proved cumbersome to administer. Although Malherbe sought to identify particular research priorities for the social sciences, most of the grants were directed to individual research projects. After the Carnegie Corporation finally withdrew its financial support to the Research Board, Frederick Keppel reflected that 'no outstanding success' had resulted from Carnegie's investment in the research grants system managed within the South African government, though some of it was of 'creditable quality'.[13]

[12] E. G. Malherbe papers, Killie Campbell Library, Durban, KCM 56980 481/1–3; National Archives, Pretoria, UOD 1954 E184/2/2 vol. 1, 'Extract from Proposed Inaugural Address to the National Research Council and Board by the Chairman, Prof. R. B. Young', 2–3; *Cape Times*, 26 July 1938 (editorial); 26 July 1938.

[13] KCM 56983 (32b), 519/2, 'Memorandum on the Activities of the Carnegie Corporation in the British Colonies and Southern Dominions', 26 January 1942, 4. This document also reveals that, between 1925 and 1942, Carnegie spent some $4.8 million in the 'southern hemisphere'. Africa (including South Africa)

In 1929, Loram spearheaded an attempt to get Carnegie to provide £10,000 for a 'Survey of the Economic and Social Conditions of Natives' as a counterpart to the Poor White Commission. This was not well received by the South African government which chose instead to embark on its own study, the Native Economic Commission of 1930–32. Keppel was acutely aware that, in South Africa, 'the whole subject of the treatment of the Natives is red hot politically' and by the end of the 1930s he had come to the conclusion that 'outside interference would accomplish more harm than good'.[14]

The liberal internationalist agenda jointly set out by Keppel, Malherbe and Loram was already sparking a reaction on the part of Afrikaner nationalist sociologists who regarded the issue of poor whiteism as far too important and sensitive a topic to be left to outside intellectuals. For politically ambitious academics like Verwoerd at Stellenbosch and Geoffrey Cronjé at the University of Pretoria, the problem of poor whiteism offered a means to lever resources towards the social upliftment of the Afrikaner *volk*. Verwoerd was clear that only through rigorous racial differentiation could the 'weaker' members of the Afrikaner nation be assisted to assume their rightful place in society. This imperative swiftly became absorbed in the broader political campaign to gain control of the state in the name of Afrikaner nationalism and apartheid.[15]

In the field of eugenics and mental testing, practitioners of race psychology like M. L. Fick and J. A. J. van Rensburg concluded that black people could not benefit from education on the same basis as whites.[16] Another new discipline where loyal Afrikaner nationalists

accounted for nearly $2 million of this, Australia $1 million, and New Zealand $700,000.

14 Keppel, 'Memorandum on the Activities of the Carnegie Corporation in the British Colonies and Southern Dominions', 6; Loram to E. G. Jansen, 3 December 1929 and 'Survey of the Economic and Social Conditions of Natives. Draft Proposals', by C. T. Loram (n.d.), Carnegie archives, Columbia University, Box III. A.A307, Research Grant Board file, 1928–31 (part 1).

15 N. Roos, 'Alcohol Panic, Social Engineering, and Some Reflections on the Management of Whites in Early Apartheid Society', *Historical Journal*, 58, 4 (2015), 1167–89.

16 M. L. Fick, *The Educability of the South African Native* (South African Council for Educational and Social Research Pretoria, 1939); J. A. J. van Rensburg, *The Learning Ability of the South African Native Compared with that of the European* (South African Council for Educational and Social Research Pretoria, 1938).

could develop theories designed to advance the interests of the *volk* was criminology. At the University of Pretoria, a crucible for Afrikaner sociology and science, W. A. Willemse and Geoffrey Cronjé identified juvenile delinquency, alcoholism, gambling and racial intermixture as deadly threats to the purity of the aspiring Afrikaner nation; reciprocally, hereditarian theories were deployed to pathologise black criminality as an insidious external threat to security and well-being.[17]

Although Malherbe's Carnegie-supported Bureau for Educational and Social Research funded such eugenic research, he was himself of a politically moderate and intellectually eclectic disposition.[18] It was largely through Malherbe's expansive personal networks that the 1934 New Education Fellowship Conference, founded by the British theosophist Beatrice Ensor, gathered in South Africa. She was influenced by the Montessori method of teaching and her peripatetic career included a period of farming fruit in the eastern Cape. This educational organisation cum international movement, formed in the midst of the First World War, preached the virtues of progressive pedagogy and humanitarianism; after the Second World War, some of its key figures went on to play important roles in the creation of UNESCO. Malherbe was reportedly offered the deputy-general directorship at UNESCO in 1950.[19]

The New Education Fellowship meeting in South Africa was a major national and international gathering. Its studiedly non-partisan approach attracted a wide range of South African intellectual leaders, including an unlikely mixture of liberals such as Malherbe, R. F. A. Hoernlé and Edgar Brookes, African political leaders such as D. D. T. Jabavu and A. B. Xuma, and Afrikaner nationalist social scientists like Verwoerd and Fick.

The month-long gathering, whose scale and ambition rivalled the joint meetings of the South African and British Associations of Science, included educationists, psychologists, policymakers, anthropologists and missionaries. Some 4,000 people attended the sessions in Cape

17 Dubow, *Scientific Racism*, 156–60; Roos, 'Alcohol, Panic, Social Engineering'.

18 E. G. Malherbe, *Never a Dull Moment* (Cape Town: Timmins, 1981), 65, 94–6.

19 J. Watras, 'The New Education Fellowship and UNESCO's Programme of Fundamental Education', *Paedagogica Historica*, 47, 1–2 (2011), 191–205; Janie Malherbe to Mr and Mrs Shepardson, 30 January 1950, Carnegie Corporation archives, Columbia University, Box III.A 209, EG Malherbe file 1932–55.

Town and Johannesburg, and the program included some 300 addresses. Among the overseas experts who travelled to South Africa were the philosopher John Dewey, the British social anthropologist Bronisław Malinowski, and the Dutch philosopher Johannes van der Leeuw, who spoke about psychoanalysis and education. Educationists included Pierre Bovet, Harold Rugg, Fred Clarke and Mabel Carney, a specialist on education and race relations.[20]

Local race relations experts like Malherbe and J. D. Rheinallt-Jones of the Institute of Race Relations both placed emphasis on the need to develop a scientific approach to educational problems – and so sought to displace the authority of missionaries and the churches.[21] The conference's overarching theme, 'Educational Adaptations in a Changing Society' spoke to the organisers' view of South Africa as a social laboratory for progressive and radical pedagogical practice and experimentation. The presence of Ensor, Carney and Clarke – all of whom had direct experience of working in South Africa – serves as a reminder of how internationalised educational science was at the time. Although the abiding ethos of the gathering was humanist and anti-authoritarian, there were strands of eugenicist thinking in discussions of local research into intelligence testing, 'African mentality' and criminology. Among the fascist sympathisers attending, Graf von Dürckheim-Montmartin from Germany – who turned out to be on a spying mission – spoke on 'The Nazi Ideal in Teacher Training'. Alarmed by the presence of the German delegation, Malherbe approached Hertzog to say that he would 'not inflict such unknown persons on the conference', but Hertzog responded that 'we must on no account offend Hitler'. Ensor defended the German visitors in the spirit of liberal intellectual exchange and the problem was contained.[22]

In their addresses, Ensor and Bovet both spoke about the possibility of maintaining a dual allegiance to country and humanity. Malherbe

[20] R. Glotzer, 'The Career of Mabel Carney: The Study of Race and Rural Development in the United States and South Africa', *International Journal of African Historical Studies*, 29, 2 (1996), 309–36.

[21] S. Krige, 'Segregation, Science and Commissions of Enquiry: The Contestation over Native Education Policy in South Africa, 1930–36', *JSAS*, 23, 3 (1997), 495–6.

[22] K. J. Brehony, 'A New Education for a New Era: The Contribution of the Conferences of the New Education Fellowship to the Disciplinary Field of Education 1921–1938', *Paedagogica Historica*, 40, 5–6 (2004), 752–3; Malherbe, *Never a Dull Moment*, 184–7.

used the opportunity to champion South Africa as a 'new country', bristling with social problems that were 'world issues'. In South Africa, these challenges were 'thrown into relief more clearly ... than in the older countries where they are often obscured by the complexities of tradition and deep underlying prejudices which parade under the name of Western Civilisation. As an experimental station and laboratory in racial and cultural relations South Africa is almost unique in the world to-day.'[23] This was reminiscent of Smuts's speech to the S2A3 a decade earlier. Similar language was being used by British policymakers in relation to its African colonies.[24]

Malherbe's contemporaneous stewardship of the Council for Educational and Social Research placed a premium on educational and social research relevant to 'race relations'. The Council supported research conducted by liberal educational psychologists, as well as upcoming social scientists like Ellen Hellmann, Simon Biesheuvel and Max Gluckman. Two black intellectuals were also given grants: B. W. Vilakazi for Zulu poetry and R. T. Caluza for African music. Research bearing on segregationist policies was regarded as of central importance. Thus, studies on comparative race intelligence pursued by Fick, Dent and van Rensburg, as well as the Afrikaner criminologist Willemse, were supported by the Council and listed among its official publications.[25]

Lightning, Astronomy and the Antarctic

In an interesting switch of priorities (perhaps mirroring its revised decision to fund Schonland's work on lightning and Ogg's on magnetism in 1932) the Carnegie Corporation allocated £10,800 in 1936 to support the creation of the Bernard Price Institute of Geophysical Research at the University of the Witwatersrand.[26] Price, a leading industrialist and benefactor, donated a similar amount. He had served as chief engineer and subsequently as general manager of the Victoria

[23] E. G. Malherbe et al., *Educational Adaptations in a Changing Society* (Cape Town: Juta, 1937), iv.
[24] Tilley, *Africa as a Living Laboratory.*
[25] E. G. Malherbe, *Educational and Social Research in South Africa* (Pretoria: S. A. Council for Educational and Social Research, 1939), 70–1.
[26] UOD 1954, 'Extract from Proposed Inaugural Address to the National Research Council and Board', 2.

Falls and Transvaal Power Company which played a vital role in the supply of energy to the Witwatersrand mines (in 1948 it was absorbed into the state-owned Electricity Supply Commission). Price also funded an Institute for Palaeontological Research (which bore his name) in 1945.[27] Opening the Institute of Geophysical Research in 1938, Smuts alluded to the poet Robert Browning as he likened Johannesburg to a place where 'meteors flash, and the stars come and go.'[28] The institute rapidly gained international attention through the work of its founding director, Basil Schonland. His interests in lightning were augmented by seismological work on rock tremors around Johannesburg and on palaeomagnetism, both of which had important practical implications for the mining industry.[29] Understanding the behaviour of lightning was significant for energy supply to the mines, since electric storms were a frequent source of power failures.

Alternative energy sources were actively under consideration by mining interests at this time. Research on oil-from-coal technology began in the 1930s when the Anglovaal mining company, founded by A. S. Hersov and S. G. Menell, acquired the rights to the Fischer-Tropsch process. The technology, developed at the Kaiser Wilhelm Institute in Berlin in the 1920s, entailed synthesising hydrocarbons from hydrogen and carbon monoxide gases which, conveniently, were readily obtainable from South Africa's abundant supply of coal. Early progress was made when H. J. van Eck joined Anglovaal as a chemical engineer in 1936. Two years later, he appointed P. Etienne Rousseau as a research engineer to work on an Anglovaal experimental shale-oil refining subsidiary. For a mixture of technical and economic reasons little advance was made with synthetic fuel, but the calculations changed after 1948 with van Eck, now head of the Industrial Development Corporation, in a powerful position to offer state assistance.[30]

[27] B. K. Murray, *Wits: The Early Years* (Johannesburg: Witwatersrand University Press, 1982).

[28] *Rand Daily Mail*, 22 October 1938.

[29] Murray, *Wits*, 290–1; A. L. Hales, 'Research at the Bernard Price Institute of Geophysical Research', *Proceedings of the Royal Society of London. Series A. Mathematical and Physical Sciences*, 258, 1292 (1960), 1–26.

[30] A. Steynberg, 'Introduction to Fischer-Tropsch Technology', in A. Steynberg and M. Dry (eds.), *Fischer-Tropsch Technology* (Amsterdam: Elsevier, 2004), 32–4.

Oil-from-coal technology provides a striking example of how industrial science soon became a core interest of the state. By contrast, pure sciences such as astronomy were not very well supported in the interwar years. Notably, the promise of the Royal Observatory in Cape Town and the Union Observatory in Johannesburg at the start of the century was not fully sustained. Astronomical activity was instead largely reliant on visiting astronomers from Europe and the United States who were keen to study the underexplored southern skies. Yale University established its Southern Station, equipped with a 26-inch photographic refractor, at the University of the Witwatersrand in 1925. The University of Leiden installed a 16-inch photographic refractor at the Union Observatory in 1938. Much of the knowledge generated by these observatories was processed abroad. The photographic plates taken in Johannesburg were shipped directly back to the United States and the Netherlands, and international collaboration was pursued very much on the terms of the visitors rather than the hosts.[31] There were few routes for South African astronomers to train and no university departments of astronomy.

In 1934, the trustees of the Radcliffe Observatory, constructed in Oxford in the 1770s, decided to sell its assets in order to build a new observatory capable of exploring the southern skies.[32] The Radcliffe Observatory was duly transferred from Oxford to Waterkloof Ridge on the outskirts of Pretoria. Technical problems with the design and installation of the large glass mirror, coupled with disruption caused by the Second World War, prevented further progress and it was only in 1948 that the telescope was commissioned. It was then the largest telescope in the southern hemisphere.[33] Bloemfontein was selected by two northern hemisphere universities as a southern outpost. Naval Hill, near the centre of town, was the site of the Lamont-Hussey Observatory, owned by the University of Michigan, and dedicated to the observation of binary stars. Harvard University operated Boyden Station north-east of the city.[34]

31 R. H. Stoy, 'Astronomy in South Africa' in Brown (ed.), *Scientific Endeavour in South Africa*, 422–3; Moore and Collins, *The Astronomy of Southern Africa*, 107ff.

32 www.gtc.ox.ac.uk.

33 I. S. Glass, 'Andrew David Thackeray at the Radcliffe Observatory', *Transactions of Royal Society of South Africa*, 64, 1 (2009), 76–8.

34 H. Haffner, 'The Boyden Observatory', *Monthly Notes of the Astronomical Society*, 15, 6 (1956), 47–8; A. D. Andrews, 'The Boyden Observatory', *Irish*

Lacking either strong state support or a significant presence in South African universities, astronomical research was also conducted by amateurs and local societies. A. W. Roberts, who researched variable stars (stellar bodies with changing luminosity) at Lovedale Institution (see Chapter 4), encouraged Alan Cousins, whose enthusiasm went back to his teenage years: in 1918 Cousins observed a nova eruption in the Aquilae constellation. Cousins became a prominent figure in the Astronomical Society of Southern Africa and subsequently gave up his career as an electrical engineer to work at the Cape Observatory, specialising in photometry (the measurement of light from objects in space).[35] In the 1930s, Arthur Bleksley pursued astronomical work on variable pulsating and binary stars while based at the Wits Department of Applied Mathematics and Astronomy. Bleksley's preferred vocation was, however, as a populariser of science and he was best known after the Second World War as a radio broadcaster and scientific educator.[36]

The idea of a South African Antarctic Research Expedition, proposed in 1919 by Ernest Goddard, an Australian-born zoology professor at Stellenbosch University, was given endorsement by the 1923 Drought Investigation Commission. The Commission's concern with environmental degradation and conservation led it to identify a need for reliable Southern Ocean meteorological data to aid with weather forecasts. Goddard was keenly interested in ancient land connections between Antarctica and the southern continents and couched his appeals in terms of national prestige and Commonwealth cooperation. Smuts, too, supported a call made by George C. Simpson, director of the Meteorological Office in London, for international collaboration in respect of weather stations in the Antarctic.[37]

Australian polar explorer George Hubert Wilkins also presented plans at this time for Antarctic science. This would include South

Astronomical Journal, 25, 2 (1998), 129–66; A. H. Jarrett, 'Boyden Observatory', *Acta Academica*, 12, 40 (1979), 70–9.

35 'A. J. W. Cousins', by D. Kilkenny, *The Observatory*, 121 (2001), 350–2.

36 D. P. Mason, 'Arthur Bleksley: Pioneer of Science Awareness in South Africa', *SAJS*, 104, 11–12 (2008), 423–7. Bleksley was an integral part of the 'Three Wise Men' on the radio quiz show 'Test the Team', together with Johannesburg astronomer W. S. Finsen.

37 Smuts, 'South Africa in Science', 14.

Africa, whose economic and strategic interests in the South Atlantic included whaling and fisheries. However, the drive to establish a sovereign South African 'sector' in the Antarctic during the interwar years was not pursued with vigour, and proposals for a South African presence in the Antarctic did not go further than a government memorandum circulated in 1935.

Two years later Reginald James, newly arrived as professor of physics at UCT, renewed public interest in the Antarctic: as a scientific member of Shackleton's 1914–16 *Endurance* expedition (for which he was awarded a polar medal), James offered a riveting personal account of the privations of those left clinging to an ice floe. From 1943 through to 1945 serious scientific plans for the exploration of Antarctica were developed by the geomorphologist, Lester C. King, at the University of Natal (Pietermaritzburg), acting in collaboration with Alex du Toit. The Geological Society of South Africa took the idea forward in 1944 and sent a deputation to lobby Smuts. Advice and support for a South African polar expedition was received from Frank Debenham at the Scott Polar Research Institute in Cambridge. These plans did not come to fruition.

For all the talk about national research funding in the 1930s, the state achieved little in developing large prestige scientific projects. As the next chapter shows, it was only in the apartheid years that a concerted effort was made by South Africa to establish a presence in the Antarctic with the construction of weather stations on Marion and Gough Islands from 1948.[38] In respect of astronomy, major state involvement only began in the 1960s and 1970s. Similarly, oil-from-coal technology, proposed in the 1930s, was only realised a generation later. Such examples all serve as a reminder that big scientific projects in South Africa, imagined in the prewar years, took a generation and more to be realised, and that science and technology often did not progress in a linear fashion.

38 S. M. E. van der Watt, 'Out in the Cold: Science and the Environment in South Africa's Involvement in the Sub-Antarctic and Antarctic in the Twentieth Century', PhD thesis, University of Stellenbosch (2012), 1, 31, 33–5, 58–60; P. Jackson, 'Meteorology and Climatology', in Brown (ed.), *A History of Scientific Endeavour*; J. F. W. Juritz, 'Professor R. W. James', www.phy.uct.ac.za; for proposals by L. C. King and the Geological Society: A. L. du Toit papers, BC 722, University of Cape Town, box I.1 'SA Antarctic Expedition 1944–7'.

Agriculture and Conservation

A partial exception to this slow-down in activity was in the fields of agriculture and conservation. The Department of Agriculture was a major beneficiary of growing state revenues in the 1930s and 1940s. In 1929–30, the budget amounted to about £1 million, of which the department earned about £176,000, mostly from the guano islands which it managed and veterinary services such as the provision of vaccines. About half of the budget was spent on veterinary services, with locust destruction and agricultural colleges the other largest items.[39] The bulk of departmental research was directed to the concerns and demands of white landowners but, especially in relation to livestock disease and conservation, the government sought to sustain a long-term foundation for agricultural production.[40] State officials and scientists devoted themselves to a wide range of environmental issues: soil erosion, the condition of the veld, rotation, water sources, invasive plants and vermin (now called problem animals). Science was invoked rhetorically and by means of direct experimentation and research in order to justify intervention. After the depression, the government attempted to apply tougher environmental regulations in respect of large settler farms as well as African smallholdings.

Delius and Schirmer argue that state conservationist policies were more coercive in relation to African than white landholders.[41] Nevertheless, vets also demanded compliance from white famers in the management of livestock diseases. Agricultural officers and foresters sometimes acted against the perceived interests of landowners in order to promote a conservationist agenda or to resolve what they considered the most urgent environmental problems. State officials thus exercised a degree of autonomy from white agricultural interests. Agrarian environmental issues were significant for the South African state and the applied natural sciences were a strong area of research for white officials, academics and popular writers. The Department of

[39] Department of Agriculture, 'Annual Report: Year Ended 30th June 1930', *Farming in South Africa*, vol. V (November 1930), 56.

[40] Beinart, *Rise of Conservation.* For general arguments about British colonial states and the history of conservation, see Richard Grove, *Green Imperialism*; Beinart and Hughes, *Environment and Empire*; and Beinart, Brown and Gilfoyle, 'Experts and Expertise'.

[41] Delius and Schirmer, 'Soil Conservation in a Racially Ordered Society'.

Agriculture was an important employer of applied scientists and a growing repository of scientific knowledge.

During the early decades of the twentieth century, successful disease control as well as sustained investment in livestock by white and black farmers resulted in spiralling livestock numbers, which peaked nationally around 1930 before the serious droughts of 1931–3. Farm animals, farm production and other agricultural statistics had been recorded alongside human numbers in the Cape colonial censuses and the national census of 1911 – evidence of their importance in South African society. More systematic annual figures, based both on farmers' returns and on dipping figures, were recorded in the annual agricultural censuses from 1918. Scientific understanding, together with increasing statistical sophistication, enabled officials to analyse and assess rising livestock numbers.

Figures showed that the total number of livestock in the country had increased rapidly from 1910 to 1930: cattle from 5.8 million to 10.6 million and sheep from 30.6 million to 48.3 million.[42] (Goats, however, declined from 11.7 million to 8 million, and ostriches from about 800,000 to a mere 32,000.) Animal numbers and 'overgrazing' were perceived by agricultural officials with scientific training to be the main cause of severe soil erosion, threatening the future of production and of the country more generally. This impression was reinforced by the drought of 1932–3 which, together with the depression, reduced sheep flocks to 35 million. The income from wool declined precipitously from an average of over £15 million in 1925–9 to little over £7 million a year. The depression hit the sheep farming areas, still politically influential, particularly hard.

The Department of Agriculture's response was in certain respects contradictory. It catered to the immediate interests of white farmers by disbursing funds under Drought Distress Acts that doubled to £720,000 in in the decade 1925–35. Together with the railways it also assisted farmers to transport livestock to areas with better pastures. Officials were adamantly opposed to the old practices of overland transhumance, which they considered to destroy veld and spread disease. During the acute drought of 1932–3, however, livestock owners put great pressure on the government, which responded by facilitating railway transhumance: an estimated 8 million head of

42 *Union Statistics for Fifty Years.*

sheep were carried by rail, probably around 20 per cent of the total in the country.[43] From 1933 to 1936, the government spent about £1.4 million on transporting livestock by train, and rail rebates on agricultural goods remained the largest item of the greatly enhanced budget till 1940.

However, there was a simultaneous view gaining strength within the department that methods of livestock farming should be changed, with fewer, better animals and, critically, with new methods of managing the pastures. Officials and more successful farmers, especially those in Karoo, articulated anxiety about denudation and soil erosion, and their concerns were echoed in the Native Affairs Department. Although these ideas had been canvassed and increasingly practised since the nineteenth century, they took their most defined and influential shape in the report of the Drought Commission of 1923.[44] The chairman, H. S. du Toit, was an experienced official committed to more scientific methods of agriculture. He and his commissioners were convinced that the major causes of environmental degradation, on which they collected extensive evidence, were livestock numbers as well as the method of livestock farming. Their views shaped debates and interventions for more than half a century.

The Commission railed against the widespread practice of bringing livestock back from the grazing pastures to a central kraal every night, in order to protect animals against predators, especially jackals, and theft. Officials agreed that kraaling was a major contributor to environmental degradation, because the daily movement of tens of millions of animals destroyed the veld through tramping. This could turn much of the country into 'a Great South African Desert uninhabitable by man'.[45] Failure to act would result in 'national suicide'. While the commissioners concentrated largely on the white-owned farms, their arguments were already being applied to the districts occupied largely by Africans. These were developed in the influential Native Economic Commission of 1932, which diagnosed 'desert conditions' in some

43 Beinart, *Rise of Conservation*, 264.

44 Union of South Africa, *Drought Investigation Commission, Interim Report*, U.G. 20-1922 (Pretoria: Government Printer); Union of South Africa, *Final Report of the Drought Investigation Commission*, U.G. 49-1923 (Pretoria: Government Printer).

45 Union of South Africa, *Final Report of the Drought Investigation Commission*, 3, 15.

African reserves.[46] As in the case of the privately funded Carnegie Commission, government inquiries could be important vehicles for national research, and the findings of these two commissions were widely publicised through newspapers, pamphlets and government publications.

Following extensive experience on privately owned farms, officials believed that animals did better if they were allowed to stay out overnight in fenced camps or paddocks. Camps had to be rotated through the year so that their vegetation was not exhausted. Fences also assisted the hunting of predators and protected livestock against them. Bounties were offered for the skins of predators: over the five years from 1919 to 1923, the Cape Provincial Council paid out over £160,000 for about 450,000 proofs of jackals, caracals, wildcats, baboons, leopards and eagles.[47] Veld conservation resulted in a mass and sustained slaughter of wild animals.

Early experiments suggested that a three-camp rotation system was adequate for most veld types and that in the summer rainfall areas it was particularly important to reserve one of these camps for the dry winter months. Ideally, camps should be demarcated and fenced around particular pasture types. Sourveld pastures, which tended to predominate in coastal areas and on higher land, were valuable in the wet spring and summer (October–March) but became rank and unpalatable to animals during the drier autumn and winter (April to September). Where possible, summer camps were demarcated around the sourveld on a farm, while winter camps were fenced around areas where sweetveld predominated. Not all farms had a mixture of such veld, and the main advice was to ensure that camps were rotated before the grazing was exhausted. 'If sheep are allowed to graze the same veld continuously', departmental scientists argued, 'good types of vegetation will automatically disappear, while inferior types will flourish'.[48] Grasses should be allowed to seed.

Research into the mineral content of different types of veld, initiated by Theiler in his diagnosis of *lamziekte*, fed into discussions about

[46] Union of South Africa, *Report of Native Economic Commission 1930–1932*, U.G. 22-1932 (Pretoria: Government Printer).
[47] Union of South Africa, Province of the Cape of Good Hope, *First and Second Reports of the Vermin Extermination Commission*, C.P. 3–1924, 6.
[48] D. C. Maree, 'The Advantages of Spare Veld', *Farming in South Africa*, VI, 62 (May 1931), 48.

rotation and recommendations around fodder and mineral supplements. Marguerite Henrici, the Swiss plant physiologist whom Theiler recruited, worked on phosphorous deficiency as well as transpiration in grasses.[49] Arriving from Basel in 1922, she was put in charge of a small research station near Vryburg in the northern Cape and went on to work in the eastern Transvaal and at Fauresmith in the southern Free State. Henrici became an authority on the nutritional content of South African grasses and Karoo bushes.

H. S. du Toit fought hard to develop the institutional basis for systematic state conservation measures; he also directed the agricultural extension services to farmers. In 1930 the government established a Soil Erosion Council and in 1932 passed a Soil Erosion Act, subsidising improvements, especially water provision, on individual farms. Dispersed water sources utilising earth dams and boreholes were essential in each grazing camp. Smuts took up the cause and in 1936 suggested that 'erosion is the biggest problem confronting the country, bigger than any politics'.[50] Dr J. C. Ross was appointed head of a dedicated division of Soil and Veld Conservation in the department in 1939, supplemented in 1941 by a Forest and Veld Conservation Act. This provided the state with powers to expropriate land and develop government schemes for erosion control, including the reservation of river catchment areas. In agriculture, too, problems identified in the 1930s and 1940s were acted upon vigorously in the postwar era when apartheid gave impetus to burgeoning developmentalist ideas with strongly interventionist implications.

The 1946 Soil Conservation Act, drawing on American legislation, laid the foundation for systematic conservation planning on private land. It funnelled resources to the white-owned farms in the shape of subsidised water storage and boreholes, mechanical works, farm planning and (along with other Acts) fencing. The management of agricultural resources, together with conservationist interventions, helped to shape the social geography of the countryside and the layout of the large commercial farms. Mechanisation after the Second World War further transformed these areas into fenced, controlled zones. By the

[49] M. D. Gunn, 'Marguerite Getrud Anna Henrici (1892–1971)', *Bothalia*, 10, 4 (1972), 503–507; M. Henrici, 'Phosphorous Content of Grassveld', *Farming in South Africa*, V, 49 (April 1930), 29–30.

[50] G. V. Jacks and R. O. Whyte, *The Rape of the Earth: A World Survey of Soil Erosion* (London: Faber and Faber, 1939), 21.

1960s about 10,000 farms had been fully planned with departmental assistance. Extension officers worked with many other landowners to implement the new practices.

Russell Thornton, who had served in the Cape Colony's Department of Agriculture, was first principal of Grootfontein College in the Karoo, a centre of experimentation in fodder supplements and rotational grazing. A committed conservationist, Thornton moved to Pretoria as national director of field and animal husbandry in 1925 and then became the first national director of agriculture in the Native Affairs Department. Touring the African reserves shortly after his appointment, he was shocked at the scale of erosion in some areas and determined on forceful intervention. He wanted massive reduction of livestock (by forced sale, for which the owners would get the proceeds) and conservation works, including dispersed earth dams. These were implemented on a large scale in Herschel district. In 1934 Thornton moved to neighbouring Lesotho, where he introduced one of the most far-reaching conservation programmes anywhere in Africa.

Following such prescriptions, the South African state passed Betterment proclamations in 1939 and 1949, which enabled intervention in the African reserves, where about 3.3 million African people lived.[51] (In the census year of 1951, soon after the National Party took power, 39 per cent of the African population was recorded in the reserves; another 35 per cent, or 3 million, on white-owned farms as tenants and workers; and 26 per cent, or 2.2 million, in the urban areas.) Betterment was tied in with comprehensive measures for segregation and African settlement under the Native Trust and Land Act passed by the United Party government in 1936.

Ostensibly, this legislation intended to make good the promises of the Natives Land Act of 1913 to extend and consolidate the African reserves by purchasing white-owned land. The United Party government saw the Act as a cornerstone of segregation and as a quid pro quo for the final removal of the African voters of the Cape from the common voters' roll. The department decided that new settlements on land bought by the Trust should be in planned rectangular villages rather than more traditional scattered homesteads or clustered wards.

51 Beinart, 'Soil Erosion, Conservationism'.

This would facilitate the introduction of rotational grazing as well as services.

F. W. Fox and D. Back, researching for the SAIMR and the Chamber of Mines, focused on malnutrition in African communities.[52] Fox, a British-trained biochemist who joined the SAIMR in 1925, argued strongly for improved rations in the mines in order to reduce scurvy.[53] The specific scientific understanding of vitamin C (ascorbic acid) was relatively new work, for which two Nobel Prizes were awarded in 1937. Transkeian officials such as the local director of agriculture, F. R. B. Thompson, were already alert to the relationship between agriculture, conservation and growing evidence of malnutrition in the rural areas. Working with such officials, Fox and Back tried to address the Chamber of Mines' concerns that new recruits to the mines were increasingly malnourished. Systematic rural social research promised to resolve such dilemmas. They too proposed fewer and better livestock, reorganisation of settlement, and expansion of health and educational services, which would be facilitated by village settlements. Similar experiments were developed in Zimbabwe; trials in the Transvaal and eastern Cape between 1936 and 1938 seemed to win some local support.

Betterment legislation envisaged planned villagisation of scattered homesteads, the separation of arable and grazing lands, cattle culling through enforced sales, and fenced camps for rotational grazing. The programme was couched in the rhetoric of scientific improvement, conservation and progress: a 'new era', as one official document put it in 1945. It was conceived before apartheid, largely by English-speaking officials, but later implemented by the Afrikaner nationalist government as part of its promotion of segregated homelands or Bantustans for Africans. The apartheid state saw Betterment, or rehabilitation, not only as a means to conserve resources and intensify African agriculture, but also as a way of stemming urbanisation. It is worth noting that they did not need to engage in so comprehensive a

52 F. W. Fox and D. Back, 'A Preliminary Survey of the Agricultural and Nutritional Problems of the Ciskei and Transkeian Territories, with Special Reference to their Bearing upon the Recruiting of Labourers for the Gold Mining Industry' (Johannesburg: unpublished report, 1938).

53 D. Wylie, *Starving on a Full Stomach: Hunger and the Triumph of Cultural Racism in Modern South Africa* (Charlottesville: University of Virginia Press, 2001), 141.

process of planning simply in order to prevent urbanisation: coercive 'influx control' by means of pass laws was the primary strategy. But in the minds of planners, conserving the soil and vegetation, as well as intensifying agriculture, would make rural areas more 'economically viable'. The programme, pursued for half a century until the 1980s, required moving, often against their will, millions of rural African people.

If state officials had set up experimental pilot schemes, they would have found that the fences were often rejected and cut, and that the villages created their own environmental pressures because of the concentration of people who required firewood and water. Especially when fences were cut, kraaling often continued, creating a new concentration of livestock in the villages.[54] Implementation became increasingly slipshod as the scale of population resettlement expanded during the heyday of apartheid in the 1950s and 1960s.[55] There was usually a minority of supporters, including some local headmen, self-conscious progressives and women, who supported the idea of village communities and the promise of services.[56] But the policy was widely opposed and triggered rural protest in Sekhukhuneland, Mpondoland and elsewhere.[57] While conservation legislation constructed a more incentive-driven framework for white landowners, the state imposed untested measures in African areas.[58]

A strong strand in the Africanist literature sees the state's attitude to soil erosion – especially where African peasant farmers were concerned – as either misinformed or as a vehicle for asserting other forms of coercive control over rural society.[59] Yet at the time, there was quite widespread consensus among white southern African scientists about the scale of environmental degradation and the kinds of intervention required. One such protagonist was Edward Roux, who lectured in

[54] P. A. McAlllister, 'Resistance to "Betterment" in the Transkei', *JSAS*, 15, 2 (1989), 346–68.

[55] F. Hendricks, 'Loose Planning and Rapid Resettlement: The Politics of Conservation and Control in Transkei, South Africa, 1950–1970', *JSAS*, 15, 2 (1989), 306–29.

[56] A. Mager, *Gender and the Making of a South African Bantustan* (Oxford: James Currey, 1991).

[57] P. Delius, *A Lion amongst the Cattle* (Oxford: James Currey, 1996).

[58] Delius and Schirmer, 'Soil Conservation in a Racially Ordered Society'.

[59] W. Beinart, 'African History and Environmental History', *African Affairs*, 99, 395 (2000), 269–302.

botany at Wits after the Second World War and who is well known for writing *Time Longer than Rope* (1948), one of the first histories of African political resistance.[60] An early member of the Communist Party (1923–36) who fell foul of ideological and personal schisms, Roux edited the trade union newspaper *Umsebenzi*, gained a master's degree at Wits and won a scholarship to Cambridge (1926–9). He gravitated to the economics lecturer Maurice Dobb, a central figure in communist networks at the university, and completed a doctorate under the supervision of the plant physiologist and ecologist F. F. Blackman.

Peder Anker shows how Roux's biology and his politics were inextricably interconnected through a view of historical and biological evolution.[61] Roux was an egalitarian ecologist and a materialist who sought to use science to overcome divisions of race and capitalism. This was quite unlike Smuts's holistic approach to ecology, which was predicated on opaque, transcendental philosophical principles and a view of ordered, evolutionary change that rendered race, in effect, insuperable. An offer from the University of Cape Town in 1937 allowed Roux to resume academic work in order to research the influence of temperature on fruit storage and, later, the vitamin content of fish liver oils, both of which had practical and commercial implications. During the Second World War, Roux was drawn to popular education. Along with publications in the *South African Journal of Science*, he wrote articles in *Bantu World* and *Trek* as well as pamphlets in the African Bookman series, illustrated by Katrine Harries. These sought to show how science, adapted for popular audiences and geared to the welfare of society, could educate and liberate black South Africans.

Roux wrote a number of pamphlets on environmental degradation and conservation aimed at a popular audience. But he too was a committed modernist and he saw no contradiction between his commitments to political freedom and agricultural progressivism. As Roux explained: 'it is clear that many of the old ways are not useful any more, and that a whole lot of new knowledge is needed. This new knowledge has been got together by "scientists" working in all parts of

60 E. Roux, *Time Longer than Rope: A History of the Black Man's Struggle for Freedom in South Africa* (London: V. Gollanz, 1948).

61 P. Anker, 'The Politics of Ecology in South Africa on the Radical Left', *Journal of the History of Biology*, 37, 2 (2004), 303–31.

the world – in Africa, in Asia, in Europe, in America. Where this new knowledge is used it makes things better . . .The knowledge found in this book ought not to belong only to a few scientists and teachers. It is knowledge which every African man and woman and every African boy and girl ought to have.'[62]

Roux explicitly encouraged African teachers and farmers to confront old ideas and prejudices, for example the tendency of to keep animals but not to consume them. Magic had to be replaced by knowledge of 'microbes and proteins and vitamins'.[63] Roux wrote a popular pamphlet, available in English, Zulu and Xhosa for African readers, which hoped that people would 'learn the new ways of farming before it is too late, before the soil has all been washed away, the grass is gone and your locations have become damaged for ever'.[64] He strongly advocated rotation and repair of dongas with stones, small dams and kikuyu grass. Roux did not differ fundamentally from the Department of Native Affairs' policy of rehabilitation but went further in advocating some kind of collective farming, assisted by agricultural officers.[65] In fact the Native Affairs Department was then cautiously promoting co-operatives, which were also enthusiastically pursued by the charismatic priest Bernard Huss in Natal Catholic mission stations.[66] A radical, but not collective, transformation of property relations, as the basis for agricultural improvement in the African communal areas, was also recommended by the National Party's Tomlinson Commission in 1955: private property on larger 'economic' plots together with Betterment programmes.

After moving to the University of the Witwatersrand, Roux was involved in grassland research at the Wits experimental farm and here he did come to rather different conclusions. In controlled circumstances, he argued, heavy grazing could increase the fertility of pastures, as long as it was stopped before soil erosion began and pastures were regularly rotated. In his inaugural lecture as professor of botany

[62] E. Roux, *Harvest and Health in Africa* (London: Thomas Nelson, 1942), 84–5.

[63] Roux, *Harvest*, 86.

[64] E. Roux, *The Cattle of Kumalo* (Cape Town: African Bookman, 1943), 28.

[65] E. Roux, 'The Native Reserves and Post-War Reconstruction' (Cape Town: Association of Scientific Workers of Southern Africa, [1944?]).

[66] P. Rich, 'Bernard Huss and the Experiment in African Cooperatives in South Africa, 1926–1948', *International Journal of African Historical Studies*, 26, 2 (1993), 297–317; M. Draper, 'In Quest of African Wilderness', USDA Forest Service Proceedings RMRS-P-27, 2003, 57–62 (on web).

in 1963, Roux directly addressed experiments in grazing management, arguing that overstocking might not be so central an issue in the African reserves where 'serious erosion' was mainly due to the 'absence of paddocking, the kraaling of the beasts at night and the long treks the cattle must make to obtain drinking water'.[67] Roux does not seem to have done detailed research in the Bantustans, but his diagnosis of the agricultural problems in African reserves – though not his politics – was strikingly similar to those of leading agricultural officials.

In sum, a scientifically based conservationist approach had profound effects on the South African countryside, helping to reorder the layout and management of large, white-owned commercial farms and also the areas reserved for Africans. However, the reorganisation of African settlements did not result immediately in investment, electrification and development, and the state did not attempt to make fundamental changes to land tenure, which remained in a regulated form of customary landholding with communal pastures. Betterment did not lead to the intensification of agriculture in the African areas; it ran in parallel with a slow decline.

In some respects, government attempts to control invasive species fed into these initiatives. By 1930 the Department of Agriculture employed about thirty entomologists – the second biggest group of scientific officers after vets. In part they worked with vets to control the many insect-borne livestock diseases and developed research on ticks, tsetse flies and scab acari. Their attention was equally devoted to scourges of fruit and crops such as citrus scale, codling moth and stem borers. In the 1930s, locusts were a particular preoccupation and the department developed a Union-wide administration to control them through poisoning. Hoppers tended to proliferate when good rains followed serious drought. In 1934–5, widespread swarms gave rise to coordinated national action that cost £1.3 million, half of the growing departmental budget.[68] Although this expenditure was not again rivalled, locust control required systematic research and, occasionally, hundreds of thousands of pounds.

Entomologist Charles Lounsbury (see Chapter 4), was enthusiastic about the use of imported insects to control other pests. By the 1920s, opuntia cactus species from the Americas had become the most

[67] E. Roux, *The Veld and the Future: Inaugural Lecture* (Johannesburg: Witwatersrand University Press, 1963), 9.

[68] *Farming in South Africa*, X (December 1935), 502–4, 526.

threatening invasive plant.[69] Initially introduced and spread as a livestock fodder in the eighteenth century, the *Opuntia ficus indica* (prickly pear) was dispersed by birds and animals consuming its fruit and seeds, and through rooting from its paddle-like cladodes. Opinion was split on the plant: it was useful for livestock in droughts and the fruit was widely eaten, especially by poorer white and black people. It was also a popular base for brandy (*witblits*) and beer (*karee* or *iqilika*). But surveys suggested that dense stands were taking over large areas of land, perhaps over a million hectares, including fertile alluvial soils; the glochids or spicules on the fruit damaged valuable livestock. While officials and wealthier farmers swung in favour of subsidised eradication in the early twentieth century, this was difficult to achieve by cutting, digging and poisoning.

Before he retired in 1927, Lounsbury reported on Australian experiments with the American cactus moth called cactoblastis as a strategy for biological control of prickly pear. Distributed in sufficient quantities, the caterpillars promised to devour the rampant prickly pear. Australian experience, as well as local tests on seventeen different commercial crops, suggested that the grubs devoted themselves entirely to opuntia species, and in the 1930s the government decided to create a breeding station. In this case, scientific work was at the heart of environmental interventions.

The project was put in the hands of William Pettey, another American entomologist in South African state service. In 1932 he travelled through infested districts, largely in the Cape Midlands, and visited Australia to study the breeding of cactoblastis. Fully absorbed in opuntia eradication for the next twenty years, Pettey established breeding stations at Graaff-Reinet and Uitenhage. When cactoblastis failed to make the same impact as in Australia, where the most invasive opuntia was a different species, he added various Latin American cochineal insects to his repertoire. These were also specialist cactus feeders; one that had long been nurtured in Mexico produced the intense red colouring exported to Europe since the sixteenth century. Breeding proved successful and by 1939 Pettey had released an estimated 30 million eggs.[70] By the late 1940s roughly 75 per cent of prickly pear had been cleared, but not the dangerous jointed cactus.

Pettey spent nearly two decades of systematic experimental work, overcoming barriers to mass breeding and effective distribution of

[69] Beinart and Wotshela, *Prickly Pear*.

[70] F. W. Pettey, 'The Biological Control of Prickly Pears in South Africa', *Union of South Africa, Department of Agriculture and Forestry, Scientific Bulletin 271* (Pretoria: Government Printer, 1948).

insects. He learnt a great deal about the life cycle, behaviour and susceptibilities of the various insects that he handled as well as the character and ecology of different species of opuntia and the experience of different countries. It was a successful exercise in patient, applied scientific work; over £1 million was probably spent on eradication and the programme continued, in different guises, until the 1960s. While scientific methods were applied to the eradication campaign, no calculation was done of the costs of prickly pear eradication for poorer rural people who used the fruit. The state engaged in prickly pear eradication primarily to serve the economic interests of white commercial livestock farmers and – in the eyes of officials – the agrarian economy as a whole. By the 1950s, officials increasingly saw eradication as a contribution to national conservation by protecting indigenous plant species.

The Coelacanth and 'Mrs Ples'

By the interwar years there was a broadly-based community of natural sciences in South Africa, dispersed around many different institutions, including universities, government departments and museums. Amateur 'citizen scientists', white and black, recorded species, took note of environmental features, and utilised plants for their medicinal properties. In the sphere of fisheries and marine sciences, the Cape government appointed John Gilchrist as marine biologist in 1895. In 1897 Gilchrist took possession of the Pieter Faure, a steam-powered boat built in Scotland, which was fitted out to his specifications and used to survey fishing stocks such as hake and sole around the extensive Cape coastal region.[71] Gilchrist was also the driving force behind the construction of a marine research aquarium established under government auspices at St James in 1902 in order to deepen understanding of local fishing stocks. He co-authored a standard text, *The Freshwater Fishes of South Africa* (1913–18).[72]

[71] Cape of Good Hope, Department of Agriculture, *Marine Investigatons in South Africa*, vol. 2 (Cape Town: Government Printer, 1904).

[72] J. D. F. Gilchrist, and W. W. Thompson, 'The Freshwater Fishes of South Africa', Annals of the South African Museum 11, 5–6 (1913) 321–464; J. D. F. Gilchrist, and W. W. Thompson, 'A Catalogue of the Sea Fishes Recorded from Natal Part 1', Annals of the Durban Museum, 1 (1916), 255–90 and 'Part 2', Annals of the Durban Museum, 1 (1917), 291–431.

Gilchrist's time was split between his responsibilities to government and his work as a curator at the South African Museum, where a dispute led him to move to the chair of zoology at the University of Cape Town. Two of his students, Cecil von Bonde and Jan Marchan, succeeded him as directors of Sea Fisheries.[73] Yet Gilchrist's energetic academic interests were not sustained following his death in 1926, with most research thereafter concentrated on practical knowledge relating to fisheries. Notwithstanding South Africa's long coastline, systematic oceanographic research only began in the 1950s with the support of the Council for Scientific and Industrial Research (CSIR). Marine science remained a relatively neglected and fragmented field until after the Second World War.[74]

The best-known scientific discovery in this field – a living coelacanth in 1938 – was the outcome of a mixture of amateur and professional curiosity, in this case a sharp-eyed museum curator and a largely self-trained ichthyologist. In December, a five-foot 'fossil fish' with four limb-like fins and heavy scales was netted by a trawler off Algoa Bay. Hendrik Goosen, skipper of the boat, alerted Marjorie Courtenay-Latimer of the East London Museum, who collected it from the harbour in the heat of the South African summer. She attempted, with limited success, to preserve the fish in formalin.

Courtenay-Latimer wrote immediately to J. L. B. Smith, a Stellenbosch- and Cambridge-trained chemist and a keen angler, at Rhodes University, to alert him to the news. She knew he was turning to the study of fishes at the time, under the guidance of John Hewitt, director of the Albany Museum. Going by Courtenay-Latimer's sketches and description, Smith concluded that the specimen was a coelacanth, a species that had been considered extinct for 65 million years and was sometimes dismissed as a 'degenerate side-line'. Smith named the species *Latimeria chalumnae* in her honour. A full announcement of the discovery was made in the South African press in February 1939, with scientific accounts carried in *The Times* and in *Nature* the following month.[75]

[73] J. H. Day, 'Marine Biology in South Africa', in Brown (ed.), *Scientific Endeavour in South Africa*, 88–90.

[74] L. van Sittert, '"The Handmaiden of Industry": Marine Science and Fisheries Development in South Africa 1895–1939', *Studies in History and Philosophy of Science*, 26, 4 (1995), 531–88.

[75] J. L. B. Smith, *Old Fourlegs: The Story of the Coelacanth* (London: Longmans Green, 1956); *The Times*, 17 March 1939; J. L. B. Smith, 'The Living Coelacanthid Fish from South Africa', *Nature*, 143, 3627 (1939), 748–50.

Smith's extraordinary claim was at first greeted with scepticism. According to his account, doubts were expressed about his suitability to conduct the research – there were suggestions he should instead send it to the British Museum – and questions were raised about the manner of its preservation. Unfortunately, the fish had been damaged in the nets and its soft tissues and gills were discarded in East London. It was imperative for further validation that another specimen be found. Smith suspected that the Mozambique Channel might yield more examples and, over the next decade, he undertook angling expeditions, offering a reward of £100 to African fishermen along the Indian Ocean coast. In 1952, Smith received word that another coelacanth had been found off the Comoros islands; it was a type of fish that was known to local fisherpeople. After frantic phone calls and lobbying, a military Dakota plane was provided by the prime minister, D. F. Malan, in order to collect the fish from the French-occupied island of Pamanzi. The chase was on to retrieve the fish before putrefaction set in.

Figure 5.1 J. L. B. Smith and the Coelacanth, 1952.
Credit: Mike Bruton; South African Institute for Aquatic Biodiversity

In a state of high excitement, Smith insisted on flying back to Cape Town so as to present Malan with his find at the prime minister's holiday retreat. In Smith's telling, his fears about Malan's strict religious beliefs and alleged Creationist sympathies proved unfounded. Malan, no doubt sensing the intense interest of the press, responded with enthusiasm. On viewing the fish, he is said to have joked: 'My, it is ugly. Do you mean to say we once looked like that?'[76] Smith named the fish *Malania anjouanae* in honour of the apartheid premier and wrote an obsequious letter of thanks, in Afrikaans, praising Malan for his honesty and strength, and assuring him that there were many Englishmen like himself who supported the government's policies of apartheid.[77]

In his published account of the discovery of *Old Fourlegs*, written as an epic of perseverance against all odds, Smith professes a strong South African scientific patriotism. Malan is here represented as a hero, in contrast to Smuts who is roundly criticised for placing his faith in foreign scientists (Smith resented the fact that Smuts once refused to see him). One sub-theme of *Old Fourlegs* is Smith's account of how he was brought up as a strong supporter of all things British but went on to abandon his prejudices against 'the Boers' in order to identify with white South Africa. His story abounds in gratuitous racism and Africa is presented as innately violent – with Mau Mau figuring as a portent of things to come.[78] Malan's support of Smith clearly paid political dividends.

By coincidence, the field of palaeoanthropology entered an exciting new phase at the same time as the discovery of the original coelacanth. In 1936, an adult *Australopithecus* cranium in perfect condition, along with baboon skulls and endocrinal casts, was found in the Sterkfontein limestone caves. This was handed to the venerable Robert Broom, now based at the Transvaal Museum. He seized on the fossil's significance, identifying a familial resemblance to Dart's Taung skull. Broom's discovery came to be known as *Plesianthropus transvaalensis*. In 1938 Broom identified a further fossil skull found by a schoolboy at Kromdraai, near Sterkfontein, as yet another type, which was named *Paranthropus robustus*.[79]

[76] Smith, *Old Fourlegs*, 172.

[77] Smith to Malan, 2 January 1953, Malan papers, 1/1/2872, Gericke Library, University of Stellenbosch. Our thanks to Lindie Koorts for this reference.

[78] Smith, *Old Fourlegs*, 4–5, 22, 66, 87–8, 94–9, 117–18, 162; Smith, 'The Living Coelacanthid Fish from South Africa'; 'The Second Coelacanth', *Nature*, 4342, (1953), 99–101. See also S.Weinberg, *A Fish Caught in Time: The Search for the Coelacanth* (London: Fourth Estate, 1999). For an informative and droll account by the navigator of the Dakota, Major General Duncan Ralston, www .dc-3.co.za.

[79] W. E. Le Gros Clark, 'Significance of the Australopithecinae, *Nature*, 157, 4000 (1946), 863–4; R. Broom and J. T. Robinson, 'Ape or Man?', *Nature*, 166, 4229 (1950), 843–4.

Rather like Broom's misidentification of *Boskop* man in 1918 (a classificatory error driven by the remorseless logic of typological distinction or the racialism of small differences) it initially appeared that the new variants, each with different estimates for their cranial capacities, sat uneasily with Raymond Dart's claims for the singular importance of *Australopithecus africanus*. Dart was in any case annoyed that the new discoveries had gone to Broom for analysis rather than the Wits Anatomy Department. But the rift was soon healed. Dart appointed Broom as a visiting lecturer and in 1946 the indefatigable eighty-one-year-old palaeontologist compiled a monograph with Gerrit W. H. Schepers entitled *The South African Fossil Ape-Men: The Australopithecinae*. Broom thus played a key role in validating Dart's original hunch about the importance of the Taung skull as a key transitional stage or 'missing link' in hominin development.

International experts now began to revisit their earlier scepticism. In a review of the developments in *Nature*, Wilfrid Le Gros Clark, the Oxford professor of anatomy, recognised the significance of the new finds, arguing that Broom and Dart had 'very considerably enhanced the prestige of South African science in the eyes of the world generally'.[80] A year later, in 1947, Broom extracted a complete cranium of *Plesianthropus*. He pronounced this the most important fossil skull ever found in the world. It came to be known with colloquial affection as 'Mrs Ples'. It is now generally regarded to have been male, and *Plesianthropus* has been subsumed into the *Australopithecus africanus* species.

At the Pan-African Congress on Prehistory held in Nairobi in 1947, Le Gros Clark clarified his views. Having just come from Johannesburg, where he inspected the australopithecine fossil evidence, he lent crucial support to the view that they were hominids rather than anthropoid apes, and he paid wholehearted tribute to the work of Dart and Broom. Sterkfontein's ongoing significance as a key international site of evolutionary origins continues to this day.[81] As in the case of the coelacanth, there was a gap between discovery of the first specimen

[80] Le Gros Clark, 'Significance of the Australopithecinae'.

[81] Le Gros Clark, 'Significance of the Australopithecinae'; 'The Pan-African Congress on Prehistory', *Nature*, 159 (1947), 216–17; P. Tobias, *Into the Past: A Memoir* (Johannesburg: Picador, 2005), chapter 8; Christa Kuljian, *Darwin's Hunch: Science, Race and the Search for Human Origins* (Auckland Park: Jacana, 2016), chapter 7.

and confirmation of the specimen's evolutionary significance. The remoteness of South Africa from Europe and the disruptions of the war may well account for the fact that broader recognition of these discoveries was deferred.

This was certainly the case in the growing field of prehistory, which, to a considerable extent, remained an amateur pursuit reliant on the 'finds' of charismatic individuals. In 1935, the government took an important step towards professionalisation when it created a Bureau of Archaeology, based at Wits University, following the passage of the Natural and Historical Monuments, Relics and Antiquities Act of 1934. The new Bureau was headed by Clarence ('Peter') van Riet Lowe. Hitherto a senior civil engineer in the Department of Public Works, his interest in archaeology was sparked by the Stone Age sites he uncovered while constructing river bridges in the Orange Free State. Lowe was made the inaugural professor of archaeology at Wits and secretary of the statutory National Monuments Commission.

The work of the Bureau gathered momentum after 1942 when the Abbé Breuil, leading French prehistorian and specialist in prehistory and rock art, was brought from Europe to South Africa by Smuts (once more prime minister) to join the staff of the Archaeological Survey.[82] Breuil was an energetic fieldworker with a gift for publicity who lent international authority to South African prehistory. His boundless enthusiasm coincided with a propensity to pronounce on finds without pause or restraint. In particular, he was strongly predisposed to speculation, based on the working assumption that the finest examples of 'Bushman art' must have been influenced or produced by foreigners.[83]

During and after the war, Breuil did a great deal to increase national and international awareness of South Africa's importance as a repository of Bushman rock art – notwithstanding (or because of) his predilection for theories rooted in assumptions around cultural and racial diffusion. Breuil's work added to the growing fascination with ideas of

[82] D. Underhill, 'A History of Stone Age Archaeological Study in South Africa', *South African Archaeological Bulletin*, 66, 193 (2011), 3–14.

[83] S. Dubow, 'Henri Breuil and the Imagination of Prehistory: "Mixing up Rubble, Trouble and Stratification"', *South African Archaeological Society Goodwin Series* (2019); S. Geroulanos, 'Polyschematic Prehistory at the Dusk of Colonialism: Internationalism, Racism, and Science in Henri Breuil and Jan Smuts', *Res: Anthropology and Aesthetics*, 69–70 (Spring–Autumn 2018), 136–57.

human origins pursued by Dart and Robert Broom, and it is perhaps not entirely coincidental that public interest in Mrs Ples occurred at just the same time that Breuil unveiled his theory of the 'White Lady' of the Brandberg. Breuil insisted that this extraordinary rock art frieze in Namibia proved a link with Minoan civilisation, though its central figure turned out to be neither a woman nor Mediterranean: the Abbé had failed to spot the figure's penis. In respect of palaeontology, the scientific and public imagination were mutually reinforcing.

Growing public awareness of the importance – and vulnerability – of South Africa's important fossil heritage led to the establishment in 1945 of a committee and a fund at Wits University, which became the Bernard Price Institute for Palaeontological Research. One account has it that, in 1945, Robert Broom gave a public lecture in which he bemoaned the lack of attention to South Africa's fossil inheritance. Bernard Price, in the audience, stood up and immediately offered to endow a new palaeontological unit to complement the Geophysical Institute that bore his name. James Kitching, who spent his Karoo childhood collecting vertebrate specimens for Broom, had recently returned from war service in Italy. Notwithstanding his lack of formal qualifications, Kitching became the Institute's first appointment and subsequently its inaugural director. He started collecting therapsid mammal-reptile fossils in the Karoo and discovered an important australopithecine fossil at Mapakansgat limeworks in 1947.[84] (In 1969–70 Kitching found tetrapod fossil skeletons in the Queen Maud mountains of Antarctica which corresponded closely to *Lystrosaurus* fauna in the Karoo, thereby providing new confirmatory evidence for Gondwanaland and continental drift.)[85] From rock art and fishes to early hominids, the decade from 1936 prompted a new wave of imagining about deep time in South Africa.

Women, Museums and Botany

Margaret Courtnay-Latimer, the co-discoverer of the first scientifically reported coelacanth, was one of a number of women who were able to find scientific positions in the first half of the twentieth century as

[84] B. S. Rubidge, 'The BPI: 50 Years of Palaeontological Activity', *Palaeontologica Africana*, 33 (1997), 1–9.

[85] J. W. Kitching et al. 'Lystrosaurus Zone (Triassic) Fauna from Antarctica', *Science*, 175, 4021 (1972), 524–7.

university education gradually opened to them. Provincial museums were an arena where they could assume leadership positions. This was an age of ambition for museums. In the Cape, substantial collections housed in buildings exuding civic pride were established in East London, Grahamstown, Kimberley and Port Elizabeth. King William's Town's Kaffrarian museum, open to black as well as white, accommodated a particularly large assemblage of zoological specimens. These were joined in 1932 by the stuffed skin of Huberta – a hippopotamus that had wandered down the east coast of the country from 1928 to 1931 in a journey that had earlier attracted as much popular interest as the coelacanth.[86]

Maria Wilman, a pioneer woman museologist, was probably the second South African woman to study at Cambridge.[87] Trained in botany and geology, she returned to work at the South African museum, where Peringuey encouraged her to research rock art. Wilman was appointed director of the McGregor museum in Kimberley in 1908 and remained in the post until 1946. She travelled widely in the area and built up collections of stone implements found by diamond diggers in the alluvial fields. Wilman published on rock art, promoted this work through the S2A3, and drew together her research in a book on *The Rock Engravings of Griqualand West and Bechuanaland, South Africa* (1933), illustrated with seventy plates and additional figures. Wilman was sufficiently highly regarded to attract funding from the Royal Society and Carnegie. At a conceptual level she is credited jointly with Neville Jones for the term 'Middle Stone Age', now used for artefacts from about 25,000–250,000 years ago; she argued in favour of a localised (rather than European) periodisation and terminology, supplementing the indigenous schema set out by van Riet Lowe and Goodwin in the late-1920s.[88] Together with Dorothea Bleek and Lucy Lloyd, Wilman made a major contribution to the

[86] L. Witz, 'The Making of an Animal Biography: Huberta's Journey into South African Natural History, 1928–1932', *Kronos* 30 (2004), 138–66; Leslie Witz, 'Hunting for Museums', *JSAS*, 41, 3 (2015), 671–85.

[87] M. Wilman, S2A3 Biographical Database of Southern African Science online.

[88] K. Weedman, 'Who's "That Girl": British, South African, and American Women as Africanist Archaeologists in Colonial Africa (1860s–1960s)', *The African Archaeological Review*, 18, 1 (2001), 1–47; Dubow, 'Henri Breuil and the Imagination of Prehistory', 37.

recording and imagining of deep human history and San mythology in South Africa.

As in the case of Wilman, Courtenay-Latimer, who did not have a university education, but effectively ran East London museum from 1931 to 1973, made her repository a notable regional centre of scientific collection and education. Even more than Wilman, she was a generalist who developed substantial and systematic botanical collections, assembling thousands of plant specimen sheets. Both women found plant species new to science, which were named after them. They were publicly spirited citizen scientists, whose approach contrasted with the more individualist Marais and Merensky (Chapter 4). Although their achievements and ideas were not always recognised, they were able to operate with considerable freedom and, like the visiting archaeologist Caton Thompson, gained authority and respect in their respective fields.

The space that opened up in botany grew from gendered nineteenth century legacies (Chapter 2), where women had made contributions as collectors, botanical artists and gardeners. In the twentieth century, it became possible to build academic careers. In 1897 following her appointment, soon after completing a doctorate at Cornell, to the Botany Department of the Huguenot College in Wellington, the American Bertha Stoneman became a key figure. She taught successive cohorts of students at this women's university college, which awarded degrees, and she served as its principal from 1921 to 1933. Stoneman was a teacher rather than researcher, but she published *Plants and Their Ways in South Africa* (1906) as well as a bilingual English-Afrikaans glossary of plant names. She was also prominent as a protagonist of botany, as a member of the S2A3, and as first president of the South African Association of University Women.

In 1902, Augusta Duthie, an early protégé, was appointed to the Botany Department at Victoria College, Stellenbosch, at the age of twenty-one. She is often referred to as the first South African-born, and solely South African-trained, professional botanist in the country. Duthie worked at Stellenbosch until 1939, served as head of department, played a role in the foundation of the university botanical garden, and made a detailed study of local plants, for which she received a doctorate. Ethel Doidge also trained at the Huguenot College with Stoneman and then received a doctorate from the University of Cape Town – possibly the first by a woman in South

Africa – on a bacterial disease of mangoes. Tropical fruit production was then taking off in suitable areas of the Transvaal, and Doidge moved to work with Pole Evans in 1908 at what became the Division of Botany in the Department of Agriculture in Pretoria. She was a highly productive researcher, rising to the position of principal plant pathologist in 1929 and publishing a standard work *The South African Fungi and Lichens* (1950). Neither Wilman, Courtenay-Latimer, Stoneman, Duthie nor Doidge married; we do not know of their individual motives, but male dominated institutions made it difficult for women to combine family life with professional careers, and some required women to resign upon marrying.

Margaret Levyns, an expert on Cape flora, felt the institutional constraints on women keenly.[89] She was a protégé of Pearson, professor of botany at UCT and founder of Kirstenbosch, who made his name studying welwitschia and cycads. On a scholarship to Cambridge, she studied with A. C. Seward, who had written on Cape flora, and found the single sex Newnham College more restrictive than UCT, where white men and women studied and socialised together. Levyns worked at UCT from 1918 to 1946. She initially concealed her marriage in 1923 because of her uncertainty as to whether she would be allowed to remain in her post; however, the University accepted it and she set a precedent in being permitted to remain. Despite completing a doctorate in 1932, publishing and becoming a member of the Royal Society, she was not promoted and was required to leave her post at the age of fifty-five – a discriminatory rule that applied only to women. There was some space for highly educated white women at UCT, but not equal opportunity.

Levyns was an innovative scientist in at least three spheres – taxonomy, the effects of fire, and phytogeography. Taxonomy, the subject of her thesis, was well-established as the primary botanical interest at the Cape and there was still a great deal of plant-finding and classification required to fill out knowledge of the hugely diverse flora. Brett Bennett notes that she was so dissatisfied with the cumbersome, nine volume *Flora Capensis* – started by William Harvey in Dublin and continued at Kew – both for teaching and general use that she

[89] B. M. Bennett, 'Margaret Levyns and the Decline of Ecological Liberalism in the Southwest Cape, 1890–1975', *South African Historical Journal*, 67, 1 (2015), 64–84; 'Margaret Levyns: The Making of a South African Botanist', www.royalsocietysa.org.za.

compiled her own *Guide to the Flora of the Cape Peninsula* in 1929.[90] Levyns probably conducted the first scientific research on the effects of fire in the fynbos biome in 1919 and continued experiments in the 1920s.[91] The results tended to confirm the views of farmers and earlier commentators that burning favoured the spread of rhinoceros bush, an agricultural pest. She and others argued that the bush became invasive, despite being indigenous, and was a secondary vegetation, caused by human disturbance, not a primary or climax vegetation. Such views reinforced scientific support for fire abolitionism at least until the 1940s, when it was gradually challenged by Henkel and Wicht – with whom she worked in the 1930s. Levyns also maintained a relatively neutral view of exotic invasive species at this time – changing only when the momentum of scientific opinion turned decisively against non-native plants in the 1960s. Her 1929 *Guide* included both indigenous and exotic plants, and her research on fire suggested that native species could also invade.

Richard Cowling and colleagues have recently coined the idea of Levyns' Law, crediting her with the first scientific analysis of the gradual decline in species diversity from the west to the east of the Cape Floristic Region.[92] This in itself required an encyclopaedic knowledge of the location of different plants, acquired over many years of field botany and taxonomic work. Levyns explored these deep historical and phytogeographical issues in publications in 1938 and 1964.[93] Many geological, climatic and landscape factors could potentially have contributed to this pattern. The eastern end of the spectrum was more typical of similar climatic zones in other parts of the world. It was the extraordinary diversity and speciation of the western parts of the Cape over a long period of time that required special explanation. She

90 Bennett, 'Margaret Levyns'.

91 Pooley, *Burning Table Mountain*, 53; B. W. van Wilgen et al., 'Ecological Research and Conservation Management in the Cape Floristic Region between 1945 and 2015 : History, Current Understanding and Future Challenges', *Transactions of the Royal Society of South Africa*, 71, 3 (2016), 207–303.

92 R. M. Cowling et al., 'Levyns' Law: Explaining the Evolution of a Remarkable Longitudinal Gradient in Cape Plant Diversity', *Transactions of the Royal Society of South Africa*, 72, 2 (2017) 184–201.

93 M. R. Levyns, 'Some Evidence Bearing on the Past History of the Cape Flora', *Transactions of the Royal Society of South Africa*, 26 (1938), 401–24; M. R. Levyns, 'Migrations and Origins of the Cape Flora', *Transactions of the Royal Society of South Africa* 37 (1964), 85–107.

pointed to climatic factors, and the great variation between winter rainfall and summer drought rather than the amount of rainfall, as well as sharp changes in altitude and aspect. Levyns also noted that this diversity was characteristic of both the fynbos and Karoo floras: 'the same urge to intensive speciation is found equally in the relatively moist conditions of the Caledon Division and in the arid stretches of Namaqualand'. Subsequent research has confirmed the positive correlation between regional-scale plant diversity, endemism, and the proportion and reliability of winter rainfall for both areas. Levyns' Law was thus defined in 2017 as: 'all clades endemic (or nearly so) to the GCFR show a westwards concentration in regional scale and a decline eastwards, concomitant with a decline in the proportion of winter rainfall'.[94] Much of the literature on plant diversity hotspots focusses on temperature and metabolic factors in the tropics. The Western Cape case, with the richest concentration globally of extratropical plant diversity and endemism, with about 11,423 species and 1,119 genera, seemed to require a different explanation.

By the 1930s and 40s, these and other South African women scientists in museums, in government and especially in universities, were making their mark as researchers, protagonists of science and as public figures in fields such as rock art, botany and palaeobotany. Levyns became the first woman to preside over the Royal Society of South Africa in 1963 – the highest recognition by the scientific community. Social medicine was also to become a focus, and nursing provided the main opportunity for women to get some training and practical skill in the healing sciences. The majority of nurses at this time were white, but opportunities were opening for black women too, who negotiated difficult roles in hierarchical hospitals at the interface between biomedical and traditional ideas about disease and healing.[95] Especially during the Second World War, when white men were required for the armed forces, women were absorbed more fully in a wider range of economic roles; training also became more accessible in the expanding post-war universities.

[94] Cowling et al., 'Levyns' Law'.

[95] S. Marks, *Divided Sisterhood: Race, Class and Gender in the South African Nursing Profession* (Basingstoke: Palgrave Macmillan, 1994); A. Digby and H. Sweet, 'Nurses as Culture Brokers in Twentieth-Century South Africa' in Waltraud Ernst (ed.), *Plural Medicine, Tradition and Modernity, 1800–2000* (London: Routledge, 2002), 113–29.

Wartime Mobilisation, Reconstruction and the Council for Scientific and Industrial Research

Over a decade from the mid-1930s it is possible to detect a thickening network of interconnected academic and state research institutions. This was in no small measure attributable to growing state investment in industrial and scientific research as well as the culture of planning and consolidation encouraged by wartime mobilisation, economic growth, and large-scale visions of post-war reconstruction. In South Africa, as in Britain, the welfare and warfare states were closely intertwined.[96] The South African experience also reflected aspects of developmental statehood which are visible too in late-colonial British Africa. There, the Colonial Welfare and Development Acts of 1940 and 1945 led to substantial amounts of money for practical scientific research, and crystallised the thinking set in train by large-scale compendia such as Lord Hailey's *African Survey* (1938) and E. B. Worthington's companion study, *Science in Africa* (1938).[97] These major interdisciplinary studies were important stimuli for Britain's intellectual and technical investment in its colonial domain, and they helped to shape the pattern of decolonisation as the old civilising mission was replaced by a new developmentalism. As Chapter 6 shows, apartheid South Africa sought to maintain its own external connections by participating in these networks of technical expertise and power.

On Smuts's return to power in 1939, he appointed Hendrik van der Bijl as director-general of war supplies. This proved to be a key moment in the mobilisation of an integrated industrial strategy; in turn, it helped to shape the post-war landscape of 'big science' in South Africa. Van der Bijl, who proudly traced his roots in South Africa back to 1688, grew up in Pretoria. He studied mathematics and physics at Victoria College (later Stellenbosch University) and then took a doctorate in physics at Leipzig (1912) where he conducted experiments on liquid ions. Prompted by Robert Millikan, van der Bijl moved to New York in 1913 to work at Western Electric (later Bell) Laboratories, where he conducted pioneering research on

96 D. Edgerton, *Warfare State: Britain, 1920–1970* (Cambridge: Cambridge University Press, 2005).

97 J. Lewis, *Empire State-building: War and Welfare in Kenya, 1925–52* (Oxford: James Currey, 2000); Tilley, *Africa as a Living Laboratory*.

thermionic valves in order to assist with the amplification of radio and telephone signals. His book, *The Thermionic Tube and Its Applications* (1920), became the standard text on the subject.[98]

At Smuts's invitation, van der Bijl returned to South Africa in 1920 in order to become the government's first scientific and technical adviser. A year previously he had written a paper on 'Scientific Research and Industrial Development' in which he advocated the adoption of scientific methods in support of secondary industrialisation. He favoured a model based on government-owned corporations run on business principles. As the first chair of ESCOM, established in 1922–3, and also as the head of the newly created ISCOR from 1928, van der Bijl was ideally positioned to shape key public utilities responsible for industrial development and transport infrastructure.[99] The outbreak of war afforded unprecedented opportunities to expand an increasingly ambitious programme of state planning, science and industrialisation with elements of social welfare too. Bill Freund sees van der Bijl as the leading force behind South Africa's wartime 'developmental state'.[100]

Vital to this programme was van der Bijl's control of the Industrial Development Corporation (IDC), set up in 1940 with £5 million in capital as a state-aided investment vehicle designed to promote economic growth and offer technical advice. Vanderbijlpark, the model industrial town south of Johannesburg which he helped to design in order to facilitate the expansion of ISCOR's steelworks immediately after the war, memorialised this half-forgotten titan of mid-century South African developmental planning.

Van der Bijl relied heavily on his deputy, Hendrik J. van Eck, a chemical engineer trained at Stellenbosch, Leipzig and Berlin, who joined ISCOR as works manager. Together, they coordinated and invested in a huge range of wartime industries, from the production of army boots and blankets to precision instruments, guns and explosives. As a result, South Africa became a key workshop and supplier to the Allies in the Mediterranean and Middle East.

98 D. J. Vermeulen, 'The Remarkable Dr. Hendrik van der Bijl', *Proceedings of the IEEE*, 86, 12 (1998), 2452.

99 B. F. J. Schonland, 'Hendrik Johannes van der Bijl 1887–1948', *Obituary Notices of Fellows of the Royal Society*, 7, 19 (1950), 26–34.

100 Freund, *Twentieth-Century South Africa*, 70.

The immediate needs of the wartime state soon morphed into conceptions of the post-war future, opening multiple 'worlds of possibilities'.[101] In the liberal and social democratic imagination, broader conceptions of welfare and citizenship were actively explored. Science in the service of a broader humanity was central to such thinking. In 1941, H. R. Burrows adapted the idiom of wartime mobilisation to outline a 'somewhat visionary' scheme for the faltering National Research Council, arguing that scientific and social research should be oriented towards the solution of longer-range problems relating to agriculture, urban poverty and 'inter racial maladjustments'.[102]

By 1942, even before the decisive battle of El Alamein where South African soldiers helped to turn the fortunes of the Allies in North Africa, a symposium on 'Science and Post-war Reconstruction' took place under the auspices of the S2A3. Smuts sent a message to say that 'a new world has to be built on the ruins of the old'.[103] Along with the government's new Social and Economic Planning Council, chaired by van Eck, Smuts looked to scientists to help with peace-time reconstruction. In keynote addresses delivered by van der Bijl and van Eck on the present and future of industry, they reiterated the importance of comprehensive planning to include the efficient mobilisation of human as well as natural resources. Van Eck argued that in any future policy 'the deciding factor' had to be 'the welfare of the greatest number in the State'.[104] Scientists, he judged, were in the best position to formulate planning and research for industrial reorganisation.

Summarising the proceedings, deputy prime minister J. H. Hofmeyr picked up on the 'human factor' and the fact that 'we are all part of one great inter-related world'.[105] In addition to van der Bijl's and van Eck's remarks on this theme, Hofmeyr mentioned Ernst Jokl's contribution on the importance of physical welfare in improving the quality of manpower, Louis Freed's criticisms of the damaging effect of

[101] S. Dubow and A. Jeeves (eds.), *South Africa's 1940s: Worlds of Possibilities* (Cape Town: Double Storey, 2005).

[102] Dubow, *Commonwealth of Knowledge*, 236.

[103] 'Message from the Prime Minister, the Rt. Hon. J. C. Smuts', to Science and Post-War Reconstruction symposium, *SAJS*, 39 (1942), 313–14.

[104] H. J. van der Bijl, 'The Industrial Aspect: Present' and H. J. van Eck, 'The Industrial Aspect: Future', *SAJS*, 39 (1942), 316–21 and 322–3, 324.

[105] J. H. Hofmeyr, 'Summary of Conclusions', *SAJS*, 39 (1942), 371–80.

compound life on black workers, as well as calls for agricultural workers to be better nourished, housed and trained.

In the field of social medicine, Dr E. H. Cluver, head of the SAIMR from 1940, produced 'striking evidence' about infant mortality and 'low physical efficiency', stressing the need to reorganise medical work so as to focus on prevention as well as treatment. Leading liberal politician, Margaret Ballinger, 'urged the necessity for the complete revision of our traditional native policy', while trades unionist A. J. Downes questioned the 'profit motive'.[106] Such ideas were foreshadowed by the pre-war Fox and Back report, discussed earlier in this chapter, while individuals in the Native Affairs Department were beginning to think along the same lines.

In Johannesburg, the SAIMR expanded its operations to support the war effort, directing a medical corps with mobile laboratories. As well as providing serum and vaccines for civilians and soldiers, the SAIMR turned to the treatment and prevention of infectious diseases. James Gear, who joined the SAIMR in 1930, and went on to direct the organisation, established a poliomyelitis research unit in 1944 to deal with a pandemic that likely spread from soldiers returning from service in the Mediterranean. A more serious epidemic broke out in 1955 which, once again, largely affected whites (Africans were believed to have greater immunity). The SAIMR laboratory released an inactivated polio vaccine in August 1955 – within months of Jonas Salk achieving this in the United States and in close consultation with the Americans.[107] Studies into human nutrition and vitamin deficiency conducted by Francis W. Fox, Joseph and Teddy Gilman, John F. Brock, Ernst Jokl, and others, were undertaken during the war years at the SAIMR or in association with it.[108] At the University of Cape Town, Edward Batson drew a direct correlation between poverty and malnutrition. Batson's Beveridge-inspired social scientific surveys of

[106] Hofmeyr, 'Summary of Conclusions'.

[107] J. Gear, 'The South African Poliomyelitis Vaccine', *South African Journal of Medicine*, 23 June 1956, 587–94; J. Gear, 'Background to the Poliomyelitis Vaccination Campaign', *South African Journal of Medicine*, 11 May 1963, 499–502.

[108] J. F. Brock, 'Some Highlights of Medical Research in South Africa', in Brown (ed.), *A History of Scientific Endeavour*, 199–200; Murray, 'History of the South African Institute for Medical Research', 389–95.

coloured households' living standards and diet bore out this point in close empirical detail.[109]

In 1939 the Wits medical school established a dispensary in Alexandra township. Attendance at the Alexandra university clinic became a compulsory requirement for the attainment of a medical degree.[110] The medical officer of health at Alexandra during the 1930s and early war years was Dr Alfred B. Xuma who became president of the ANC in 1940.[111] He represented a parallel, but linked, impulse to promote social medicine. Xuma had taken his first degree in agriculture in Minnesota and worked with an entomologist before completing a medical degree in Chicago. He practised as a doctor in Sophiatown, Johannesburg, from 1928 and he addressed the joint British and S2A3 conference in 1929, perhaps the first black scientist to do so. He was also part of the 1934 Education Conference.

The training of African medical personnel was already a key issue for the mines, and elements in the state saw the need to address pressing issues of disease in the rural areas. In 1927 the Department of Native Affairs appointed a committee under C. T. Loram which recommended full (though segregated) medical training for black students at the University of the Witwatersrand.[112] Xuma, working with Loram, argued strongly for this development, both on the grounds of equal opportunity and because he believed that African doctors would be more successful than whites in introducing scientific medical practice to black people.

Although Wits refused to admit black students to its existing medical courses, it was willing to start a segregated facility. The university received a number of applications from 'Native or Indian candidates ... which if the applicants were not coloured, would undoubtedly secure them admission to and indeed in some cases, partial exemption from the medical course'.[113] The Rockefeller Foundation's offer in 1931 of a donation for a separate black medical

[109] Wylie, *Starving on a Full Stomach*, 106, 117–8, 150.

[110] Murray, *Wits. The Early Years*, 249–50.

[111] Gish, *Alfred B. Xuma*; P. Limb (ed.), *A. B. Xuma: Autobiography and Selected Works* (Cape Town: van Riebeeck Society, Second Series, no. 43, 2012).

[112] K. A. Shapiro, 'Doctors or Medical Aids: The Debate over the Training of Black Medical Personnel for the Rural Black Population in South Africa in the 1920s and 1930s', *JSAS*, 13, 2 (1987), 234–55.

[113] Shapiro, 'Doctors or Medical Aids', 244, 253.

school was turned down by Hertzog's government. Three Indian students were eventually admitted to Wits in 1938 and the first African students in 1941. The government, together with the mining industry, instead pursued the route of training Africans as medical aides. Xuma and others were strongly critical of this 'auxiliary medical scheme', which was in keeping with segregationist ideology. The University of Natal established a separate black medical school after the war (see Chapter 6).

In the 1930s, Xuma articulated consistent concern for socio-medical issues, condemning the conditions in mining compounds, arguing for urgent measures to deal with miners' phthisis, and linking tuberculosis in the African population with 'bad socio-economic conditions'.[114] In 1937 he visited the United States again, took a London diploma in public health, and sought funding for scientific laboratories and scholarships at Fort Hare. In 1940 he read a paper to the Pretoria Joint Council, attended by Wits medical students, which criticised the government's failure to collect health statistics, meaning that causes of death in the urban African population and the scale of infant mortality were inadequately recorded. He argued that 'so-called African native health' was wrongly considered to be 'something distinct and apart from European health.' The root cause of poor standard of African health was economic inequality as well as segregationist legislation like the Land Acts. Xuma's analysis of African health, distilled from his Alexandra experience and his racially mixed practice in Sophiatown, drew on and fed into a wider concern about social medicine.

Diana Wylie illustrates the growing concern with African health. She glosses the reformist interventions at this time as 'scientific paternalism' and also draws attention to the biological racism that conditioned many to see the health status of whites and blacks as existing in different spheres. It is true that the mining industry's concerns with inadequate nutrition was grounded first and foremost in pragmatic calculations about economic profit and efficiency. Racism, founded in biological ideas about degeneration, formed part of this attitudinal context. Many medical experts saw Africans as disease-carrying bodies rather than as patients with individual health needs. Yet it would be mistaken to underestimate the broader shift at this time amongst a

[114] Wylie, *Starving on a Full Stomach*, 116.

critical mass of health workers towards viewing black and white South Africans as constituting part of the same body politic.

For some young medical students and social welfare practitioners, such as Gertrude Kark, working with black patients in the clinics and hospitals of Johannesburg and Cape Town, as well as in rural outposts, proved transformative. Strategies of disease prevention and welfare necessitated new and often radical ideas. Wylie attributes the source of some of the most inspiring thought about social medicine in South Africa to the Swiss-born, American-based, Henry Sigerist. In the course of a four-month visit, Sigerist, who had recently published *Socialized Medicine in the Soviet Union* (1937), lectured widely in South Africa. He delivered a farewell radio broadcast in which he appealed to enlightened national self-interest by reminding his audience that a sick black population was not only economically disadvantageous, but also a threat to the health of whites: 'You are a nation, not of two, but of ten million inhabitants.'[115]

In 1942 an Association of Scientific Workers of Southern Africa was formed. Use of the term 'workers' in its title was indicative of a more democratic outlook. Its membership, listed as 800 that year, included established scientists, those with 'some scientific qualification', and students too. The organisation's stated objectives were 'the application of science and the scientific method to all problems concerning the welfare of society' as well as the promotion of scientific understanding by the general public. The Johannesburg branch arranged film evenings, lectures, and a symposium on 'Feeding the People' in 1944. In 1945, branch members in Cape Town organised a symposium on 'African Housing' and a biannual *Bulletin* was launched.[116]

The emphasis on interconnected dimensions of health and social welfare, and the willingness of significant sections of the scientific public to acknowledge that South Africa's social problems crossed boundaries of race, were indicative of a broader acceptance of the need for state intervention. These sentiments were reflected in Smuts's announcement, in a widely publicised address to the Institute of Race

[115] Sigerist was given an honorary degree by the University of the Witwatersrand in 1939. Wylie, *Starving on a Full Stomach*, 150.

[116] A. L. du Toit papers, BC722, University of Cape Town, box M4, 'The South African Association for the Advancement of Science. Report on the Association of Scientific Workers of Southern Africa', 13 February 1946; C2.8, 'Proposed Scientific Bulletin', 1945.

Relations that year, that 'Isolation has gone and segregation has fallen on evil days, too'.[117] Smuts's address was conditioned by the need to gain at least tacit political support amongst blacks (at a time of growing internal divisions between whites) as well as recognition that the country's wartime political and economic needs necessitated greater participation of blacks in the domestic economy. The inclusion of blacks in state pension and disability grants, albeit on an unequal basis, was one signal of this shift. Although Smuts manifestly failed to follow through on his reformist promises, one consequence was to embolden liberal and left-wing idealists, technocrats and scientists, to explore more equitable social policies for the future.

Social medicine was an area where new approaches to health were explored with considerable enthusiasm and some initial success. Sidney and Emily Kark, pioneers of a health centre built at Pholela in rural Zululand in 1940, developed a radical integrated approach to community-based health care. This took account of societal conditions as well as biomedical factors and combined practical service, training and epidemiological research into preventable infectious diseases, as well as new thinking about nutrition. Sidney Kark, who was involved in a 1938 Bantu Nutrition Survey, was clear that soil science and erosion was one of the key factors in malnutrition. The community healthcare initiative was supported by the government Department of Health, and especially by its dynamic secretary, George Gale. The Pholela idea was subsequently extended in 1946 to the overcrowded working-class Coloured area of Grassy Park, Cape Town, where Nancy Barlow and Jacob Henson adapted the rural community approach to urban conditions.

The Natal and Cape experiments attracted the interest of John Ryle, professor of social medicine at Oxford, who visited South Africa in 1947–8, by which stage forty-four health centres had been established. These experiments expressed a broad shift towards the racially inclusive, comprehensive national health service envisaged by Henry Gluckman, chair of the landmark 1942–4 National Health Services Commission. Smuts's failure to act on its recommendations was one example of the way in which he stymied the radical proposals that he himself had helped to set in train.[118]

117 J. van der Poel (ed.), *Selections from the Smuts Papers, vol. VI*, doc. 556, 'Address to the South African Institute of Race Relations', 21/1/42, 336.

118 A. Jeeves, 'Delivering Primary Care in Impoverished Urban and Rural Communities: The Institute of Family and Community Health in the 1940s', in

Figure 5.2 General Smuts on Table Mountain, 1947.
Credit: Keystone Pictures USA/ZUMAPRESS.com/Alamy Live News

By contrast, Smuts gave consistent and wholehearted support to the South African Council for Scientific and Industrial Research (CSIR), established in 1945. The origins of the organisation go back to 1942 when Smuts invited Basil Schonland to sketch out a scientific organisation modelled along the lines of similar institutions in Britain and the Dominions. Smuts knew of Schonland through personal and natural history networks. Schonland's father, Selmar, was the founder of the botany department at Rhodes University and co-translator of August Weismann's *Essays upon Heredity* into English; his mother, Flora, was the daughter of the botanist Peter MacOwan.

Schonland completed an undergraduate degree at Cambridge in 1915, and a doctorate at the Cavendish Laboratory (1920–22), working with leading physicists Ernest Rutherford, C. T. R. Wilson, and J. A. Crowther. At Cambridge he met a Newnham history student, Ismay Craib, and followed her back to South Africa and to a lectureship in physics at the University of Cape Town. In the 1930s, Schonland worked on the phenomenon of lightning discharges on the Witwatersrand, using novel photographic techniques to record patterns of flashes and strokes. The study of atmospheric lightning had

Dubow and Jeeves (eds.), *South Africa's 1940s*; Marks, 'South Africa's Early Experiment'; S. Marks, 'Reflections on the 1944 National Health Services Commission: A Response to Bill Freund and Anne Digby on the Gluckman Commission', *South African Historical Journal*, 66, 1 (2014), 169–87.

important theoretical as well as practical implications, not least because electricity transmission was vulnerable to electrical storms. This research brought Schonland into contact with Bernard Price of the Victoria Falls Power Company. It was judged sufficiently important to merit his election to a fellowship of the Royal Society in 1938.[119]

As head of the new Bernard Price Institute, Schonland took responsibility in 1939 for developing a system of radar, at great speed and in utmost secrecy, in order to defend the South African coastline from German attack. This was achieved, working from first principles, by a team of radio engineers, including G. R. Bozzoli, later the vice-chancellor of Wits University, and gifted inventor Trevor Wadley. A prototype was ready for testing within three months and the system was soon made operational with South African troops in East Africa and around the southern African coastline.[120] From 1941, Schonland was on active service in Britain, working with John Cockroft on radar before moving on to the Army Operational Research Group. By 1944 he was serving as General Montgomery's official scientific adviser with the rank of brigadier. After VE Day Schonland returned to become Smuts's scientific adviser and he mapped out a plan for the CSIR, which was established by statute, in 1945.

The CSIR was modelled on similar bodies in Britain, Canada, Australia and the United States (though the Indian CSIR, constituted in 1942, was not mentioned in the planning documents). As founding president and chief executive, Schonland insisted that it be constituted as an autonomous agency bearing direct responsibility to the prime minister: he wished to avoid it becoming ensnared by the civil service.[121] The CSIR absorbed the functions of the existing state grant-giving bodies (the Research Grant Board and the National Research Council and Board) and positioned South Africa to enter the post-war era of applied science operating on a national scale. An early statement of its intention, citing Franklin Delano Roosevelt, proclaimed the role

[119] T. E. Alllibone, 'Basil Ferdinand Jamieson Schonland 1896–1972', *Biographical Memoirs of Fellows of the Royal Society*, 19 (1973), 629–53.

[120] B. Austin, 'On the Development of Radar in South Africa and Its Use in the Second World War', *URSI Radio Science Bulletin*, 358 (2016), 69–81. B. Austin, *Schonland: Scientist and Soldier* (Bristol: Institute of Physics, 2001).

[121] Austin, *Schonland*, 305–6; Schonland papers, Churchill College, Cambridge, SCHO 1/2, Schonland to Hancock, 2 January 1957.

of science in advancing new 'frontiers of the mind', with particular emphasis on the pursuit of economic well-being and health.[122]

The core of the CSIR comprised a number of dedicated national laboratories: the National Physical Laboratory, National Chemical Research Laboratory, and National Building Research Institute. These were soon followed by the National Institute for Personnel Research and the Telecommunications Research Laboratory. The South African Bureau of Standards was constituted separately in 1945, tasked with testing and maintaining quality control over commodities and industrial processes. Beyond the CSIR's immediate remit were a number of existing specialist research institutes and laboratories dedicated to veterinary science (Onderstepoort), medicine, mining, agriculture and fisheries. The CSIR maintained links to universities by making available grants to academics and medical researchers. Initially housed in temporary accommodation in central Pretoria, it moved to a dedicated new campus to the east of the city in landscaped parkland known as Scientia.[123]

The creation of the CSIR marked a new moment for scientific research in South Africa and proved a high point for Smutsian South Africanism. Schonland later reflected that for all his personal engagement in the project, Smuts was not comfortable with laboratory-based science and had little 'understanding of the technical world of 1946 or of modern physics'.[124] As a progressive-minded internationalist, Schonland believed that scientific research should cater to the greater good in a non-political manner. Schonland also revealed himself to be an astute political operator, not least in appointing, as director of the National Physical Laboratory, Stefan Meiring Naudé, a respected Afrikaner research scientist and technocrat who identified with, but was not captive to, the rising nationalist movement. In seeking to persuade Naudé to join him, Schonland remarked that if he did not accept the offer, he would appoint someone 'from overseas' (likely an Englishman) instead.[125]

122 *The South African Council for Scientific and Industrial Research. Objects and Policy of the C.S.I.R. (An Initial Statement)* (Pretoria: CSIR, 1945), 3–5.

123 D. G. Kingwill, *The CSIR: The First Forty Years* (Pretoria: Scientia Printers, CSIR, 1990), 287.

124 Schonland papers, Churchill College, Cambridge, SCHO 1/2, Schonland to Hancock, 2 January 1957.

125 Austin, *Schonland*, 314.

Other key appointments showed the influence of personal, wartime and Commonwealth bonds. William Rapson, a New Zealand- and Oxford-educated chemist, came to the CSIR via the University of Cape Town to head up the Chemical Research Laboratory. Rapson later became vice-president of the CSIR before working for the Chamber of Mines. Frank Hewitt, recruited by Schonland to direct the Telecommunications Laboratory, was a fellow Grahamstown-born electrical engineer, working alongside Schonland on lightning research and then on the development of radar. Hewitt went on to help in developing the tellurometer for the CSIR, a patented device for measuring distances, which proved to be an international commercial success (see Chapter 6).[126]

The National Institute for Personnel Research (NIPR), headed by Simon Biesheuvel from its creation in 1946, was another agency which owed its creation to wartime experience. Smuts dismissed the aptitude testing programme conducted by Biesheuvel for the South African Air Force as 'pretty good nonsense', but Schonland prevailed and the development of aptitude tests provided lucrative income streams for the NIPR from large industries, the military and the mines.[127] Under Biesheuvel's leadership, the NIPR developed group intelligence tests and training designed to improve the efficiency of black mineworkers, for example in the selection of 'boss boys'. The NIPR developed a 'General Adaptability Battery' of five tests for the purposes of screening. It also became heavily involved in cross-cultural mental testing. Biesheuvel, born in Rotterdam, trained as a psychologist at the University of Cape Town before taking a lectureship at Wits under Professor I. D. MacCrone. Both were outspoken critics of racialised assumptions in mental testing. Nevertheless, the NIPR played host to many psychometric research programmes, which did as much to reinforce racial and cultural stereotypes as to challenge them.[128]

Alternative Scientific Traditions on the Left and Radical Right

The moderation and scientific centrism epitomised by the likes of Schonland, van der Bijl and Biesheuvel did not survive intact under

126 Kingwill, *CSIR*, 323; Obituary, 'F. J. Hewitt (1919–2007)', *SAJS*, 104, 208 (2008), 414.

127 Schonland papers, Churchill College, Cambridge, SCHO 1/2, Schonland to Hancock, 2 January 1957.

128 Dubow, *Scientific Racism in Modern South Africa*, 236–42.

apartheid, but it left enduring institutional legacies. Several competing scientific imaginaries were evident during and after the war. On the political left there was a thin but committed tradition of radical liberal and Marxist scientific thinking that can be traced back to at least the 1920s. This coalesced in Johannesburg and Cape Town, emerging most prominently in the 1940s when communism became more visible as a result of wartime alliances against Fascism. Socialists were committed to a multiracial or non-racial future and some measure of economic redistribution.

Wulf Sachs, a medical doctor in Johannesburg and pioneer of psychoanalysis, was one such intellectual. In 1937 he wrote a remarkable account of his attempt to psychoanalyse John Chavafambira, a Zimbabwean-born diviner-healer who practised his trade in the slumyards of Doornfontein, Johannesburg. Sachs's *Black Hamlet* was written in the form of a gripping novel, but Sachs also had a serious purpose in seeking to test the universal applicability of Freudian analysis. From the 1930s he was becoming increasingly persuaded that the mental structures and states of Africans and whites were fundamentally the same.

In a later reworking of this subject, *Black Anger* (1947), Sachs came to the more radical conclusion that Chavafambira's mental stress was caused not only by the Hamlet syndrome, but also by profound socio-economic and political inequalities.[129] Here he was closely in tune with progressive medical thinking in Johannesburg. Sachs's interest in ethnopsychiatry was prompted by earlier research conducted on black patients in the Pretoria mental hospital who were suffering from schizophrenia. This led him to conclude that 'the delusions, and hallucinations of the insane native were in structure, in origin and, partly in content, similar to those of the European'.[130] In effect, Sachs was challenging very widespread eugenic-based assumptions that there were basic differences in mental capacity and functioning between whites and blacks.

A little earlier, in Cape Town, biologist Lancelot Hogben, professor of zoology from 1927 to 1930, became sensitised to the prevailing climate of racism. This experience fed into the strongly anti-eugenic views that he developed as professor of social biology at the London

129 W. Sachs, *Black Hamlet*, reissued with introductions by Saul Dubow and Jacqueline Rose (Baltimore MD: Johns Hopkins, 1996), first published 1937.

130 W. Sachs, 'The Insane Native: An Introduction to a Psychological Study', *SAJS*, 30 (1933), 710, 713.

School of Economics. Hogben's passionate advocacy of experimental zoology at the University of Cape Town immediately brought him into conflict with marine biologist Cecil von Bonde, successor to Gilchrist, who removed the department's library to the aquarium at St James. Hogben tossed most of Gilchrist's type specimens out of the window, declaring 'this is not zoology'.[131] A mercurial and provocative character who enjoyed causing uproar – he alienated some of his closest associates – Hogben easily made enemies.

As a committed opponent of segregation, and an uncompromising exponent of scientific materialism, Hogben castigated Smuts's metaphysical theories of holism when the British and South African Associations met in South Africa in 1929. Together with his wife, Enid Charles, a feminist statistician and zoologist, regular Saturday night parties were held at their Rondebosch home, Xenopus. The house was named after the South African clawed frog they studied, which became a test animal of choice in developmental biology – thus spawning the xenopus pregnancy test. These social gatherings attracted a lively bohemian set of intellectuals, artists and left-wing activists.[132] Within the extended circle were Ruth Schechter, a friend and devotee of Olive Schreiner; Benjamin Farrington, an Irish republican and classicist; the Swiss linguist Frederick Bodmer (author of *The Loom of Language*); and historian Jean van der Poel, who translated the *Communist Manifesto* into Afrikaans before writing a key text on the *Jameson Raid* (1951), which clearly implicated Rhodes in the imbroglio. Van der Poel went on to work with W. K. Hancock on the *Smuts Papers*.

Edward Roux, mentioned above in connection with botany and scientific agriculture, gravitated to Hogben's circle. He was a committed scientific rationalist, as well as a convinced communist, who met his future wife, Winifred, at a Xenopus gathering. Roux was much taken by Hogben's iconoclasm and his predilection for 'happily shocking the local philistines as well as some quite nice people in Cape Town'.[133] After the Hogbens' departure for London in 1930, the left-wing salon moved to the house of Cissie and Abdul Hamid

131 J. H. Day, 'Marine Biology', 92.

132 J. B. Gurdon and N. Hopwood, 'The Introduction of *Xenopus laevis* into Developmental Biology: Of Empire, Pregnancy Testing and Ribosomal Genes', *International Journal of Developmental Biology*, 44 (2000), 43–50.

133 E. and W. Roux, *Rebel Pity: The Life of Eddie Roux* (London: Rex Collings, 1970), 75.

Gool in District Six. Here, black political activists, like the Trotskyite I. B. Tabata, gathered along with whites to exchange ideas.[134]

The intellectual culture of the 1940s allowed other forms of radical scientific culture to gain a foothold. In Johannesburg and Durban, as we have seen, social and community medicine provided networks where welfare merged into forms of socialism. A government-appointed Social Security Committee, which reported in 1943, proposed basic universal entitlements for blacks as well as whites (albeit not without discriminatory elements). At the University of Cape Town, sociologist Edward Batson pioneered the concept of the 'poverty datum line', a measure of destitution that extended concerns about 'poor whites' to include Coloured, Indian and African people. Batson's work derived more from British Fabianism than from radical political traditions, and his motivation was reformist and meliorist. Yet, as Grace Davie has shown, the principle of the poverty datum line could be, and was, appropriated to support iconic wartime protests like the 1943–4 Alexandra bus boycotts, which drew the ANC (and Nelson Mandela) into a new awareness of the possibilities for, and necessity of, popular mobilisation.[135]

The ANC provided a linked platform that focussed especially on socio-medical ideas that fed into their political programme. Dr A. B. Xuma was a key conduit for these links. In the early decades of the twentieth century, formal medical training was only available to black South Africans outside the country. By 1940, over thirty had completed their qualifications, mostly in Glasgow, Edinburgh and London, and returned to practice in South Africa.[136]

They faced formidable barriers in the public sector. Provincial authorities were generally opposed to employing black doctors, they received lower salaries, and endured hostility from white doctors and nurses. Most went into private practice, either in the African reserves or in townships, and some were particularly successful with a multi-

134 B. Hirson, *The Cape Town Intellectuals: Ruth Schechter and Her Circle, 1907–1954* (Cape Town: The Merlin Press, 2001), chapter 7.

135 J. Seekings, 'Visions, Hopes and Views about the Future: The Radical Moment of South African Welfare Reform', in Dubow and Jeeves (eds.), *South Africa's 1940s*; G. Davie, *Poverty Knowledge in South Africa: A Social History of Human Science, 1855–2005* (Cambridge: Cambridge University Press, 2015), chapter 3.

136 Digby, 'Early Black Doctors'.

racial clientele. A number became leading political activists: Abdullah Abdurahman was a long serving Cape Town councillor and James Moroka followed Xuma as president of the ANC (1949–52). Modiri Molema served as President of the African Races Association of Glasgow in 1919, wrote books on history, and in 1927 won a court case against white nurses who objected 'to being ordered about by any native doctor'.[137] He was involved in regional politics in the Northern Cape and Botswana, and worked as treasurer of the ANC in the 1950s. Both Molema and Xuma were committed to the primacy of biomedicine, although they also distinguished between witchcraft and the more positive role of herbalists, and in Xuma's case wrote about 'The African Concept of Disease'.[138]

While Xuma focussed on broader political issues in his speeches, his presidential address to the ANC in 1941 emphasised that African people, as the poorest, should have full rights to social welfare as equal citizens, and benefit from 'full extension of public health and preventative health measures', on the 'basis of ability and training and recognised professional rates and conditions'.[139] These ideas made their way into the ANC's pathbreaking 1943 *Africans' Claims* document, demanding equal citizenship and human rights, including free medical and health services.[140] Not for the first time in South Africa, the language of political and human rights followed from practical understandings of inequality and injustice.[141]

The early ANC doctors firmly connected disease with poverty and inserted this social democratic perspective into the major documents and rhetoric of the movement – a legacy that was carried in part into the Freedom Charter (1955). As ANC president, Xuma was criticised for his caution in relation to mass action, such as the Alexandra bus boycott (1943), and the more militant Africanist ideologies propagated by the ANC Youth League (1944). The latter in turn focussed on the political realm and less on the specifics of a scientifically based programme for health and welfare. In the mid-1940s both Xuma and the

137 Marks, *Divided Sisterhood*, 60. 138 Digby, *Diversity and Division.*

139 South African History Online, Dr A. B. Xuma, ANC Presidential Address, 14 December 1941.

140 Padayachee and van Niekerk, *Shadow of Liberation.*

141 S. Dubow, *South Africa's Struggle for Human Rights* (Auckland Park: Jacana, 2012).

Youth League focussed on the critical shortage of housing provision in Johannesburg.[142]

Less visible in the 1930s and early 1940s, but of great importance in the apartheid era, were attempts to organise Afrikaner nationalist science along political lines. Most accounts of the rise of Afrikaner nationalism at this time concentrate on agencies such as churches, business, intellectuals and social activists. Yet science and technology were also spheres that began to command attention as a rising generation of Afrikaner scientists became determined to serve the interests of *volk* and state, as well as to realise personal ambitions.

The young Afrikaner generation of the 1930s and 1940s were far more stridently nationalist than their predecessors, for whom Afrikanerdom was a rather more fluid identity. Eugene Marais, E. G. Malherbe, van der Bijl and van Eck for example, were all from Afrikaner backgrounds, but with diverse political loyalties. Marais certainly identified with Afrikaner culture and his poetry became important to the Afrikaans language movement, yet he was an elusive individualist, whose psychological torment led him to commit suicide on the farm Pelindaba (which later featured in South Africa's nuclear enrichment programme).

Malherbe's father's church at Luckhoff in the Free State was destroyed during the South African War, its piano, pulpit and pews chopped up for firewood by the British occupiers.[143] Yet, unlike many who experienced such treatment, he was not embittered and, to the fury of many Afrikaner nationalists, became a committed advocate of bilingual education and a close ally of Smuts during the Second World War. In his capacity as wartime director of military intelligence, Malherbe exposed the secret networks of the Afrikaner Broederbond. As a youth, van der Bijl had seen rifles being collected to repel the Jameson Raid and he later witnessed the British occupation of Pretoria. His father chose to send him to be schooled at the Cape. Like Malherbe, van der Bijl did not seek to redress the wrongs of the past by joining the radical nationalists, but saw science and technology as the best way to serve the purposes of nation-building.

142 E. Sisulu, *Walter and Albertina Sisulu: In Our Lifetime* (Cape Town: David Philip, 2002).

143 Malherbe, *Never a Dull Moment*.

By contrast H. S. du Toit, mentioned above in connection with agricultural improvement and conservation, fought in the South African War and was considered a '*bittereinder*'. In 1902 he travelled to Mexico and the United States, where he was impressed by Theodore Roosevelt's progressivism; in Russia, he met Tolstoy. On his return to South Africa, du Toit joined the Department of Agriculture and devoted himself initially to experimental dry farming in the western Transvaal. As a follower of General Botha, he opposed the Afrikaner Rebellion in 1914, and as a government agronomist he worked with both Afrikaner and English-speaking farmers, but in the late 1920s he gravitated to Hertzog's nationalists.[144]

Other young Afrikaner scientists who came to maturity in the 1920s and 1930s attached themselves to radical Afrikaner nationalism. Prominent among them were A. J. A. Roux, a future leader in atomic energy; meteorologist T. E. W. Schumann; geneticist Gerhardus Eloff; Netherlands-born mining engineer and palaeontologist E. C. N. van Hoepen; and experimental physicist Meiring Naudé, later president of the CSIR. All played prominent roles in the Suid-Afrikaanse Akademie vir Wetenskap en Kuns (South African Academy for Science and the Arts), which did a great deal to promote Afrikaner scientists and science from the early 1940s.

The mobilisation of Afrikaner scientists and practitioners was part of a broader imperative to take control of the cities – a movement that was powerfully symbolised in the emotional 1938 centenary celebrations of the 'Great Trek'. In 1937, while a student at Wits in the centre of industrial Johannesburg, Roux formed the Afrikaanse Vereniging vir Wetenskap en Wysbegeerte (Society for Science and Philosophy). He also edited its journal, *Die Denker* (The Thinker). The Vereniging was short-lived but its ambitions were realised more fully in the Suid-Afrikaanse Akademie vir Wetenskap en Tegniek (South African Academy for Science and Technology). J. K. Marais served as the first chairman, with Roux acting as secretary and Piet Meyer, of the Afrikaner Broederbond and FAK (Federation of Afrikaner Cultural Organisations), maintaining close oversight.[145]

These initiatives prompted the prestigious Suid-Afrikaanse Akademie voor Taal, Letteren en Kuns, originally formed in 1909, to take notice. The august Akademie was hitherto almost entirely focused

144 Beinart, *Rise of Conservation*, chapter 7.

145 C. A. Engelbrecht, 'Natuurwetenskap, Tegniek en die Akademie', in *Akademie 75* (Pretoria: Suid-Afrikaanse Akademie vir Wetenskap en Kuns, 1984), 132–3.

on literary matters, although some attention was paid to scientific affairs in the 1920s. Van Hoepen was deputed to open negotiations with a view to absorbing the new scientific academy within the ambit of its established cultural equivalent. In 1941–2 the Akademie carried through a major renaming and restructuring process by which it became known as the Akademie vir Wetenskap en Kuns (Academy for Science and the Arts). In effect, Roux's student organisation was now integrated into the older Academy with science granted equal standing. A faculty of natural science and technology was duly established, chaired by J. K. Marais (1941–7), and a dedicated scientific journal, *Tegnikon*, launched in 1948. Specialised subcommittees catered to subjects such as physics, medicine, engineering, agriculture, economics and biology.

The formation of a national institution to advance Afrikaner science represented a conscious effort to achieve parity with English-speakers. Some of the new Afrikaner scientists were politically on the radical right. Sociologist Cronjé and biologist Eloff both advanced eugenic arguments in support of apartheid while playing prominent roles in the Akademie. The organisation took a broad view of Afrikaner accomplishment and professional ascendancy in support of the *volksbeweging* (people's movement). Its desire to promote Afrikaner nationalist science posed intellectual as well as organisational difficulties. Turning Afrikaans into a scientific and technical language was one obstacle. This challenge was energetically pursued by the Akademie, which, in 1944 produced its first example of a specialised medical vocabulary in Afrikaans, followed by a word list for physics in 1948, and then ones for meteorology, palaeontology, pharmacy, botany and mathematics.[146] The Christian-nationalist insistence that science should be consistent with religion exerted constraints on the pursuit of certain forms of endeavour, especially where this bore on matters of evolution and creation. At Stellenbosch, Robert Broom's lectures on evolution had earlier caused considerable consternation, resulting in a celebrated debate with J. I. Marais, theologian and critic of Darwinism, who became the first chancellor of Stellenbosch University.[147]

[146] Engelbrecht, 'Natuurwetenskap', 137.

[147] G. H. Findlay, *Dr Robert Broom, F.R.S.* (Cape Town: A. A. Balkema 1972), 26. Broom was a committed believer in evolution, though he was also a spiritualist who believed that evolution had a higher guiding force behind it.

In the late 1920s and early 1930s, evolutionism was vigorously debated in the journal, *Tydskrif vir Wetenskap en Kuns*, with contributions from the theologian H. G. Stoker on theism and science and from palaeontologist van Hoepen (whose interests in fossil ammonites and *Equus capensis*, an early horse, could not have been pursued without an acceptance of some version of evolution).[148] Prime minister Malan's joke about the coelacanth being an ugly forebear of humankind revealed that his own theological orientation was more broad-minded than expected. This open-mindedness aroused consternation in sections of the nationalist movement.[149]

Resistance to evolutionist theory remained palpable in theological circles throughout the apartheid period. Palaeontological exhibits at the Transvaal Museum, Pretoria, in 1952, and again in 1963, as well as at the Bloemfontein Museum in 1955, sparked public controversy as to whether evolution should be presented as a theory rather than established fact.[150] Yet, as the following chapter shows, this did not constrain the apartheid state and its supporters from embracing the instruments of modernity to pursue techno-nationalist objectives. The lessons of the Second World War, namely that state-sponsored research and intervention could be powerful instruments for transformation – as well as a source of patronage and power – were not lost on apartheid's ideologists, who became highly committed technical as well as social engineers. Some scientific initiatives begun but barely pursued in the 1930s, ranging from Antarctic research to agricultural Betterment, were taken up in the apartheid era and given fresh backing by an ever more interventionist and technicist-minded bureaucracy.

148 For a flavour of such debates, see *Tydskrif vir Wetenskap en Kuns* 8 (1929–30), 9 (1930–1) and 10 (1931–2). From 1920 to the early 1940s, van Hoepen published on fossil ammonites, Karoo dinosaurs and *Equus capensis*, a pleistocene Zebra, both through the auspices of the Bloemfontein museum and the *Tydskrif vir Wetenskap*.

149 Thanks to Lindie Koorts for this observation.

150 Thanks to Becky Sigmon and Jurie van den Heever for confirming these incidents. See also J. F. Durand, 'Paleontologie en die Suid-Afrikaanse Samelewing Vandag', *Tydskrif vir Geesteswetenskappe*, 53, 3 (2013), 338–9.

6 *The Republic of Science, 1948–90*

Apartheid Science: Reset or Rupture?

The surprise electoral defeat of the United Party in 1948 marked an abrupt end to old-style racial segregation and the era of Smutsian South Africanism. In the field of scientific cooperation, the future of well-established links with the Commonwealth and associated international organisations were uncertain. Yet, as Freund points out, 'The institutional foundations of an industrial development state had been laid by 1948' and these were not so easily disrupted.[1] Under prime minister D. F. Malan, the new apartheid regime was fiercely nationalistic, often defensive, avowedly anti-cosmopolitan, and committed to breaking the bonds of British imperialism. But Malan's worldview was not as insular as his critics averred. In the fields of science and technology, the apartheid government sought to adapt rather than disrupt industrial development; it took considerable care to cement international links, especially where economic and strategic interests were at play.

The imperative to avoid isolation was augmented by post-war economic expansion and state-led 'big science' in areas like nuclear physics and industrial chemistry where high investment and sophisticated skills were required. Less well-known areas of state involvement include radio astronomy, such as the HartRAO facility in Hartebeesthoek, which was used to track American and French satellites. By the 1960s South Africa's old commonwealth of knowledge was transmuting into an autarkic republic of science and technology. This represented a new technopolitics of power, a process in which the progressive scientific developmentalism of the wartime years became increasingly focused on the security and defence of white South Africa.

[1] Freund, *Twentieth-Century South Africa*, 68.

The full meaning and intent of apartheid may not have been clear when the Nationalists won the 1948 election, but it was immediately evident that dimensions of post-war reconstruction envisaging a more equitable, common society would no longer command state support. Smuts's political defeat was compounded by the deaths in late 1948 of J. H. Hofmeyr and H. J. van der Bijl, both his trusted political intimates and champions of scientific progress. Bernard Price, benefactor of Wits, also died in 1948. With their passing, the Smutsian project of Commonwealth-oriented scientific internationalism and broad South Africanism was eclipsed.

Even so, the shift was not immediate. Consider, for instance, the career of Basil Schonland, now the leading representative of liberal internationalist, anglophile science. Schonland envisioned a major role for nuclear physics in South Africa, drawing on his close associations with the Cavendish Laboratory in Cambridge and wartime connections. He enjoyed a friendship with Edward Appleton, recently the recipient of the Nobel Prize in physics, and John Cockroft, a key figure in the British nuclear establishment. Cockroft worked with the Chalk River Laboratories in Canada, which formed part of the Manhattan Project. He was aware of South Africa's uranium resources, which, from 1943, also attracted American interest. In an exchange of letters with Cockroft bearing on the creation of the CSIR in February 1945, Schonland expressed a fear that 'politically ... we are heading for a republic in the British Commonwealth but I don't know how long it will take'.[2] He sought an interview with Nationalist opposition leader D. F. Malan to ask whether, in the event of a change of government, the new administration would 'support the CSIR as a national institution'. Schonland was discomforted by Malan's insistence that the new research body would henceforth have to adopt '*'n Suid-Afrikaanse kleur*' (South African colouring) and that it should dedicate itself exclusively to the interests of South Africa.

Schonland remained at the CSIR until 1950, but he retained the directorship of the Bernard Price Institute in order to keep open his options as an academic research scientist.[3] Schonland felt he could not count on building an internationally competitive physics research institute under Malan and he was being sounded out about becoming director of the National Physical Laboratory in London. Other senior

[2] Austin, *Schonland*, 308, 311, 345. [3] Austin, *Schonland*, 303–13.

positions were offered in England, including the office of chief scientist of the Ministry of Supply. In 1954, Schonland agreed to join the Harwell Atomic Energy Research Establishment as deputy director to Cockroft. Four years later he became its director.[4]

There was little outcry over Schonland's resignation from the CSIR – as there had been when Major-General Evered Poole was passed over as the army's chief of staff in 1948. Poole's failure to get the top post is often cited as indicative of the rapid capture by Afrikaners of the higher reaches of the state and was seen as a portent of things to come. Schonland's departure from South Africa was not under duress. Rather, it was a calculated response on his part – in which personal ambition played no small part – to a markedly changed political climate where liberal English-speakers, with degrees from Cambridge or Oxford, feared that their careers might be jeopardised. Afrikaners, by contrast, often preferred to attend German and Dutch universities and even if they accepted Rhodes Scholarships to Oxford, they 'almost invariably' completed their research elsewhere.[5]

In 1948, a group of physicists met at the CSIR to talk about forming a South African Institute of Physics. Of those present, Schonland was keenest to maintain an affiliation with equivalent British and Dominion organisations. A further meeting, chaired by Meiring Naudé, took place at the University of the Witwatersrand in 1949. Tellingly, the Interim Committee chose to defer the question of affiliation. The institutional initiative stalled until it was revived, in 1953, by two junior scientists at the University of the Orange Free State, a bastion of Afrikaner nationalism. The formal launch of the South African Institute of Physics took place in Pretoria in 1955.

The 102 founder members of the South African Institute of Physics included a mixture of Afrikaans and English names, with a fair representation from universities across the country. Among the English-speaking founder members was UCT's celebrated Reginald W. James, who taught two of South Africa's three Nobel laureates in science: Aaron Klug (chemistry) and Allan Cormack (medicine). Cormack, a specialist in X-ray crystallography before he switched to nuclear physics at Cambridge, was also a founder member. So too were

[4] Austin, *Schonland*, 353–4, 393–4.
[5] P. C. Wagener, 'The Founding of the SAIP' in P. R. de Kock and H. Moraal (eds.), *Physics in South Africa* (Pretoria: South African Institute of Physics, 2014), 4.

the British Radcliffe Observatory astronomers Michael Feast and David Thackeray. Naudé, now supremo at the CSIR, was installed as the first president of the Institute of Physics. His deputy was the Wits professor of physics, Frank Nabarro, who had worked closely with Schonland during the war on the explosive effects of shells. Though a strong supporter of the SAIP, Nabarro never became president. Close perusal of the executive officer lists indicates that although the ethnic balance was carefully preserved, the office of presidency remained heavily dominated by Afrikaners or Afrikaans-medium universities until the election of Friedel Sellschop, the leading Wits nuclear physicist, in 1995.[6]

In the humanities and social sciences, E. G. Malherbe gained notoriety among Afrikaner nationalists for his championship of bilingualism, his efforts to suppress the Afrikaner Broederbond during the war, and his close connections to Smuts. In 1950 – the year in which he was approached to join UNESCO – he lost his position on the governing committee of the National Council for Social Research, which was overhauled to include experts sympathetic to the new apartheid government. Malherbe left government service entirely to take up the principalship of the University of Natal. Under his energetic leadership, in 1951 the university established the first medical school for black students. Although segregated, it was not a creature of apartheid, and it opened up new spaces for discussion and debate: Steve Biko, the founder of Black Consciousness, was a student there, as was William Makgoba.

The medical school's first dean, George Gale, came to Natal because he felt that his initiatives in government service were being stymied. Gale did not intend the school to perpetuate racial divisions and brought to it a more liberal worldview. Indeed, the Institute of Family and Community Health, which Gale started with Sidney and Emily Kark in Durban, was regarded by politicians with the greatest suspicion as an incubus of socialist and non-racial ideas. By 1957, Gale, as well as the Karks, and social epidemiologist John Cassel had all left South Africa for the United States or Israel. They took with them ideas and experience that subsequently fed into international

[6] Wagener, 'The Founding of the SAIP' and H. Moraal, 'The South African Institute of Physics', in De Kock and Moraal (eds.), *Physics in South Africa*, 5–9, 13–14; R. Cherry, 'Allan MacLeod Cormack', www.phy.uct.ac.za.

models of community health care championed by bodies like the World Health Organisation (WHO).[7]

In 1949 South Africa was still sufficiently accepted by the international community to host the African Regional Scientific Conference. Roughly 100 delegates were drawn from the European colonial powers, the settler colonies, and supra-national scientific and technical affiliates of the United Nations, such as UNESCO, the WHO, and the Food and Agriculture Organisation (FAO). The meeting was opened by Prime Minister Malan in the Wits Great Hall.[8] His address was polite and pedestrian, quite unlike the expansive thinking that characterised Smuts's ruminations about Africa's scientific place in the world. Malan reminded delegates that there was no 'English' or 'French' way of doing science, laying emphasis instead on the need for international collaboration to solve practical problems in Africa.[9]

These were not idle words. Malan was keen to maintain a degree of continuity in the transition to apartheid and reluctant to squander overseas links. He was initially reluctant to accept Schonland's resignation from the CSIR. When Schonland's friend and confidant, Cockroft, came to South Africa at the invitation of the CSIR in 1949 and delivered a lecture at Wits on the promise of atomic energy, the unstated reason was to reopen discussions with the South African government about supplying Britain with uranium.[10] In nuclear affairs and in other areas of strategic and economic cooperation, the post-1948 period was more a matter of resetting than rupturing long-term relationships.

Scientific Cooperation and International Links

The 1949 African Regional Scientific Conference, held at Wits, was chaired by veterinarian P. J. du Toit, who took over from Schonland as head of the CSIR in 1950; Donald Sole, later a key diplomatic figure in

[7] Marks, 'South Africa's Early Experiment', 452–9; T. Brown and E. Fee, 'Sidney Kark and John Cassel: Social Medicine Pioneers and South African Emigrés', *American Journal of Public Health*, 92, 11 (2002), 1744–5; Jeeves, 'Delivering Primary Care', 103–4.

[8] Austin, *Schonland*, 386.

[9] *African Regional Scientific Conference Johannesburg 1949, vol. 1: Proceedings and Resolutions* (Pretoria: Government Printer, 1949), 23–5.

[10] Austin, *Schonland*, 337, 374, 382, 385–6.

South Africa's 'outward policy' under apartheid, headed the secretariat. The proceedings featured a disproportionate number of South Africans and the local press indulged in stereotypes about Africa as the 'dark' and 'backward' continent. In many respects the record of the conference's proceedings reflected the ethos of late-colonial scientific interventionism. This was obliquely observed by the Kenyan anthropologist and palaeontologist Louis Leakey, who, for reasons not unrelated to his own personal ambition, stressed his own Africanist credentials. He informed colleagues that he spoke 'as one who was born and brought up among an African tribe' whereas South African-trained anthropologists were 'unlikely to be acceptable' to Africans in other parts of the continent.[11]

Two years earlier, Leakey had been organising secretary of the first Pan-African Congress on Prehistory that met in Nairobi in 1947. Smuts made a plane available in order to ensure a substantial South African delegation in Kenya. But the next Congress, scheduled to meet in South Africa in 1951, was moved at short notice to Algiers. It was reported in the anthropological journal *Man* that the original invitation, issued by Smuts, had now been rescinded by the Malan government. The withdrawal of state financial support and the possibility that visas might not be approved for some visitors (presumably Africans) were further factors in the decision to shift the venue.[12] Whatever the precise sequence and cause, the cancellation of the 1951 Congress was a form of mutual disengagement – well before anti-apartheid academic boycotts had been conceived.

Malan was nevertheless keen to maintain most existing strategic, scientific and financial links. In 1949 he launched an 'African Charter' focussed on economic and technical development. The thrust of this

[11] *African Regional Scientific Conference Johannesburg 1949, vol. II* (Pretoria: Government Printer, 1949), 349; 'Minutes of African Regional Scientific Conference. Section E&D, 21 October 1949, Wits Africana Collection, pam G.1718; Ruth First papers, Institute of Commonwealth Studies, London, 'Scrapbook RF/1/4/1/1: 1949–50' undated and unattributed clipping, 'South Africans "Highly Suspect"', very likely *The Guardian*, c. 1949.

[12] S. Cole, J. D. Clark and O. Davis (eds.), 'The Second Pan-African Congress on Prehistory', *South African Archaeological Bulletin*, 7, 28 (1952), 150; A. J. H. Goodwin, 'Second Pan-African Congress on Prehistory, 1951', *South African Archaeological Bulletin*, 3, 9 (1948), 16; *Man*, 51 (July 1951), 97; L. S. B. Leakey and S. Cole (eds.), *Proceedings of the Pan-African Congress on Prehistory, 1947* (Oxford: Blackwell, 1952), 3.

Charter was the need to cement South Africa's position as a bulwark of anti-communism and to preserve the country's regional dominance. In some ways this was consonant with earlier initiatives led by Smuts. But where Smuts was more than happy to sponsor curiosity-driven science concerned with human evolution and palaeontology, as in pan-African anthropology, the approach of the apartheid government was more narrowly technicist and state-directed.

The apartheid government was especially keen to participate in the Scientific Council for Africa South of the Sahara (CSA). Its creation in 1950 arose out of the 1949 African Regional Scientific Conference. Established by agreement between the governments of Britain, France, Belgium and Portugal, as well as Southern Rhodesia and South Africa, the CSA was designed to facilitate joint scientific meetings and technical cooperation in Africa. The government's External Affairs Department lent strong backing to the initiative and a key planning meeting was held at South Africa House in London. E. B. Worthington, formerly participant in Lord Hailey's *Survey* and scientific head of the British East Africa High Commission, served as the CSA's first secretary-general. His landmark book, *Science in the Development of Africa* (1958), was published under the auspices of this organisation.[13]

South Africa's willingness to back the CSA was reciprocated when the CSIR's Petrus du Toit became the CSA's inaugural chair. Strikingly, du Toit retained this position until September 1960, the year of Macmillan's 'Wind of Change' speech and the Sharpeville–Langa killings. Under his direction, the CSA organised three or four large meetings each year in Africa, tackling problems such as soil, forestry, nutrition and education. There were, in addition, thirty to forty smaller annual meetings dedicated to the discussion of issues ranging from water treatment and roads to worker absenteeism. Administrative functions were mainly based in London until 1959, when the secretary-general moved to Lagos.

In terms of technical scientific cooperation and institutional collaboration, the achievements of the CSA were considerable. Helen Tilley characterises the Council as an 'inter-imperial coordinating body led by scientific experts'.[14] But she also notes that as early as 1949 British

[13] Kingwill, *The CSIR*, 64–5.
[14] Tilley, *Africa as a Living Laboratory*, 74, 324.

officials expressed disquiet at the prospect of future meetings being held in apartheid South Africa. The fact that a South African could have served as the chair until 1960 is nevertheless indicative of the willingness of late-colonial powers to keep apartheid South Africa in the fold – and also to benefit from its organisational strength and financial resources. South African involvement continued until 1965 when the CSA, increasingly seen as a residue of colonialism by newly independent African countries, fell into abeyance.[15]

South Africa continued to take a lead in the Southern African Regional Commission for the Conservation and Utilisation of the Soil (SARCCUS), which was established in Pretoria in 1950 as one of four regional African organisations. By the 1960s the other African regional commissions had faded from view, but SARCCUS continued to hold regular meetings in affiliate countries (mostly, but not exclusively, settler colonies) on subjects such as land use, pastures, water supply, soil conservation, animal husbandry and nature conservation.[16] A 1972 report compiled by G. M. E. Leistner for the South African Institute of International Affairs on 'Accelerated Development in Southern Africa' expressed broad satisfaction with the workings and achievements of SARCCUS, viewing it as a model of 'cooperation' in the region.[17]

The government was likewise keen to maintain international status in the field of wildlife conservation. In 1950, the newly created International Union for the Protection of Nature admitted thirty-five members, including South Africa, represented by C. H. Taljaard, a diplomat based in Brussels. Several South African wildlife bodies joined the International Union, including the Kruger National Park and the Natal Parks, Game and Fish Preservation Board. In Jane Carruthers's view, South Africa's influence 'appeared to be growing' through the 1950s. It was strongly represented at the meeting in Bukavu, Belgian Congo, in 1953, and the National Parks Board's enthusiastic new supremo, Rocco Knobel, joined the executive

15 Kingwill, *The CSIR*, 64–6.

16 E. D. Adler, 'Inter-territorial Technical Cooperation in Southern Africa with Special Reference to SARCCUS', *SAJS*, 67, 3 (1971), 140–4.

17 G. M. E. Leistner, 'Co-operation for Development in Southern Africa', paper delivered to conference on 'Accelerated Development in Southern Africa', DEV/72/C2/2, South African Institute of International Affairs, 1972, Jan Smuts House, Johannesburg, 4–5.

committee in 1958. Apartheid undoubtedly complicated South Africa's connections, yet it was not expelled from the International Union. In 1958 the National Parks Board launched a bilingual in-house journal, *Koedoe*, to publish research on nature conservation, biology and mammalogy. The journal signalled an increasing commitment to scientific conservationism; it still exists.[18]

The Antarctic

The Malan government's desire to maintain transcontinental scientific links was also strongly evident in respect of the Antarctic. South African involvement in this region revolved around fishing rights, geopolitical interests, and the influence of southern weather systems on meteorological predictions, rainfall in particular. These concerns had been debated in the 1920s and 1930s (see Chapter 4). Britain controlled, somewhat reluctantly, the Marion and Prince Edward Islands, 2,200 km south of Cape Town and in 1947 negotiated for a South African takeover.[19] Early in 1948 they were occupied by a South African naval detachment, resulting in the establishment of a meteorological station at 'Transvaal Cove' (named in honour of the landbound Boer republic). The highest peak visible from the Marion Island station was named after Jan Smuts. Later, this was topped by naming a higher mountain after apartheid state president C. R. Swart (who also happened to be tall in stature).[20] Scientific research was the stated objective of the Antarctic initiative, but security was a significant factor as well. The impetus for occupation was sustained by anxiety, expressed by the Defence Force in 1951, that the Soviet Union might occupy the islands and use them as a base for missiles.

Fear of *omsingeling* (encirclement) as well as a desire to remain part of international scientific associations and geopolitical treaty systems, persuaded successive apartheid administrations to take a more active role in the Antarctic. South Africa joined in the Commonwealth Trans-

18 Carruthers, *National Park Science*, 136–43.

19 L.-M. van der Watt and S. Swart, 'Falling off the Map: South Africa, Antarctica and Empire, c. 1919–59', *Journal of Imperial and Commonwealth History*, 43, 2 (2015), 267–91.

20 O. Langenegger and W. J. Verwoerd, 'Topographical Survey', in E. M. van Zinderen Bakker et al. (eds.), *Marion and Prince Edward Islands* (Cape Town: A. A. Balkema, 1971), 38.

Antarctic Expedition of 1955–8, led by Vivian Fuchs, in which Sir Edmund Hillary, in charge of the New Zealand party, reprised his Everest role with a reckless 'dash to the pole'. South Africa also participated in the International Geophysical Year of 1957–8 where Antarctic research featured prominently. The country's role in the Geophysical Year was coordinated by the director of the Weather Bureau, T. E. W. Schumann, with more than £150,000 provided by the treasury and the CSIR. These funds went to support collaborative programmes in geomagnetism, physical oceanography, geodesy and geology.[21]

South African participation emerged out of its membership of the International Council of Scientific Unions (ICSU), which dated back to 1931. This organisation comprised around 100 member states and sought to uphold the principle of the universality of science. In the period from 1963 to 1985 some forty ICSU scientific meetings were held in South Africa. In 1970, the South African government hosted the International Union of Geological Sciences' Second Symposium on Gondwana Stratigraphy and Palaeontology, which was organized by the CSIR. Half of the 180 delegates were international visitors.[22]

As head of the Weather Bureau from 1933, Schumann worked to develop its scientific capacity so as to better predict droughts and floods. But as well as weather and climate, he was deeply interested in geostrategy. Highly politicised and well connected, Schumann was closely engaged in international negotiations over the Antarctic. He was also a founding member of the South African Institute of Physics and served as head of the Akademie vir Wetenskap en Kuns (1952–5). In the 1960s Schumann became part of a hard-right international group of eugenicists and anti-egalitarians who, reacting against what they perceived as UNESCO 'equalitarianism', promoted apartheid South Africa internationally as a bastion of anti-communism and a standard-bearer of global white supremacy.[23] Schumann's Spenglerian tract, *The Abdication of the White Man* (1963), registered anger at the West's failure to defend white supremacy against socialism and

21 Kingwill, *The CSIR*, 67–8.

22 P. L. Robinson, 'Second Symposium on Gondwana Stratigraphy and Palaeontology, South Africa, I970', *Journal of the Geological Society*, 127 (1971), 189–91.

23 S. Dubow, 'Racial Irredentism, Ethnogenesis, and White Supremacy in High-Apartheid South Africa', *Kronos*, 41, 1(2015), 236–64.

communism. Awarded an honorary degree in science by Stellenbosch University in 1960, Schumann wrote a resentful letter to the university in 1977 objecting to its decision to admit black students.[24]

In respect of South Africa's rights in the Antarctic, Schumann was strongly supported by Eric Louw, South Africa's minister of external affairs, who was notorious for his pugnacious and undiplomatic responses to criticisms of apartheid at the UN. Schumann and Louw viewed national involvement in the Southern Ocean as part of South Africa's wider geopolitical role in defending 'the West'. This was significantly different in tone from the 1930s enthusiasm for joint Commowealth engagement in Antarctica. The great white spaces of the Antarctic frontier happened to be the only remaining area of international engagement where the apartheid government could avoid dealing with black or Asian (with the exception of the Japanese) representatives.[25]

One important outcome of the Geophysical Year was the Antarctic Treaty of 1959; South Africa was one of a dozen signatories and this boosted its scientific and strategic interest in the continent. J. J. 'Hannes' la Grange, a meteorologist from the South African Weather Bureau, served as a scientific member of the Fuchs–Hillary expedition. He thus became the first South African to reach the South Pole in January 1958 and was awarded the Polar Medal in a ceremony conducted by Queen Elizabeth II.[26] Brought up in the small Karoo town of Ladismith, la Grange fitted the part of Afrikaner Antarctic pioneer perfectly. He had gained earlier experience in 1950–2 on Marion Island, where a wooden structure erected at Transvaal Cove was jokingly named 'La Grange Villa'. Fuchs described la Grange as a 'blind date' who turned out to be 'as solid in character as he was tall and heavy of build'.[27] Symbolism was important in the process of naming and claiming. When la Grange reached the South Pole, he left

24 T. E. W. Schumann to Rector, Stellenbosch University, 15 June 1977, Stellenbosch University Archives, 'Eredoktorsgraad (D.Sc.): Dr T. E. W. Schumann'.

25 P. Roberts, C. Dodds and L.-M. van der Watt, '"But Why Do You Go There?" Norway and South Africa in the Antarctic during the 1950s', in S. Sörlin (ed.), *Science, Geopolitics and Culture in the Polar Region* (Farnham: Ashgate, 2013), 79–110.

26 J. J. la Grange, 'Obituary', *Polar Record*, 37, 203 (October 2003), 369–70.

27 V. Fuchs and E. Hillary, *The Crossing of Antarctica: The Commonwealth Trans-Antarctic Expedition 1955–58* (London: Cassel, 1958), 10.

a green flag embroidered with a protea on one side and a springbok on the other.[28]

In 1959, la Grange was selected to lead the National Antarctic Expedition at the personal invitation of Prime Minister Hendrik Verwoerd, who is said to have telephoned him to say, 'South Africa needs you'. As a result of these efforts a permanent South African Antarctic Base (SANAE) was established at Queen Maud Land during 1960, replacing the existing Norwegian station. The Weather Bureau's monthly *Nuusbrief/News Letter* carried regular reports from the Antarctic, mostly in Afrikaans, which combined items of scientific interest with more informal reports. An illustrative example was la Grange's item on sleigh dogs, 'Die Brakkies [mongrels] van Sanniespos', which observed that the team responded to orders in Afrikaans to start ('*trek*') and halt ('*hokaai*').[29] The reports in the *Nuusbrief* suggest a clubby atmosphere in which masculine camaraderie and dry humour were important to maintaining a sense of fraternity.

In 1962 a relief vessel, the *RSA*, was commissioned to supply the country's Antarctic footholds. The process of laying scientific and strategic claim to the Antarctic was presented as a South African and, more particularly, an Afrikaner triumph. La Grange and his team were honoured with newly minted Antarctica medals on their return home to Cape Town in 1962; English-speaking South African newspapers featured cartoons with jokes highlighting black and white penguins on an otherwise pristine white landscape.[30] Such exploits exemplify Klaus Dodds's observation that the Antarctic serves as 'a kind of fantasy space for white European men, in particular, to perform "firsts" and record their "achievements" in the name of national and individual power'.[31]

In the 1970s and 1980s the increasingly isolated apartheid government jealously guarded its status as a signatory to the Antarctic Treaty,

[28] Fuchs and Hillary, *The Crossing of Antarctica*, 254.

[29] J. J. la Grange, 'Die Brakkies van Sanniespos', *Nuusbrief*, 142, January 1961, 7.

[30] L.-M. van der Watt, 'Out in the Cold: Science and the Environment in the Sub-Antarctic and Antarctic in the Twentieth Century', DPhil thesis, Stellenbosch University (2012), 120–1, 123–5, 131; D. Pinnock, *The Woman Who Lived in a Tree and Other Perfect Strangers* (Auckland Park: Jacana, 2009), 134–9.

[31] K. Dodds, *The Antarctic: A Very Short Introduction* (Oxford: Oxford University Press, 2012), 41, 63.

which kept open international diplomatic and scientific channels that were otherwise closed. In 1983 and 1986, a group of non-aligned countries, in which Malaysia played a prominent role, sought to open up the 'question of Antarctica' at the United Nations. Pressure to remove or suspend South Africa from continued participation in the Antarctic Treaty figured prominently. South African diplomats, led by Brand Fourie, doggedly resisted such pressure. Some of Fourie's objections were couched in environmental terms. In the context of growing awareness of the fragility of the earth's ecosystems, and efforts to agree international treaties designed to protect Antarctic marine animals and this 'common heritage of mankind', Fourie argued that mineral exploitation of the Antarctic risked pollution and that South Africa's waters and marine life were especially vulnerable.[32]

The fact that South Africa was rich in rare and strategic metals such as platinum provided a means for the country to assume the high moral ground while protecting itself from competition. It was for this reason that Maarten de Wit, in a controversial book published in 1985, argued that platinum mining of the Dufek Massif near the Weddell Sea (part of Gondwanaland and geologically closely related to the platinum-rich South African Bushveld) might be economically viable.[33] Exploiting such resources would be costly but it would have the merit of reducing the world's heavy reliance on valuable South African platinum-group metals. By these means, additional pressure would be placed on the apartheid regime.

The end of the Cold War, coupled with South Africa's re-entry into the global community after 1990, facilitated a new approach to Antarctica based not only on the shared geological fact of Gondwana but also on conservation and biodiversity.[34] One of the conspicuous shifts in Antarctic science from the apartheid to the post-apartheid era is the way in which exploration has been reimagined. In the 1960s and 1970s, the SANAE ethos was highly masculinised and considerably Afrikanerised. In 1996, a new Antarctic base, SANAE IV, was opened

[32] L.-M. van der Watt, 'Return to Gondwanaland: South Africa, Antarctica, Minerals and Apartheid', *Polar Journal*, 3, 1 (2013), 72–93.

[33] M. de Wit et al., 'Boycott of South African Science', *Nature*, 330, 31 December 1987, 688; M. de Wit, *Minerals and Mining in Antarctica: Science and Technology, Economics and Politics* (Oxford: Clarendon Press 1985); van der Watt, 'Return to Gondwanaland', 88.

[34] van der Watt, 'Out in the Cold'.

on Queen Maud Land with several ANC government ministers in attendance. Accounts of the new base laid stress on a more racially and gender-diverse scientific fraternity. The language of 'ecosystems', 'biodiversity' and 'global warming' now described the Antarctic project, and resource management replaced concerns with geopolitical security. In 1995 the Prince Edward Islands were designated a Special Nature Reserve.

South Africa's long scientific involvement in the Antarctic was thus bracketed by the internationalist-minded governments of Smuts and Mandela but, paradoxically, received strongest support from an apartheid government attempting to counter international isolation. Its Antarctic initiative was celebrated as a national success, the only African country with an active research interest in the white continent during the era of the Cold War. Even more, it was a victory for Afrikanerdom: T. E. W. Schumann was awarded the Akademie vir Wetenskap's first Havenga Prize in 1947, while la Grange became a recipient in 1958. Verwoerd's own son, Wilhelm, a professor of geology at the University of Stellenbosch, was a strong protagonist of South African Antarctic research. He was awarded the Havenga Prize in 1987. In post-apartheid South Africa, one of the continuing justifications for maintaining an Antarctic presence is that it is the only country in Africa equipped to do so.

Akademie vir Wetenskap en Kuns, the Industrial and Development Corporation and the CSIR

The Akademie's Havenga Prize was the scientific counterpart to the literary Hertzog Prize, established in 1916, and was thus indicative of Afrikaner nationalism's growing interest in scientific and technological achievements. A tell-tale shift was the Akademie's success in persuading Malan to rename the Department of Education as the Department of Education, Arts and Science in 1949. Also noteworthy was the joint meeting between the Akademie and the S2A3 in 1950 – a clear signal of formal parity between Afrikaans- and English-speaking scientific institutions.[35]

[35] 'Die Pad van did Akademie (1909–1984)' and C. A. Engelbrecht, 'Natuurwetenskap, Tegniek en die Akademie', in *Akademie 75* (Pretoria: Suid-Afrikaanse Akademie vir Wetenskap en Kuns, 1984), 6–8, 132–3, 135–6, 139, 150.

Over the course of the 1950s the commitment of the Akademie vir Wetenskap en Kuns to science and technology (the two were invariably paired) became increasingly clear. Growth in the Akademie's engineering section was especially rapid, expanding from 35 members in 1942 to 180 in 1958, with several branches around the country. One of the motivating factors in this development was the creation of new engineering faculties in Afrikaans-speaking universities. Another was to maintain parity between engineers and scientists in state employ. The Akademie's A. J. A. Roux and R. L. Straszacker (leading figures in ESCOM and the Atomic Energy Board) chaired government commissions in 1952 and 1966 on policies towards engineering. Roux was awarded the Havenga Prize in 1959, as was fellow nuclear physicist Wally Grant in 1966. In a message of congratulations on the 50th anniversary of the Akademie's foundation in 1959, Verwoerd commended the organisation's creation of a scientific faculty as a 'definite, forward step'. Roux delivered an address on the 'consolidation of the Afrikaner in the natural sciences'.[36]

By now the National Party government was clearly committed to scientific endeavour under the aegis of the state. As shown in Chapter 5, H. J. van Eck, protégé of van der Bijl, gained experience in ESCOM and ISCOR during the war. He also played an important role in Smuts's wartime scientific establishment as chair of the Social and Economic Planning Council (1942–52). At this stage, van Eck had a reputation as a progressive: he was even seen by some as a figure who might be sympathetic to the formation of a centrist post-war political party to be led by J. H. Hofmeyr. The state corporations formed by van der Bijl and van Eck were initially treated with some suspicion by the Nationalists on account of their links to the despised Smuts regime and the mining houses; van der Bijl enjoyed close connections with industrialists like Oppenheimer and Menell.[37]

[36] 'Boodksap van Sy Edele Dr. H. F. Verwoerd', *Feesalbum 1909–1959* (Pretoria: van Schaik, 1959), 17; A. J. A. Roux, 'Die Konsolideering van die Afrikaner in die Natuurwetenskappe' in *Feesalbum*, 53–4; Engelbrecht, 'Natuurwetenskap, Tegniek en die Akademie', 46–7.

[37] N. L. Clark. *Manufacturing Apartheid: State Corporations in South Africa* (New Haven CT: Yale University Press, 1994); S. Sparks, 'Between "Artificial Economics" and the "Discipline of the Market": SASOL from Parastatal to Privatisation', *JSAS*, 42, 4 (2016), 716.

After he took over at the Industrial Development Corporation (IDC) in 1948, van Eck began to steer the organisation in ways adapted to the new strictures of the apartheid era. Under his direction, the IDC acted as a semi-autonomous development agency, with very broad scope to interpret 'industry', loaning money to national enterprise schemes while paying close attention to its own profitability. The IDC's first loan, issued in 1941, was to a Mrs Greyvensteyn, producer of 'Ouma' rusks.[38] Now the IDC was able to use a range of schemes such as export credits, low-interest loans, structured financing, and protection for import substitution to build up large new parastatal enterprises. Like van der Bijl, van Eck had a very clear view that modern state-led industrial development was key to South Africa's future prosperity.[39] His presence can be found in virtually every major state initiative affecting science and industrial policy in the early apartheid era.

The IDC's two most important strategic initiatives in the early 1950s were the establishment of a South African Coal, Oil and Gas Corporation (SASOL), which utilised the German Fischer-Tropsch process to produce oil from coal, and the Phosphate Development Corporation (FOSKOR) producing fertiliser.[40] The SASOL plant was built adjacent to Sigma colliery and together they gave birth to the new model town of Sasolburg in the northern Free State. FOSKOR emerged out of the exploratory work by Hans Merensky (see Chapter 4) near Phalaborwa. Concerned by South Africa's dependence on imports of phosphates, and also by the interest shown by the Anglo American Corporation in gaining control over the Phalaborwa phosphate deposits, the apartheid government declared these a national interest and established FOSKOR as an IDC-controlled parastatal in 1952.

The venture took little account of the interests and rights of people living in the vicinity, and the reversal of African fortunes in the Phalaborwa area was particularly striking. Phalaborwa had been the site of ancient iron and copper manufacturing. Archaeological evidence of vertical shafts and smelting furnaces worked by the

38 A. P. Cartwright, *The Industrial Development Corporation of South Africa Limited* (Johannesburg: Industrial Development Corporation, 1971), 13–14.

39 H. J. van Eck, *Some Aspects of the South African Industrial Revolution* (Johannesburg: South African Institute of Race Relations, 1951); H. T. Andrews et al., *South Africa in the Sixties* (Cape Town: South Africa Foundation, 2nd edn, 1965), 135.

40 Cartwright, *Industrial Development Corporation*, 23–4.

BaPhalaborwa people had long been known. Modern mining in the 1960s displaced Africans but kept them conveniently close – through the local configuration of the apartheid Bantustan system – as a ready source of unskilled labour.[41]

Both parastatals tapped into South Africa's abundant and cheap natural resources in the quest for economic self-sufficiency. They also both depended on, and created new demands for, expertise in physics, chemistry, and engineering. In so doing, they gave a significant boost to South Africa's technical capacity. Demand for petrol soared as the number of vehicles quadrupled from 1,107,000 in 1960 to 4,677,000 in 1989.[42] By this time, SASOL produced about 40 per cent of oil needs. FOSKOR became a major supplier of fertiliser in the years in which South African arable agriculture mechanised and expanded exponentially. By 1948, farmers were already spending £3 million a year on phosphate fertilisers.[43] The number of tractors in the country increased from 6,000 in 1937 to 119,000 in 1960, and the gross value of arable production rose from about £33 million to £200 million.[44]

In the 1960s and 1970s the IDC became further involved in promoting strategic and capital-intensive industries. These included aluminium smelting, synthetic rubber, computers, aircraft manufacturing (Atlas), automotive parts, coastal oil exploration (SOEKOR), military hardware (ARMSCOR), hydroelectric schemes and nuclear energy.[45] As well as concentrating industrial and scientific power in large-scale enterprises, the IDC's promotion of industrial decentralisation and border industries producing textiles, wood products, and foodstuffs (sited adjacent to the ethnic homelands, where they benefited from cheap labour and favourable tax regimes) was intimately connected to the Verwoerdian conception of establishing nominally independent Bantustans.

If the IDC formed the centrepiece of the state's industrialisation strategy, the CSIR was its counterpart in fundamental and applied

41 E. H. Roux et al., 'Phosphate in South Africa', *Journal of the South African Institute of Mining and Metallurgy*, 89, 5 (1989), 129–39; N. J. van der Merwe and R. T. K. Scully, 'The Phalaborwa Story: Archaeological and Ethnographic Investigation of a South African Iron Age Group', *World Archaeology*, 3, 2 (1971), 178–96.

42 Beinart, *Twentieth-Century South Africa*, 355.

43 *Farming in South Africa*, vol. 23, 269 (August 1948), 531.

44 *Union Statistics for 50 Years*, I-53.

45 Carwright, *Industrial Development Corporation*.

research. After the 1948 election the CSIR suffered a temporary decline in status when its direct line of reporting shifted from the Prime Minister's Office to the Ministry of Economic Affairs (whose title made no reference to science or technology). There was a perception that the government was seeking to constrain the CSIR for its own ends and a suspicion that senior new appointments, as elsewhere in the civil service, were being made for political purposes.[46] The CSIR's prospects were soon revived by a combination of factors: sustained economic growth, the international shift towards 'big' science, and the growing willingness of the modernising apartheid state to support scientific and technological advance.[47]

In 1952 P. J. du Toit was succeeded by his deputy, Meiring Naudé. As founding director of the CSIR's National Physical Laboratory, Naudé benefited from having been prepared for the role by his immediate predecessors. Born in the small Karoo town of De Doorns in 1904, Naudé took an MSc at Stellenbosch, turned down a Rhodes Scholarship, and proceeded instead in 1926 to Germany where he attended courses given by Nobel laureates like Einstein, Laue, Hertz and Schrödinger. Naudé graduated from Berlin in 1928 and worked for two years on X-rays in Chicago with Robert Millikan and Arthur Compton before returning to South Africa to take up a senior lectureship in 1930 at the University of Cape Town, where Schonland was a colleague. Naudé gained international recognition at this time for discovering a number of isotopes and moved to Stellenbosch as professor of experimental physics. He was singled out by Schonland to head up the National Physical Laboratory in 1946. During his five years in this capacity, Naudé expanded its remit to include optics, spectrochemistry, national measuring standards (metrology), X-ray crystallography, radioactivity, geophysics, and atmospheric physics.[48]

As a leading scientific representative of the Afrikaner ascendancy, Naudé was a credible successor to van der Bijl and Schonland. Like them, he insisted on excellence in all senior appointments, regardless of official pressures. Promising recruits were carefully dispatched to overseas centres for further training. Naudé was ideally positioned to drive the expansion of the CSIR at a time of opening scientific frontiers.

[46] N. Basson, *Passage to Progress: The CSIR's Journey of Change 1945–1995* (Johannesburg: Jonathan Ball, 1996), 21–2.

[47] Kingwill, *The CSIR*, 17–18.

[48] SA Physics MS, 54, 192–3.

Particle physics, radio astronomy, jet aircraft, and space exploration attracted fresh talent and funding. He also appreciated the need to expand applied research and supported scientific work on water treatment, road-building and surfacing, mechanical engineering, construction, nutrition, telecommunications, engineering and defence.[49] In its annual report for 1965, marking two decades since its establishment, the CSIR began with a confident pronouncement: 'South Africa has taken its place among the scientifically oriented countries of the world.'[50]

Naudé's intellectual stature and personal confidence allowed him to lead the CSIR and the Institute of Physics in ways that transcended – without necessarily erasing – political, institutional, and ethnic rivalries within the white establishment. Whereas Schonland preferred Johannesburg as the site for the Physical Laboratory, Naudé insisted on placing it in Pretoria, the heart of government and growing centre of South Africa's military-industrial complex.[51] Naudé's broad acceptability to Afrikaner and English-speaking colleagues is indicated by his receipt of honorary doctorates from eight South African universities. He served as president of the Royal Society (1960–1) as well as the Akademie vir Wetenskap en Kuns (1967–9) and the South African Institute of Physics. He remained close to the state and personally supervised aspects of the CSIR's defence division. On his retirement from the CSIR, Naudé became scientific adviser to Prime Minister Vorster.

Under Naudé's presidency (1952–72) the CSIR's staff expanded from 685 to more than 4,000 and the number of its national institutes grew from the original five to thirteen. Basic science continued to be done – journal publications and monographs were encouraged – and this was augmented by industrial contract work. From 1951 to 1971 income generated by contract work increased from 4.3 per cent of the total institutional funding to 37.4 per cent (allowing for inflation). Patented inventions in variable frequency crystal control for radios and work on optical laser technology led to products that attracted international commercial interest.[52]

A notable practical invention was the tellurometer. In 1954 the South African Trigonometrical Survey requested an instrument

[49] *CSIR Annual Report 1965* (Pretoria, 1966), 9–12.
[50] *CSIR Annual Report 1965* (Pretoria, 1966), 1.
[51] Austin, *Schonland*, 314–15, 397–8; M. McDowell, *The National Metrology Laboratory of South Africa* (Pretoria: CSIR, 1997), 29.
[52] Kingwill, *The CSIR*; Basson, *Passage to Progress*, 25.

capable of more accurate long-distance measurement, and Trevor Wadley, a Durban-educated physicist and veteran of radar experimentation during the war, came up with a prototype which functioned by measuring the frequency of reflected microwaves between two points. Beacons 50 km apart were accurately measured to within about 5 cm. Wadley demonstrated the capacity of the tellurometer in 1957 to an international audience in Cape Town, attracting global interest. By 1960 a new model was developed for hydrographic measurement. With investment by British electronics firm Plessey, a company was set up in Cape Town, with subsidiaries in Britain, Canada, Australia and the United States. In 1966, more accurate instruments were produced for the US Army followed by the British defence ministry in 1973. Lighter devices for civilian use were marketed, including instruments for extreme conditions such as the Antarctic – where it was tested by Hannes la Grange. In its different guises, the tellurometer revolutionised geodetic surveying by making available a cheap and accurate means of distance measurement, so replacing old-style theodolites.[53]

Figure 6.1 Trevor Wadley demonstrating his Tellurometer.
Credit: South African Institute of Electrical Engineers, Historical Section

[53] B. Sturman and A. Wright, 'The History of Tellurometer', FIG Working Week, Stockholm, 2008, www.fig.net/resources/proceedings; Bjørn Geirr Harsson, 'The Tellurometer, from Dr Wadley to MRA7', *Survey Review*, 42, 316 (2010), 207–8.

The CSIR also took an interest in, but did not originate, the 'dolos', an innovative twisted 'H' formation interlocking concrete block which was first used to protect the East London harbour breakwater. Credit for the practical invention has been given to harbour engineer Eric Merrifield, though it was draughtsman Aubrey Kruger who came up with the innovative design in 1963. Kruger's family has claimed that its design was based on a common thorn, 'the dubbeltjie', and the name 'dolos' refers to the bones thrown by African sangomas and diviners and a game of the same name played by rural children. It was a genuinely indigenous invention and its use soon spread internationally, though neither Merrifield nor Kruger benefited, since they believed the intellectual property to reside with their state employers.[54]

Nicoline Basson's account of the CSIR in the 1960s and early 1970s portrays the organisation as something of a 'research utopia' at this time. Salaries and working conditions were controlled by the civil service but she describes the atmosphere of the modern Pretoria campus, Scientia, as more like a university with a strong sense of collegiality, a good library – and no teaching responsibilities. There was considerable autonomy in respect of selection of research topics: 'the sheer pleasure of cracking tough problems … was multiplied by the challenge of building up a world-class research establishment in a relatively insignificant country'.[55] This account plays down the extent to which the organisation's technical and research orientation was increasingly geared towards self-sufficiency and white survival. From the 1970s, in response to international trade sanctions, pure science increasingly gave way to strategically focused and contract-oriented research.[56] South Africa's declining economic performance from the middle of the decade, the government's growing interest in market-based reforms, and the increasing orientation towards military and security research, all conditioned this shift.

A key inflection point came in 1986–7 when C. F. Garbers, president of the CSIR, announced a major policy change in a document titled 'Strategy for the Future Course of the CSIR'. 'Outputs' rather than 'inputs' would henceforth define research priorities with 'successful

[54] T. Hutson, 'Shipping: Real Deal on a Home-Grown Success Story', *The Mercury*, 10 August 2016. Thanks to Jonathan Cane for advice.
[55] Basson, *Passage to Progress*, 31–2.
[56] McDowell, *The National Metrology Laboratory*, 32.

implementation of R&D as its business'. The CSIR's annual parliamentary grant would no longer increase automatically with inflation, performance indicators would be used as a management tool, and activity was to be measured according to rigorous cost-benefit criteria. It was envisaged that large research facilities including astronomy and nuclear physics would continue to be supported, yet the major emphasis would now be on technology innovation with business spin-offs.[57] Many leading CSIR scientists left the organisation at this time for the private sector or the universities.

One notable CSIR project that partly succeeded in bridging curiosity-led fundamental science and market-driven imperatives was in the field of battery technology. The spike in oil prices in the mid-1970s, coupled with international embargoes on South African importation of oil, sharpened requirements for alternative energy sources. Michael Thackeray (son of astronomer David Thackeray) joined the CSIR in 1973 from the University of Cape Town. He teamed up with Johan Coetzer, a structural chemist based in the CSIR's X-ray crystallography division from 1968. In line with the CSIR's preference for private sector cooperation, Anglo American Corporation and De Beers funded Thackeray's and Coetzer's research programmes into battery storage technology from 1977. One promising development was the Zebra battery (code name for Zeolite Battery Research in Africa) which attracted investment by Daimler-Benz. Yet, anti-apartheid pressures meant that collaboration with British and German research institutions and corporations had to be, in Thackeray's words, 'kept below the radar screen'.[58] Zebra technology nevertheless proved workable and in 1986 a company was spun off from the CSIR under Coetzer's direction producing batteries to power electric vehicles. Zebra technology was used in Mercedes buses at the 1992 Olympic games in Barcelona. However, impatient with what it considered to be slow further progress in this technology, Anglo American and Daimler-Benz abandoned their Zebra investments in 1999.

[57] C. F. Garbers, president CSIR, 'Strategy for the Future Course of the CSIR', 1987, South African Astronomical Observatory Archives, Observatory, Cape Town; Basson, *Passage to Progress*, chapter 6.

[58] 'Energy Storage for a Cleaner Planet. Fellow's seminar by Michael Thackeray', Stellenbosch Institute for Advanced Studies, 7 May 2018, stias.ac.za/2018/05; 'Battery Pioneers: Johan Coetzer', www.batteriesinternational.com/2016/09/22.

Thackeray persisted with lithium technology at the CSIR with Coetzer's assistance, albeit without corporate funding. A spell in Oxford in 1981–2 working with Nobel laureate John Goodenough, provided Thackeray with fresh ideas about how to produce safe cost-effective lithium-manganese cells operating at room temperatures. An American patent followed. Thackeray returned to the CSIR at the end of 1982 and continued to work on layered lithium electrode technology until 1994, when he moved to the United States to work at the well-resourced Argonne National Laboratory. There, major progress in lithium-ion battery technology was achieved by a large team in electrochemical storage team led by Thackeray. The CSIR's decision, in 1992, to cut its funding of battery technology on the grounds that there was insufficient local market demand – apparently unaware of Sony's introduction of lithium-ion batteries into consumer products the previous year – was a crucial factor in ending Thackeray's twenty-year association. Thackeray was left to wonder what might have happened had the CSIR and Anglo American persisted with their investment.[59] While it is impossible to answer this hypothetical question definitively, it is hard to avoid the conclusion that apartheid isolation as well as the quest for immediate solutions contributed to the initial quest for rechargeable batteries as well as constraining the development of a commercially successful homegrown product.

Defence and Security

National security was a growing concern for the CSIR from 1962 with the creation of a Defence Research Council, led by the minister of defence and the president of the CSIR. A National Institute for Rocket Research and Development followed in 1963 and the National Institute for Defence Research in 1965. There were further high-level military links through van Eck, who was at this time a member of the CSIR Council as well as chair of the IDC and of the Munitions Production Board. Under the directorship of T. J. Hugo, the National Institute for Defence Research worked closely with the military on technology such as guided missiles, aircraft and computers. In

[59] M. Thackeray, 'Twenty Golden Years of Battery R&D at CSIR, 1974–1994', *South African Journal of Chemistry*, 64 (2011), 65; M. Thackeray, *Running with Lithium* (Bloomington: Archway, 2019).

its annual report for 1965, the CSIR noted that half of its research income was now derived from 'direct services and contract work' with NASA, which relied on the CSIR's radio astronomy facility at Hartebeesthoek to help track American satellites, contributing 13 per cent of the CSIR's income.[60] The elements of a military-industrial complex, centred on Pretoria, were by now in place. And this was to expand over the following two decades.

From the 1970s, and particularly after the Soweto uprisings in 1976, concerted efforts were made to integrate military, political and research developments. This was especially the case after the minister of defence, P. W. Botha, succeeded John Vorster as prime minister in 1978. In that year key CSIR facilities and around 350 staff were moved to an independent subsidiary of ARMSCOR, the precursor of Denel, which used CSIR know-how to develop guided weapons systems, including short-range air-to-air missiles.[61] The CSIR also provided intellectual feeder fuel for South Africa's nuclear programme. Key figures in the development of the atomic bomb, including A. J. A. Roux, Wally Grant, and Wynand de Villiers, were all products of the CSIR.[62] Company annual reports, including those of the CSIR, suggest ever closer integration between government parastatals, Afrikaans universities (Pretoria in particular), the CSIR, and the military.

The CSIR's status as a semi-autonomous research organisation working in the national interest brought it into the orbit of American foreign policy. Following the launch of the Russian Sputnik programme in 1957, South Africa moved to support the United States' ventures into space. In 1958 a Minitrack radio system operated by the CSIR began to monitor US satellites. A separate satellite tracking station north of Pretoria was signed with the French space agency (CNES) in 1966 and remained operative until 1980. More ambitiously, the CSIR operated a Deep Space Instrumentation Facility from 1961 on behalf of NASA. Here it used a 26-metre parabolic antenna situated at Hartebeesthoek, north-west of Johannesburg, to track satellites and space missions. South Africa expected diplomatic protection and continued military cooperation as a quid pro quo. The Deep Space initiative was described in the mining house journal *Optima* in a

[60] *CSIR Annual Report 1965* (Pretoria, 1966), 4–5.
[61] SA Physics, 478–9; Kingwill, *CSIR*, 25.
[62] N. von Wielligh and L. von Wielligh-Steyn, *Die Bom: Suid-Afrika se Kernwapenprogram* (Pretoria: Litera, 2014), 98–101.

boosterish article titled 'South Africa's Vast Natural Laboratory Serves World Science'.[63]

In 1973 the South African Astronomical Observatory, working in collaboration with the CSIR, opened a major new observatory at Sutherland in the Karoo housing three reflecting telescopes. The new facility absorbed the Royal Observatory in Cape Town and the Radcliffe Observatory in Pretoria, both with strong British connections and associations. The government's Republic Observatory (run by the CSIR since 1964) was shut, despite strong representations from Johannesburg astronomers.

The idea of integrating South Africa's highly dispersed observatory system in one central location, with excellent optical viewing potential, made good scientific sense. It also marked a signal success for the CSIR, which, since the early 1960s, was increasingly unhappy about the lack of national control over the country's observatories. In 1970 a secret agreement was struck with the British Science Research Council to consolidate optical observatory work at Sutherland. This was a joint enterprise. British–South African links were strengthened through the appointment of Sir Richard van der Riet Woolley as the SAAO's inaugural director from 1971. A long-serving British Astronomer Royal and leading establishment figure, Woolley had studied in Cape Town (with Schonland among others), where his father had been a naval officer and his mother the daughter of a Simonstown magistrate.

At the opening ceremony of the SAAO a smiling Margaret Thatcher, then British secretary of state for science and technology, sat next to a gruff Prime Minister John Vorster, who stressed the desirability of international and bilateral cooperation in efforts to further knowledge of the southern skies. At a time when Vorster's policy of 'détente' in southern Africa was beginning to stall, and when British-South African links were coming under new pressure from the anti-apartheid movement, the observatory at Sutherland became a visible demonstration of ongoing scientific goodwill and interchange. It was also a demonstration of the South African government's desire to consolidate its control of astronomical activity in the country.[64]

[63] D. G. Kingwill, 'South Africa's Vast Natural Laboratory Serves World Science', *Optima*, 15, 1 (1965), 24–6; Kingwill, *The CSIR*, 188.

[64] S. Dubow, '200 Years of Astronomy in South Africa: From the Royal Observatory to the "Big Bang" of the Square Kilometre Array', *JSAS*, 45, 4 (2019), 663–687.

International collaboration in astrophysics continued through and beyond the apartheid era, as was the case in Antarctic research. But these pursuits were not immune to external criticism as the world's attention focused increasingly on apartheid. The Deep Space agreement at Hartebeesthoek came to an end in 1974, three years after visiting anti-apartheid and civil rights campaigner Charles Diggs called for NASA to terminate its relationship with South Africa. Diggs, who became a public hate figure for many white South Africans, hailed NASA's announcement of a phased withdrawal as 'a major victory' against apartheid. The Hartebeesthoek facility was thereafter converted to a radio astronomical observatory, HartRAO. This observatory developed expertise which later fed into South Africa's massive post-apartheid bid to host the international Square Kilometre Array (see Chapter 7).[65]

The decline in state support for fundamental science in the late-apartheid era was outweighed by growing demand for military technology, making safety and security the largest source of external grant income for the CSIR by 1990.[66] A complex web of interrelated state-owned and private sector companies proliferated with ARMSCOR at its heart. One of its subsidiaries, Kentron, split off from the CSIR in 1978 to take a lead in missile and guided weapons technology. A close affiliate, Denel Aviation, produced hardware like the Rooivalk attack helicopter. This potent weapon was adapted from the French Super Puma. Its design owed much to the CSIR's Computational Aerodynamics Group. Defencetek, an offshoot of the CSIR, offered high security devices like the ACL-20 perimeter fence utilising infrared beams governed by 'neural networks' that purported to distinguish between real and false alarms. Even when defence budgets were slashed in 1990 and local demand for military products diminished, the South African armaments industry was able to reorient its business model to take advantage of international export opportunities.[67]

65 Kingwill, *The CSIR*, 182–5; *Survey of Race Relations in South Africa 1972* (Johannesburg: South African Institute of Race Relations, 1973), 224; *Rand Daily Mail*, 19 July 1973.

66 Basson, *Passage to Progress*, fig. 117; 'Strategy for the Future Course of the CSIR', 13 November 1986, 2, in SAAO archives, Observatory, Cape Town.

67 D. Henk, 'Guns and Butter: Reframing South Africa's Arms Industry', *African Security Review*, 13, 3 (2004), 1–33.

Nuclear Power and Militarisation

In 1948 an Atomic Energy Board was given parliamentary approval, providing for regulation of the uranium industry and for supporting research. India passed an Atomic Energy Act in the same year: the impetus to become part of the nuclear establishment in the case of both countries was their desire to be seen as technically advanced, modern, regional powers.[68] In 1955, leaders of the CSIR, ESCOM and the Industrial Development Corporation (Naudé, Hattingh and van Eck, respectively) were deputed to investigate the application of atomic energy, representing South Africa at the UN's First Geneva Conference on 'Peaceful Uses of Atomic Energy'.[69]

Recent attention to the story of South Africa's nuclear weapons production tends to obscure the larger history of its involvement in nuclear science or, alternatively, to collapse these into each other. Yet it is important to separate these strands and not to presume a direct line from nuclear physics to nuclear weapons. Research into nuclear physics was pursued as early as 1922 at the University of Cape Town.[70] The production of uranium from 1944 to 1945 was initially geared to satisfying international strategic obligations rather than meeting domestic needs, namely, facilitating the Manhattan Project and maintaining links with the British nuclear programme .[71] Smuts, Schonland, and the Chamber of Mines, were closely involved in this pre-apartheid initiative. It is possible to demarcate distinct phases in nuclear physics leading up to the production of weaponry: civil uses pre-1970; uranium enrichment from 1970 to 1978; and bomb-making post-1978.[72] In practice, these shaded into one another. The precise trajectory of development could not easily be traced, even by those who were involved in the programme, which proceeded strictly on a need-to-know basis.

South Africa's ample supplies of uranium, a by-product of gold mining and a highly valuable commodity, provided an entrée into the nuclear community as well as inviting experimentation with production, sales and enrichment. In Gabrielle Hecht's words, 'Uranium

[68] G. Hecht, *Being Nuclear: Africans and the Global Nuclear Trade* (Johannesburg: Wits University Press, 2012), 29.
[69] Kingwill, *The CSIR*, 22. [70] *Physics in South Africa*, 41–2.
[71] Austin, *Schonland*, 356.
[72] A.-M van Wyk, 'South Africa's Nuclear Programme and the Cold War', *History Compass*, 8, 7 (2010), 562, offers this periodisation.

reserves gave apartheid South Africa a material role in the "defense of the West".'[73] The country's first uranium plant, opened on the West Rand in 1952, was hailed by Prime Minister Malan both as proof of the country's advanced industrial capabilities and as a contribution to 'the cause of the Western Powers'.[74] National security in this sense was coextensive with South Africa's keen participation in the Korean War. National pride was also a significant spur to action. South African metallurgists resented the notion that they were incapable of testing and processing the country's uranium resources themselves; by 1952, the Government Metallurgical Laboratory could justifiably claim to be expert in this field.

The first phase of nuclear research and experimentation privileged civil projects: energy production, radionuclides for biomedicine and industry, and even the possibility of nuclear explosions for mining and construction. Young South African scientists were drawn to atomic physics for reasons of intellectual excitement and because of the opportunities that were opening up.

In 1965, Friedel Sellschop, director of the Schonland Research Centre for Nuclear Sciences at Wits University, organised a collaborative international experiment together with the American scientist Frederick Reines of the Case Institute of Technology, Cleveland. It took place 3,000 metres underground in the bowels of a deep-level gold mine on the East Rand. With support from the US Atomic Energy Commission, a laboratory was hacked out of solid rock. In February, detectors recorded a muon travelling horizontally, the first naturally occurring high-energy neutrino ever detected. The existence of this elusive atomic particle had first been predicted by Wolfgang Pauli in 1930.[75] This achievement in particle physics research was a minor triumph for South African nuclear science as well as testimony to Sellschop's success in developing a leading laboratory from scratch. It was a boon for Wits and confirmed, too, that the university's long association with the mining industry could continue to benefit both parties.

[73] Hecht, *Being Nuclear*, 22.

[74] P. N. Edwards and G. Hecht, 'History and the Technopolitics of Identity: The Case of Apartheid South Africa', *JSAS*, 36, 3 (2010), 622–3.

[75] *Physics in South Africa*, 76–8; F. Reines and J. P. F. Sellschop, 'Neutrinos from the Atmosphere and Beyond', *Scientific American*, 2, 214 (1966), 40–8.

Theoretical physics in South African universities remained fragmented and dispersed in the post-war decade. The first large particle accelerator in the country was a 16 MV cyclotron built by the CSIR at its specially designed National Physics Research Laboratory in Pretoria. Government approval for this major undertaking was granted in 1950 but it took until the end of the decade before the first deuteron beams were extracted. The cyclotron was locally designed and based on Swedish technology with assistance from German scientists who joined the South African group.[76] Developed in the 1930s, cyclotrons spun accelerated charged particles between powerful magnetic fields to produce high-energy beams. These were used for experimental nuclear physics and in medicine, including the diagnosis and treatment of cancerous tumours. Prior to the commissioning of South Africa's cyclotron, iodine-131 and cobalt-60 radioactive isotopes for medical purposes had to be imported. In an effort to save on costs, the CSIR's Physical Research Laboratory designed canisters fitted to the wing tips of planes for the transportation of radioactive isotopes rather than carrying heavy lead-shielded containers in the cabin.[77]

Greater national integration – and cost-sharing – in nuclear physics began in the 1960s with a collaborative nuclear programme between the universities of Cape Town and Stellenbosch at the Southern Universities Nuclear Institute at Faure, using a 5.5 MV van de Graaff accelerator. Within a few years there were relatively small accelerators based at the universities of the Witwatersrand, Potchefstroom (now North West) and Pretoria. In 1979 the Faure facility was brought together with the CSIR's Pretoria accelerator and rebranded the National Accelerator Centre. Faure gained a large 200 MV cyclotron, which became fully operational by 1986. Its objectives were to undertake applied and basic particle research as well as to produce nuclear isotopes for use in medicine and industry. The cyclotron was an expensive but well-engineered 'gold-plated' facility, which provided a model for cooperation between different universities and scientists, including the training of postgraduates from the University of the Western Cape.[78] This research infrastructure had little or no direct influence on nuclear weapons production, though it did provide

[76] *Physics in South Africa*, 4–5, 264–7. [77] Kingwill, *The CSIR*, 142–3.

[78] *Physics in South Africa*, 264ff; *CSIR Annual Report 1986* (Pretoria, 1987), 14; G. F. R. Ellis, *Science Research Policy in South Africa: A Discussion Document* (Rondebosch: Royal Society of South Africa, 1994), 83.

training for experts associated with the Atomic Energy Corporation who were recruited into the military programme.

Nuclear interchange with Britain and the United States in respect of personnel and material was barely affected by apartheid policies in the 1950s. The British Atomic Energy Authority signed a collaborative agreement with the South African Atomic Energy Board in 1957, just two years after the Simonstown Agreement transferred command of the naval base from Britain to South Africa. There were also close links with the United States which, in the same year, signed a secret agreement with South Africa in terms of the Eisenhower administration's 'Atoms for Peace' initiative. The Kennedy administration's arms embargo, somewhat reluctantly imposed in 1963, did not affect nuclear collaboration. By one estimate, nearly ninety South Africans were trained at US facilities like Argonne and Oak Ridge. South Africa also enjoyed close links with West German nuclear facilities after 1955.

In 1959 A. J. A. Roux left the CSIR, of which he was vice-president, in order to join the Atomic Energy Board. Roux's ambition to establish a national nuclear research centre independent of the CSIR received government approval the same year.[79] This opened the way for the Atomic Energy Board to pursue covert research on nuclear enrichment with potential military applications. The National Nuclear Centre was opened at Pelindaba (taken to be isiZulu for 'we don't talk about this anymore') west of Pretoria, in 1963. SAFARI-1 (the South African Fundamental Atomic Research Installation) was manufactured by the United States in terms of the 1957 bilateral agreement and went critical in 1965. This facility ran on American-supplied enriched uranium (U_{235}), the level required for weapons development, but was initially designed for research rather than military purposes.[80]

The Atomic Energy Board moved rapidly to develop its own enrichment capacity. One argument was that nuclear power reactors could

[79] H. E. Purkitt and S. F. Burgess, *South Africa's Weapons of Mass Destruction* (Bloomington: Indiana University Press, 2005), 35–6; Kingwill, *The CSIR*, 22–3.

[80] D. Fig, 'Sanctions and the Nuclear Industry', in N. C. Crawford and A. Klotz, *How Sanctions Work: Lessons from South Africa* (Basingstoke: Palgrave Macmillan, 1999), 80–1; W. Stumpf, 'South Africa's Nuclear Weapons Program: From Deterrence to Dismantlement', *Arms Control Today* (December 1995/January 1996), 3. The name Pelindaba, precedes the nuclear facility. It was the unofficial name given to the farm by its owner, the Afrikaner journalist and historian Gustav Preller, and it was here that Eugene Marais committed suicide in 1936.

run more efficiently on enriched uranium. Gas centrifuges required high technical expertise and gaseous diffusion methods demanded costly capital investment. Prime Minister Verwoerd gave his full support for an independent route to enrichment, agreeing to supply funds within days of being approached by Roux. The method chosen was a stationary centrifuge fed by uranium hexafluoride and hydrogen gases. Much of the original scientific design was conducted by the multi-talented Pretoria University physicist, chemist and engineer Pierre Haarhoff, who became manager of the Uranium Enrichment Corporation in 1971.[81]

Haarhoff was responsible for designing the bespoke Helikon aerodynamic vortex tube. At once charismatic and secretive, Haarhoff took four doctorates (including a 1976 Stellenbosch dissertation on the principles of uranium enrichment). He also published short stories in Afrikaans, and reputedly made a fortune as a part-time estate agent near Hartbeespoort Dam. For these multiple accomplishments, he won both the Havenga Prize for science and the Eugene Marais Prize for literature. In 1977, Haarhoff returned to Pretoria University as professor of mechanical engineering, where he remained until his death in 1990.[82] Wally Grant, an unconventionally trained engineer who had worked on jet turbines during the war (under the British pioneer Frank Whittle), developed the engineering of the compressor and provided overall direction. Another key figure in the enrichment programme was the Stellenbosch- and American-trained physicist J. W. L. de Villiers.[83]

The novelty of this locally developed method of isotope separation was exaggerated as a national triumph: 'It is all the work of South Africans.'[84] Subsequent capture of documents in Bonn by the ANC pointed to close collaboration with West German scientists on the use of jet nozzles to facilitate enrichment. Franz Josef Strauss, who headed the German Federal nuclear programme, made several visits to Pretoria at this time. Key South African scientists trained at Karlsruhe from 1969–70 and Erwin W. Becker, pioneer of the German jet-nozzle process, inspected the South African facilities in 1974 (and apparently did not think it would work).[85] South Africa's enrichment programme

[81] Von Wielligh and Von Wielligh-Steyn, *Die Bom*, 114.
[82] www.stellenboschwriters.com/haarhoff.
[83] Von Wielligh and Von Wielligh-Steyn, *Die Bom*, 99–101, 115.
[84] Edwards and Hecht, 'History and the Technopolitics of Identity', 635–6.
[85] Fig, 'Sanctions and the Nuclear Industry', 81–3; Stumpf, 'South Africa's Nuclear Weapons Program', 3.

is thus best thought of as a creative adaptation to local conditions of the German-designed Becker process, using a mix of theoretical physics and practical engineering prowess. It demonstrated the determination of the state and nuclear scientists to ensure that science and technology were transferred rapidly.

Uranium enrichment entailed very high energy requirements. As Gabrielle Hecht shows, the mining industry remained unconvinced that there was a commercial market for South African-processed hexafluoride, but the Atomic Energy Board, under Roux's leadership, chose to ignore the question of cost, stressing national prestige and self-sufficiency. At the IDC, van Eck played a key role in persuading the government to support an indigenous enrichment programme.[86] The Sharpeville crisis and growing international condemnation of apartheid created the impetus for autonomous nuclear capability and, in the short term, it drew South Africa closer to its Western allies. Enrichment was perceived as both a necessary response to international isolation and a statement of the country's exceptional global standing.

In 1970, Vorster announced to the world in 1970 that South Africa was about to embark on its own programme of enrichment at the 'Y-plant' at Valindaba (a Zulu-derived name taken to mean 'we don't talk about this at all' and so a coded statement of security and defiance). 'South African scientists', he boasted, 'have thus again added to the prestige of their country in no uncertain terms. In the past they have made lasting contributions to science, but perhaps the achievement that I am announcing to-day is unequalled in the history of our country.'[87] In 1967, a ten-year deal was renewed with the United States for enriched uranium imports under International Atomic Energy Agency (IAEA) protocols. In 1965 South Africa signed an agreement to ship 10 tons of uranium yellowcake to Israel (without any mention of the IAEA), thereby initiating a reciprocal nuclear relationship that bypassed their existing Western patrons.[88]

86 Hecht, *Being Nuclear*, 90–4; Von Wielligh and Von Wielligh-Steyn, *Die Bom*, 118.

87 A. R. Newby-Fraser, *Chain Reaction: Twenty Years of Nuclear Research and Development in South Africa* (Pretoria: Atomic Energy Board, 1979), chapter 4.

88 S. Polakow-Suransky, *The Unspoken Alliance: Israel's Secret Relationship with Apartheid South Africa* (Auckland Park: Jacana, 2002), 43.

After 1974, US–South African relations came under increased strain (this was also the moment when American satellite tracking came to an end) and the apartheid government now made a determined, secret decision to proceed with nuclear weaponisation. It was also the moment at which the CSIR began battery research, prompted by anxieties about energy supplies. American suspension of enriched uranium exports in 1975, and the political crisis created by South African military intervention in Angola in the same year, were important trigger points. The Soweto uprising in 1976 and the loss by South Africa of its coveted position on the IAEA Board of Governors were important contributory factors. The decision to produce a bomb was part of a more general shift towards militarisation by 'securocrats' whose hold on government grew rapidly from the late-1970s.

By 1977 South Africa was ready to 'cold test' a nuclear device in the Kalahari Desert using depleted uranium (sufficient supplies of enriched uranium were as yet unavailable). However, concerted political pressure by the United States led to the abandonment of these plans.[89] At this stage, South Africa's technological capacities were handicapped by the lack of an appropriate delivery system. This helps to explain why defence minister, Botha, is said to have brokered a secret agreement with Israeli prime minister Peres in 1975, code-named 'Chalet', by which Israel would supply South Africa with missiles armed with nuclear warheads. No such enabled missiles were delivered, but as soon as Botha became prime minister in 1978, the government began to implement the plan to produce functional nuclear weapons.[90]

It took South Africa some twenty years from the commissioning of the Pelindaba plant to the point where it produced the first viable nuclear device. This was assembled at the top-secret ARMSCOR facility, the Kentron-Sirkel. Opening the facility in May 1981, Prime Minister P. W. Botha, the third successive apartheid leader to have had direct involvement in the atomic programme – and the most closely aligned to the military – took visible pride in the project. He compared it to the achievements at Los Alamos and spoke menacingly of the bomb as an instance of turning ploughshares into swords. On the same day, foreign minister 'Pik' Botha informed President Reagan

[89] Or Rabinowitz, *Bargaining on Nuclear Tests: Washington and Its Cold War Deals* (Oxford: Oxford University Press, 2014), 114–15.
[90] Polakow-Suransky, *The Unspoken Alliance*, 83.

that South Africa had the capacity to make a nuclear bomb. A year later the first completed nuclear weapon was presented to Botha as a Strangelovian 'Christmas gift'.[91]

The first six nuclear devices were completed at a relatively slow pace (consonant with the capacity to enrich uranium) over the rest of the 1980s. The relatively extended time to construct these devices may partly be accounted for by the fact that the Y-plant at Valindaba only began to produce 80 per cent enriched material in 1978. In 1979, a massive explosion and release of gas set back operations for two years.[92] According to Waldo Stumpf, a senior official at the Atomic Energy Corporation, only a very small proportion of those involved in nuclear research – perhaps 250 at any one time and 1,000 in total – were ever directly drawn into weapons production.[93] Much of the bomb-making activity was centred round Pretoria, where the university, ARMSCOR (and its affiliate companies), the CSIR and government departments were grouped. The circulation of individuals between these various agencies, and careful selection of personnel based on tacit knowledge and political affiliations, kept the circle tight and mutually sustaining.

Secret knowledge and military codes meant that those in the know were a privileged elite. This has important implications for understanding the hierarchies of knowledge and power in Botha's late-apartheid securocrat-dominated presidency. Growing isolation fed a powerful strain of can-do, survivalist nationalism that affected scientists, politicians and the wider public alike. Yet the drift towards weaponisation was incremental, with the pace forced only from 1978. It reflected shifting power-politics within the apartheid deep state. The role of bluff and counter-bluff, disclosure and non-disclosure, was a major factor in diplomatic and strategic positioning, and the nuclear threat had to be backed up by credible capacity.

As in the case of Israel, rumours, disclosures, and official denials of nuclear capability were important aspects of high-level bargaining and negotiation. The apartheid bomb was seen as a valuable bargaining chip by Botha's securocrats, and as a weapon of last resort over which

91 Von Wielligh and Von Wielligh-Steyn, *Die Bom*, 169, 173; 'PW se Bom', *Rapport*, 26 July 2014.

92 Stumpf, 'South Africa's Nuclear Weapons Program', 3–4.

93 Stumpf, 'South Africa's Nuclear Weapons Program', 4; Von Wielligh and Von Wielligh-Steyn, *Die Bom*, 121.

a select few were able to exercise authority. In the 1970s and 1980s, conflicts grew within the ruling elite over the future of apartheid. Control of nuclear weapons production helped Botha's military-based faction to retain an advantage over reformers who advocated political solutions.

The secrecy surrounding Pelindaba and Valindaba fed the perception that South Africa could take care of itself. The very names given to these facilities contained a implicit threat, which was all the more insidious because of the use of isiZulu to suggest a warlike state of readiness. Use of locally developed technology to enhance uranium emphasised self-reliance. So too, did the decision to construct high-pressure water reactors at Koeberg – the only nuclear power-generating plant in Africa, based on French technology. The production of low-enriched uranium at Pelindaba's Z-plant to power Koeberg was itself a major technological achievement, quite distinct from the Valindaba enrichment process. Taken together, these initiatives formed part of a greater claim to national technological and scientific prowess: its ideological valency was as important as, if not more important than, the strategic utility of possessing a bomb.

SASOL and Oil Supply

The creation of a petrochemical company dedicated to transforming coal into oil preceded the pursuit of a nuclear industry. Here, too, technology initially borrowed from Germany and the United States was adapted to local conditions. Just as low-cost electricity rendered energy-hungry jet nozzles viable for uranium enrichment, so abundant supplies of coal, mined relatively cheaply by black workers, made oil-from-coal technology viable over the long term. The mantra of self-sufficiency prompted by the international oil embargo of the 1970s gave SASOL an important strategic reason for existence. It also allowed for the diversion of substantial government funding to subsidise investment and operating costs. Stephen Sparks's account of SASOL demonstrates how important this prestige project was to the Afrikaner state and scientists.[94]

[94] S. J. Sparks, 'Apartheid Modern: South Africa's Oil from Coal Project and the History of a Company Town', DPhil thesis, University of Michigan (2012).

Van Eck and managing director Etienne Rousseau decided that the best way forward was to combine the process developed in Nazi Germany (which had also been starved of oil supplies) together with a new American process pioneered by the M. W. Kellogg Corporation. This technology required rapid local innovation by SASOL to become operational in 1954 and to produce petrol a year later. Key funding was secured in secret talks with the World Bank which, in a striking mirror of the Marxist critique of apartheid as a system of cheap labour exploitation, was assured that oil-from-coal production was economically viable because of the exceptionally low costs of black labour in South Africa. Similar arguments about 'South African conditions' were advanced to persuade sceptical overseas investors that uranium enrichment was cost-effective.[95]

Rousseau was a resolute Afrikaner nationalist and, in line with official government policy, he positioned the venture in terms of a strategic alliance with the West against communism. Yet he saw himself as an entrepreneur rather than a state official and did not disdain working with private domestic industries such as the chemicals giant AECI. He prioritised skills over Afrikanerisation and fully understood that oil-from-coal production involved 'artificial economics' with heavy state support.[96] It was only in the era of growing isolation after Sharpeville that SASOL became fully emblematic of South African autarky and resilience, in ways that recall the ideological appeal of oil-from-coal production in fascist Germany and Spain.

Early technical problems with the Kellogg synthesis unit prompted a search for innovative local solutions and by the mid-1960s it was claimed that 'Sasol Leads the World in Catalyst Research'.[97] SASOL took pride in the technical abilities demonstrated by South African scientists and engineers, their ingenuity, and their willingness to work long hours in order to solve problems.[98] The dusty Free State veld was tamed by industry and transformed into an orderly company town with attractive, landscaped precincts and curved roads. Max

[95] Sparks, 'Apartheid Modern', 45; A. P. Steynberg 'Chapter 1: Introduction to Fischer-Tropsch Technology', *Studies in Surface Science and Catalysis*, 152 (2004), 163, 356; Hecht, *Being Nuclear*, 93.

[96] Sparks, 'Apartheid Modern', 427; Sparks, 'Between "Artificial Economics" and the "Discipline of the Market"', 712–17, 721–2.

[97] Sparks, 'Apartheid Modern', 72.

[98] *Sasol, 50 Years of Innovation* (SASOL, [2000?]), 42–5.

Kirchhofer, a Swiss-born, Johannesburg-based town planner, was selected in 1951 to design this model garden city; the arborealist Aart Jurriaanse selected and grew the trees. As the company town developed, it was allocated its own bespoke car licence plate: 'OIL'. Street names honoured men like van Eck, Havenga and Eric Louw along with German scientists Fischer and Tropsch; the Etienne Rousseau Theatre opened in 1975.

Key to Sasolburg's self-image was the idea that the town was not merely a preserve of the scientific and engineering elite: it was defined by the community of 'Sasolites', ordinary white employees who joined the new, expanding industrial town.[99] Black employees, many of whom continued to live on surrounding farms, were not encompassed in this model community. Black clerical workers, housed in a neighbouring township known as Zamdela, were to some extent incorporated into the paternalist environment of the company town, but they nonetheless glossed the name of the township as meaning 'we have given up hope'.[100] They commuted to Sasolburg along a road named after apartheid's principal defender at the United Nations, Eric Louw.

SASOL registered a small profit for the first time in 1960, but in commercial terms it was the combination of cheap coal and rising fuel costs after the 1973 oil crisis, compounded by the 1979 Iranian revolution (which brought a temporary end to crude oil importation from South Africa's largest supplier), that gave SASOL a real chance of success. Two giant oil-from-coal plants were built in 1976 at Secunda on the coalfields of Mpumalanga. Technical achievements and improvements now positioned SASOL to license its own technology to overseas purchasers. As well as motor fuel, the petrochemical process included by-products such as fertilisers, a full range of hydrocarbons, ethylene for plastics, waxes and solvents. After 1979, SASOL was listed on the Johannesburg Stock Exchange and so became an early experiment in part-privatisation.[101]

Extensive advertising was used to distinguish SASOL petrol, marketed as 'Trek' – supposedly a premier product and a patriotic

[99] Sparks, 'Apartheid Modern', 64ff; chapter 3; Steynberg 'Introduction to Fischer-Tropsch Technology', 36–7.

[100] Sparks, 'Apartheid Modern', chapter 5; *Sasol, 50 Years of Innovation*, 45. Thanks to Stephen Sparks for clarification on this point.

[101] Andrews et al., *South Africa in the Sixties*, 125–6; Steynberg 'Introduction to Fischer-Tropsch Technology', 38–9.

consumer choice. Its logo featured a zebra (or quagga) in full flight. The SASOL plant came to feature on the obverse side of the two-rand note, and the 25th anniversary of SASOL in 1975 was marked by a commemorative stamp.[102] This imagery was not lost on the liberation movement: in June 1980 a special unit of Umkhonto we Sizwe attacked the Sasolburg refinery as part of its campaign of armed propaganda. The symbolic value of South Africa's growing self-sufficiency in energy also helps to explain why MK chose to attack the Koeberg nuclear facility in 1982, effectively delaying its completion by 18 months. Rodney Wilkinson, who laid the four limpet mines, stole plans while working at the Koeberg plant and smuggled them to the ANC abroad.

Techno-Nationalism

In a discussion about the future of university scientific research, held at the University of Pretoria just months after the Sharpeville massacre in 1960, one of the participants observed: 'The power of a nation (volk) has always depended on the level of technological progress in its military, agricultural, and industrial spheres.'[103] Far from old stereotypes about Afrikanerdom being backward-looking and resistant to modernity, technological advancement was encouraged by the state and became central to its self-image.

Apartheid was heavily reliant on technology to control the body politic. Keith Breckenridge argues that South Africa became a 'biometric state' over the course of the twentieth century.[104] He traces this process back to the beginning of the century when Galtonian-influenced programmes to fingerprint African workers were devised and implemented. The Population Registration Act of 1950, a foundational piece of apartheid legislation, underpinned race classification and social engineering. During the 1950s and 1960s, registration proceeded through laborious paperwork to satisfy the objective of categorising all South Africans within one of four racial types. But, from the late 1960s, a new, phase in the centralisation of population

[102] Sparks, 'Apartheid Modern', 80–1.

[103] H. L. de Waal, 'Natuurwetenskaplike navorsing aans ons Universiteite – Waarheen?', 'Die Suid-Afrikaanse Akademie vir Wetenskap en Kuns. Simposium gehou in the Universiteit van Pretoria, 20 October 1960' Gubbins Collection, Wits University, pam.4950 (emphasis in original).

[104] Breckenridge, *Biometric State*.

registration was inaugurated: every white, Coloured and Indian person was issued with a 'Book of Life' intended to record not just race and identity, but extensive personal details and certificates. In quest of 'unrestrained panopticism', the state required that Africans were issued with, and had to carry, separate documents.

In pursuit of this plan, the grand demographic operation moved in 1972 from the modest offices of the Census Bureau to a thirty-one-storey glass-clad modernist building in the heart of Pretoria, known as Civitas. This prominent external display of architectural power was complemented by the powers of surveillance implicit in access to the centralised digital records of South African citizens. A large staff of clerks and data processors inputted information on new IBM computers in the basement. This initiative came to be seen as a measure of how eager the South African state was to demonstrate that it was at the forefront of modern technology, specifically by embracing 'the world of the computer', as the minister of the interior put it.[105] At the CSIR, which installed its first digital computer, code-named ZEBRA, in 1958, computing was also becoming a focus of interest. The organisation sought to take the lead as the national centre of computing, and from 1967 an IBM 360 mainframe platform was installed and regularly updated.

Afrikaner technocrats also used the state to modernise infrastructure. The rolling out of a network of national roads, increased telecommunications capacity, electrification of white-owned farming districts by means of high-voltage power lines, and the erection of large grain silos towering over agricultural hamlets, were all visible examples of state-led modernisation. Large-scale dams both for irrigation and electricity were an ambitious element in planning.

Elsewhere in Africa, grand trophy projects such as the Aswan High Dam and the Akosombo hydroelectric scheme on the Volta River in Ghana became defining monuments to late-colonial and newly independent statecraft. In southern Africa, the Kariba Dam, constructed on the Zambezi River between 1955 and 1959, emerged from the confluence of late-colonial and settler nationalist power during the failed attempt to bring both territories together as the Central African

[105] K. Breckenridge, 'The Book of Life: The South African Population Register and the Invention of Racial Descent, 1950–1980', *Kronos*, 40 (November 2014), 234, 235.

Federation. As Julia Tischler demonstrates, this enormous scheme was a 'twin project of nation-building and modernisation' whose construction brought to the fore asymmetries of race and power affecting those who built the dam, those who benefited from it, and those who were displaced by it.[106] While conceived in part as a transnational invitation to share hydroelectric power, Kariba was also realised as a show of political power.

Further down the Zambezi, late-colonial technical developmentalism was also on display in the Cahora Bassa hydroelectric scheme.[107] As with Kariba (whose construction prompted the Portuguese decision to build their own dam), patriotic colonial pride was widely in evidence. The Salazar regime portrayed Cahora Bassa as a tangible statement of Portuguese determination to remain in Africa. Intense anti-colonial armed resistance to Portuguese rule in Mozambique meant that the dam was also a form of strategic defence: its precise location was influenced by efforts to constrain the movement of guerrillas across the Zambezi River. Given the relative weakness of Portuguese economic and political power, the dam was heavily reliant on South African corporate start-up capital and demand for hydroelectricity: in 1969, after four years of discussions, an agreement was signed between Lisbon, Pretoria, ESCOM, and a consortium of fifteen companies, under which electricity would be sold to South Africa at a low price in exchange for military support. By 1976 a high-voltage transmission line was able to deliver power over 1,400 km to a converter station near Pretoria. The technopolitics of the 'deterritorialised' Cahora Bassa scheme was therefore closely bound up with apartheid South Africa's determination to remain the regional hegemon.

Both the Cahora Bassa and the Kariba dams were international enterprises to the extent that they relied on overseas funding and technical expertise. This was also true of the Orange River Project in South Africa, which was developed by a French–South African

[106] J. Tischler, *Light and Power for a Multiracial Nation: The Kariba Dam Scheme in the Central African Federation* (Basingstoke: Palgrave Macmillan, 2013), 12; J. McGregor, *Crossing the Zambezi: The Politics of Landscape on a Central African Frontier* (Oxford: James Currey, 2009).

[107] K. Middlemas, *Cabora Bassa: Engineering and Politics in Southern Africa* (London: Weidenfeld and Nicholson, 1975); A F. Isaacman and B. S. Isaacman, *Dams, Displacement, and the Delusion of Development: Cahora Bassa and Its Legacies in Mozambique, 1965–2007* (Athens: Ohio University Press, 2014).

consortium. Yet, in the case of this prestige scheme, funding and technical knowledge were under direct state control, and the water and power generated were entirely geared to national requirements. The Irrigation Department was responsible for all aspects of the planning, construction and operation of the Verwoerd (now Gariep) and P. K. le Roux (now Vanderkloof) dams, which formed the core of the Orange River Project.[108]

Figure 6.2 Gariep (formerly Verwoerd) Dam, 1970s.
Credit: Eskom

Since the foundation of the Cape's Department of Irrigation in the early twentieth century, expansive hydraulic schemes had gradually been realised for irrigation and urban water supply. For nearly a century these had included ideas to canalise the Orange River for the benefit of the eastern Cape.[109] In 1929, following a period of intense drought, the government director of irrigation, Alfred Dale Lewis, cautiously considered a scheme to divert the Orange River into the

[108] W. Alexander, 'The Orange River Project', Part 3, *Civil Engineering*, 13, 8 (August 2005), 8–11.

[109] A. R. Turton et al., *A Hydropolitical History of South Africa's River Basins*, WRC Report No. 1220/1/04, 2004, 168–70, 119ff, 168–70, 396. www.wrc.org.za

Sundays River. But this was ruled out by the expense involved in satisfying the irrigation requirements of farmers in a relatively marginal area of the country. Apartheid strategic and ideological considerations tipped the balance and the project was duly reconfigured to suit several purposes: irrigation in the northern as well as the eastern Cape; urban and industrial water supply; and hydroelectric power. These were vital requirements for the future in a rapidly urbanising society. The Orange River Project also gave national morale a boost, emphasising Afrikaner power, with Verwoerd's full backing, and saying it 'with concrete'.[110]

Work began in 1963 and three years later Vorster, the new prime minister, triggered explosions at Oranjekrag to start the work on the huge curvilinear concrete wall of the Hendrik Verwoerd Dam, which was sealed in 1970 and began to generate power a year later. A second hydroelectric dam 130 km downstream at Vanderkloof followed in 1977. The scheme involved diverting water through 82 km of concrete tunnels – then among the longest concrete-lined arteries in the world – to the Great Fish and Sundays rivers in the eastern Cape.[111] One of the external consultants was Alan Muir Wood, who was involved in major tunnelling projects in England, including the Heathrow Airport Cargo Tunnel and the Channel Tunnel. He introduced the latest civil engineering techniques, again demonstrating the receptivity of the South African state to innovation.

The flow of engineering talent was not one-way only. Jack Zunz, who trained in civil engineering at Wits, left the country for London in 1950 and found work with the Ove Arup engineering company in London. Zunz returned to Johannesburg on behalf of Arup to build the iconic Albert Hertzog (now Sentech) telecommunications tower in the suburb of Brixton (though his disapproval of apartheid caused him to leave the country again in 1961). He played a significant role in solving technical problems with the folding plates of the Sydney Opera House, and he successfully brought this project to fruition after the Danish architect Utzon withdrew in 1966. Zunz ended his career as head of Ove Arup.[112]

110 Christie, *Electricity, Industry and Class*, 167–8.
111 W. Alexander, 'The Orange River Project', Parts 1–3, *Civil Engineering*, 13, 6 (2005), 4–7; 13, 7 (2005), 4–5; 13, 8 (2005), 8–11; H. Olivier, *Great Dams in Southern Africa* (Cape Town and New York: Purnell, 1977), chapter 14.
112 Obituary, Sir Zack Zunz by Elain Harwood, *The Guardian*, 2 January 2019.

Conceived and delivered in the era of high apartheid, the Orange River Project was a dramatic illustration of the confluence of social and water engineering led by a state with a strong sense of purpose and the means to achieve its objectives. In tune with apartheid priorities, these irrigation works primarily benefited white landholders, and it provided the infrastructure for a sophisticated irrigated grape industry along the Gariep river. But the state also invested in dams and irrigation projects in the homeland areas. As in the case of rehabilitation schemes and decentralised industries, Bantustan policy aimed to create economically viable African statelets that might reverse the pattern of urbanisation to 'white' South Africa and justify political balkanisation. During the apartheid era, the extent of land under irrigation in the homelands increased from less than 10,000 ha to about 50,000 ha.

Wildlife, National Parks and Scientific Management

South Africans were also significant innovators in the field of wildlife management, based on an increasingly scientific approach, during the apartheid era. As in other spheres, Afrikaners became more dominant at the national level in the National Parks Board. But wildlife conservation was a devolved project, with provincial agencies, especially the Natal Parks Board, and private landholders playing important roles. Jane Carruthers argues that the National Parks Act of 1926 was promoted primarily by anglophone South Africans, imbued with growing empire-wide ideas of preservation.[113] Kruger's name was used for the flagship Kruger National Park because the pre-Union Republic had first reserved land for wildlife in the area. At a time of growing Afrikaner nationalism in the 1920s, this project was seen as a means to unite white South Africans in a common cause. Under its first warden, James Stevenson-Hamilton, preservation was the priority, and tourism gradually expanded from the 1930s. Other new national parks initially focused on saving particular species: Addo for the small surviving herd of elephants in the Eastern Cape; and the Kalahari Gemsbok, Bontebok and Mountain Zebra parks, named after these rare species.[114]

[113] J. Carruthers, *The Kruger National Park: A Social and Political History* (Pietermaritzburg: University of Natal Press, 1995).

[114] J. Skotnes-Brown, 'Domestication, Degeneration, and the Establishment of the Addo National Elephant Park in South Africa', *The Historical Journal* (online, 2020).

As in the case of soil conservation, these initiatives were heralded by government propaganda and pamphlets. The South African Railways played a major role in promoting popular tours and visual material. Photographic books, novels, tourist guides, and children's stories all contributed to the popular promotion of zoological knowledge and the parks.[115] Internal tourism was underpinned by the growth of private car ownership: by 1960, white South Africans probably ranked fourth globally in per capita vehicle ownership.[116] Visitor numbers to Kruger grew to about 100,000 in 1960, 400,000 in 1980 and nearly a million by 1998.[117]

Kruger authorities initially allowed some African people to continue living in the area demarcated for the park but removed them by the 1950s.[118] As many critiques of 'fortress conservation' now argue, human settlement had helped to shape the natural world that the parks inherited.[119] National parks in Europe generally allowed some restricted human settlement as part of park management. However, South African scientific conceptions in the post-war period prioritised the restoration of nature without human inhabitants and also involved the reintroduction of predators.

Stevenson-Hamilton's legacy was challenged in a government commission chaired by P. W. Hoek in 1952, which heard extensive evidence of disorganisation in Kruger.[120] This led to reforms in management, accompanied by increasing Afrikaner dominance in the parks board as a whole. The leading Transvaal nationalist politician and future prime minister, J. G. Strijdom, installed F. H. Odendaal, a lawyer and close business partner with impeccable National Party connections (but with no experience in wildlife management) as chair of the National Parks Board. He was joined a year later by Rocco

115 E. Teversham, 'Representations and Perceptions of the Kruger National Park and the Manyeleti Game Reserve, 1926–2010', DPhil thesis, University of Oxford (2014).

116 Beinart, *Twentieth-Century South Africa*, 182.

117 S. Freitag-Ronaldson and L. C. Foxcroft, 'Anthropogenic Influences at the Ecosystem Level' in J. T. du Toit, K. H. Rogers and H. C. Biggs (eds.), *The Kruger Experience* (Washington DC: Island Press, 2003), 402.

118 D. Bunn, 'An Unnatural State: Tourism, Water and Wildlife Photography in the Early Kruger National Park' in W. Beinart and J. McGregor (eds.), *Social History and African Environments* (Oxford: James Currey, 2003), 199–218.

119 The term is from Dan Brockington, *Fortress Conservation: The Preservation of the Mkomazi Game Reserve, Tanzania* (Oxford: James Currey, 2002).

120 Carruthers, *National Park Science*, 51–6, 126–31, 141.

Knobel, a social worker and committed Dutch Reformed Church elder, as director of National Parks.

A. M. Brynard was the first scientist to be appointed as director of Kruger in 1960, followed in 1970 by U. de V. Pienaar, who oversaw a rapid expansion in visitor numbers, facilities and research. The mandate of the national parks was explicitly modified to expand from the idea of 'protect and preserve' in order to include scientific work.[121] From the 1950s, strategies for preservation focused less on specific animal species and more on typical habitats or biomes.[122] Scientists, influenced by ecological thinking, were starting to view environments through a more complex, inclusive lens rather than concentrating on endangered mammals. Both national and provincial authorities used scientific ecological mapping at a national scale to ensure that samples of the most distinctive environments were reserved. Ideas about ecology and biodiversity (a term coined in the 1980s) increasingly underpinned the justification for expanding protected land.

A new dynamic also developed around provincial parks, and a cadre of professional conservation officers emerged to staff them. Douglas Hey trained as a zoologist and in the 1940s worked in trout hatcheries in Jonkershoek near Stellenbosch, Cape.[123] This in turn alerted him to the health of rivers, many dammed for agriculture and urban water supplies and often polluted. Hey helped to promote legislation that conserved riverine environments. He was also involved in experiments to restock South African rivers with fish to control mosquitoes, vectors of malaria, and water snails, vectors of bilharzia. In 1952 the Cape Provincial Council set up a provincial Department of Nature Conservation with the express task of enforcing new controls on private farms and extending reserved areas. Hey became the director, and the province started to purchase farms systematically, focusing initially on a large coastal zone in the Overberg called De Hoop. Watershed and forestry reserves were added to the portfolio. By the end of the twentieth century, the Western Cape province had reserved

121 Carruthers, *National Park Science*, 165

122 J. Carruthers, 'Conservation and Wildlife Management in South African National Parks 1930s–1960s', *Journal of the History of Biology*, 41, 2 (2008), 203–36.

123 D. Hey, *A Nature Conservationist Looks Back* (Cape Town: Cape Nature Conservation, 1995).

around 700,000 ha of land, considerably more than the national parks in the province.

In Natal, the Hluhluwe and Umfolozi reserves (now Hluhlwe-iMfolozi), initially proclaimed in 1895, protected most of the surviving white rhino. Anglophone Natal conservationists fought to maintain provincial control, and their authority was institutionalised in a separate Natal Parks, Game and Fish Preservation Board (NPB) in 1947. Under Jack Vincent, a British ornithologist with field experience in many parts of Africa, the NPB developed a committed cadre of game rangers and became perhaps the most innovative of the provincial wildlife conservation bodies. Rhinos continued to be a flagship project, but officers also tried to develop control over rare plants such as cycads, and expanded their authority over the Drakensberg (Kahlamba) watershed.

A key scientific advance in the late 1950s, animal immobilisation, facilitated new strategies of management, including translocation and veterinary treatment of wild animals. Toni Harthoorn, a British-trained vet of Dutch origins who worked at the new University of East Africa in Kampala, pioneered the adoption of an American dart syringe.[124] During his training, he had worked with animal anaesthesia for livestock and he experimented in Uganda on African buffalo. While his main duties involved training a new generation of African vets specialising in livestock, he became increasingly committed to the growing ethos of wildlife conservation. Harthoorn later wrote: 'with ever increasing certainty I saw in the great wild herds of Africa the last chance for mankind to redeem itself ... Above all, we needed knowledge of the ecology of Africa. This then became my consuming interest.'[125]

Elephants were a particular danger for people and crops at the interface of parks and settlements in the unfenced reserves of East Africa. Harthoorn was opposed to culling, which seemed to undermine the argument for protection: translocation seemed a far more humane option. He modified the American rifles so they became more powerful

124 A. M. Harthoorn, *The Flying Syringe: Ten Years of Immobilising Wild Animals in Africa* (London: Geoffrey Bles, 1970); A. M. Harthoorn, *The Chemical Capture of Animals: A Guide to the Chemical Restraint of Wild and Captive Animals* (London: Bailliere Tindall, 1976); Ian Player, *The White Rhino Saga* (London: Collins, 1972).

125 Harthoorn, *The Flying Syringe*, 16.

and switched to a stronger drug called Flaxedil, which also had the advantage of an antidote.[126] In 1959 he was asked to assist in removing rhino from islands created in the Zambezi by the Kariba Dam. Rhinos were immobilised in the Operation Noah rescue operation, which, he wrote, 'was an outstanding success and a blueprint for other similar operations'.[127] It was filmed and gained wide publicity in the Western world because of the care and expense taken over removing animals. More recently, critics have contrasted this attention on animals with the callous neglect of displaced people.[128]

While white rhinos had been shot out of most of their ecological range in Africa, numbers in Umfolozi had grown by 1959 to 567 with 86 calves, and the NPB believed that they were putting pressure on available habitat.[129] Ian Player became senior ranger in 1958 with immediate responsibility for managing this problem; Harthoorn was called in to experiment with immobilisation.[130] The first three rhinos darted in 1960 died, and they had difficulty in getting close enough to shoot the syringe effectively. Player turned to Maqubu Ntombela, who had worked in the Zululand parks since the 1920s, combining his stalking skills with the latest equipment. A dedicated local team refined techniques and Harthoorn later introduced a new morphine-based drug.

After initial failures, this new Operation Rhino became 'streamlined and highly efficient'.[131] By 1962 they had moved fifty-five animals, mostly to Kruger, where the last white rhino had been shot in 1896. Further innovations enabled them to move over 1,000 in the 1960s to many other southern and eastern African parks, as well as private game reserves such as Mala Mala and zoos worldwide. The white rhino saga was profitable for the NPB, with favourable international media coverage, including a feature film, *Rhino!* (1963). Harthoorn married a vet, Sue Hart, who had begun to develop a wildlife practice near Kruger, and together they became major conservationist figures in Kenya and South Africa as well as the models for the lead actors in the long-running American TV series *Daktari*.[132]

126 Harthoorn, *The Flying Syringe*, 49.
127 Harthoorn, *The Flying Syringe*, 91.
128 Tischler, *Light and Power*.
129 Harthoorn, *The Flying Syringe*, 95.
130 Player, *The White Rhino Saga*; G. Linscott, *Into the River of Life: A Biography of Ian Player* (Johannesburg: Jonathan Ball, 2013).
131 Harthoorn, *The Flying Syringe*, 108.
132 W. Beinart, 'The Renaturing of African Animals: Film and Literature in the 1950s and 1960s' *Kronos*, 27 (November 2001), 201–26.

Player was not specifically engaged in scientific work, though his team kept good records.[133] He became a powerful proponent of the idea of wilderness and the need to connect conservation to an African and spiritual ethos.[134] With Maqubu Ntombela, he started long-distance walks through the Zululand reserves in order to immerse his clientele in the experience of nature. He also took the unusual step of promoting local knowledge, realising that support from local people and from the KwaZulu homeland government was essential for the long-term success of this enterprise. Ntombela was given an opportunity to narrate and guide tours; Player wrote this up in a column in the *Natal Mercury* and later in a book entitled *Zulu Wilderness*.[135] He also formed an association with Mangosuthu Buthelezi, chief minister of KwaZulu, who was committed to expanding protected areas and saw environmental knowledge as part of a broader Zulu heritage.

Player conceptualised Ntombela's knowledge around the idea of reading the landscape – a metaphor that Anders Sparrman had articulated two centuries before in the 1770s (see Chapter 1). A central element involved the ability to read signs on the ground. He illustrated Magqubu's understanding, as well as the theatre of his presentation, by means of dramatic narrative reconstructions of the behaviour of predators and their kills.[136] Through wilderness walks, schools and conferences, these pluralistic forms of conservation knowledge gained global exposure.

The managers of Kruger carefully enumerated species and believed that elephants, once so rare, were multiplying from an estimated 1,186 in the first survey in 1960 to 6,585 in a 1967 helicopter count.[137] Elephants were by then so numerous that there was limited scope for translocating large numbers: twenty-seven were sold to a dealer who

133 H. Wels, *Securing Wilderness Landscapes in South Africa: Nick Steele, Private Wildlife Conservancies and Saving Rhinos* (Leiden: Brill, 2015).

134 M. Draper, 'Zen and the Art of Garden Province Maintenance: The Soft Intimacy of Hard Men in the Wilderness of KwaZulu-Natal, South Africa, 1952–1997', *JSAS*, 24, 4 (1998), 801–28.

135 I. Player, *Zululand Wilderness: Shadow and Soul* (Cape Town: David Philip, 1997). In this text the spelling is Magqubu.

136 Linscott, *Into the River of Life*, 263–4.

137 I. J. Whyte, R. van Aarde and S. L. Pimm, 'Kruger's Elephant Population: Its Size and Consequences for Ecosystem Heterogeneity' in Du Toit, Rogers and Biggs (eds.), *The Kruger Experience*, 332–48.

transported them to California.[138] The issue of elephant culling was intensely debated in eastern and southern African in the 1960s. On the advice of a British scientist, elephants were culled in Uganda, but in Kenya the first African parks director, Perez Olindo, decided against because of adverse publicity and the threat to the idea of national parks as a haven for animals.[139] David Sheldrick, the influential warden of Tsavo, believed that the exclusion of people and livestock from the park had allowed vegetation to thicken and that elephants were needed to control it. Opponents of culling argued that variable rainfall, fire and drought affected vegetation in unpredictable ways and that culling did not adequately allow for this. Such views were often articulated in later decades as a critique of equilibrium theories in ecology. But in Kenya, drought losses of elephants were significant and in the longer term, Tsavo's elephant population declined – largely due to poaching – from an estimated 35,000 in 1970 to about 6,000 in 1990.

In Kruger, by contrast, Brynard was committed to scientific ideas of carrying capacity. Botanical surveys revealed significant damage to Kruger trees and aloes, especially valuable species such as the marula, which seemed to coincide with the sharp rise in elephant numbers in the 1960s. Brynard aimed to keep elephant numbers at around 7,000.[140] Systematic culling started in 1967 and was sustained until 1994, by which time 14,629 elephants had been killed and 1,313 translocated. The policy was further justified by more rigid boundaries and successful policing after the completion of fencing and it was extended to other species. Critics pointed to cruelty, adverse publicity, and ecological unpredictability, but culling was quietly maintained with the visual record minimised. Elephant meat was made available largely to black employees and tusks could be sold until 1989 when CITES banned trade in ivory. Policies were reinforced by consensus within the circle of Afrikaner scientists, based around Kruger and the University of Pretoria, where a course in wildlife management was introduced in 1965. Pienaar, warden till 1987, argued that for the sake of the animals, the vegetation and tourism, Kruger could not have starving elephants. Culling on this scale successfully held elephant numbers stable, with a peak of 8,768 in 1983.

[138] Carruthers, *National Park Science*, 303.
[139] J. Schauer, 'The Elephant Problem: Science, Bureaucracy, and Kenya's National Parks, 1955 to 1975', *African Studies Review*, 15, 1 (2015), 177–98.
[140] Whyte, van Aarde and Pimm, 'Kruger's Elephant Population', 338–9.

From the 1980s, however, ideas about ecological history and management shifted towards disequilibrium theory. An increasing number of scientists questioned the necessity for such intensive management and also the provision of so many additional water sources. They were supported by a growing lobby for animal rights and changing global policies by CITES, which removed any income from culling. In 1991, as the political transition took shape, Robbie Robinson, an English-speaking warden, originally a marine biologist, was appointed and convened scientific conferences and public consultations, in keeping with the wider spirit of debate and inclusion.[141] Elephant culling was suspended in 1994 and, in this sense, elephants too were beneficiaries of the transition from apartheid – or at least they escaped the muscular managerial science that underpinned tight control of their numbers. It was accepted that they should find their own level and by 2015 Kruger housed 17,000 elephants.[142] These increases were, however, to some degree catered for by expansion of the range through the opening of fences on the western boundary to include adjacent private game reserves, and also a loosening of park boundaries with protected areas in Mozambique.

Kruger remained racially segregated till the 1980s, though as Jacob Dlamini shows, some black South Africans did in fact manage to circumvent the whites-only signage and enjoy the main Park over many years. Apartheid planners provided a small adjacent area, called Manyeleti, where black South Africans could visit for game viewing and stay overnight.[143] In the hubristic age of high-apartheid, this involved the removal of an African community. In 1985 Manyeleti was incorporated into the homeland of Gazankulu whose leader, Hudson Ntsanwisi, subscribed to the conservationist project. In the post-apartheid era the area was subject to a successful land claim under the Restitution Act of 1994. It is now being run as an elite private wildlife reserve under a co-management agreement with the displaced community.

141 Carruthers, *National Park Science*, 402.

142 S. M. Ferreira, C. Greaver and C. Simms, 'Elephant Population Growth in Kruger National Park, South Africa, under a Landscape Management Approach', *Koedoe*, 59, 1 (2017).

143 Teversham, 'Representations and Perceptions'. Cf. Dlamini, *Safari Nation*, 142–51, 155–62, 202–5

South African conservation strategies in the apartheid era did not always gain international approval in that they were prototypical examples of fortress conservation, with military approaches to protection and relatively little space for community involvement. Culling in national parks, and the breeding of wild animals on private land for hunting rather than conservation, also attracted criticism. The overall outcome was a rapid rise in the number of wildlife, including some endangered species, and an increasingly strong commitment to the protection of unique and highly diverse habitats, which few other countries could rival. In this case, an authoritarian state demonstrated a capacity for some important advances in scientific conservation and left a valuable, if still contested, heritage for post-apartheid society. Wildlife remained significant in the national imagination.

Apartheid Science and Science Under Apartheid

The apartheid government demonstrated considerable commitment to science and technology. It pursued ambitious, even hubristic projects, reimagining South Africa not only in the light of Afrikaner priorities, but also as a powerful modern nation resisting isolation and international moral condemnation. The first decade after 1948 was one of transition as Afrikaner scientists were inserted into many key positions. However, the state also attempted with some success to maintain international contacts, including technical and agricultural programmes in Africa, as well as high-profile Antarctic venturing. Sharpeville in 1960 was a turning point, and the period of high apartheid that followed saw state-led expansion in science and technology. Vigorous economic growth until the mid-1970s helped to provide the necessary resources for investment in the military, nuclear and energy industries.

In earlier chapters we distinguished between colonial science and science in the colonised world. The material presented here suggests that a distinction might also be drawn between *apartheid* science and science *under apartheid*: the former refers to the role of state-sponsored science in support of apartheid; the latter concerns scientific endeavour that took place in the times of apartheid but which was only indirectly subject to state priorities. There were close overlaps, particularly where patriotic nationalism featured, but because apartheid was never fully a totalitarian system of government, the state did not command all scientific activities. The nuclear weapons programme, for example, was clearly directly related to the apartheid state, but it was not an automatic outcome of the prior theoretical work on which it depended, nor the civilian applications that proceeded in parallel. The Natal Parks Board conservation initiatives, while dependent on exclusive and racialised reservation of land, retained a degree of autonomy from the central state. Research into energy storage in batteries was spurred by the 1973 oil crisis and the OPEC embargo on oil exports, but it was primarily intended for civilian and commercial use. A distinction between apartheid science and science under apartheid is therefore valuable in understanding apartheid as a system and scientific initiatives in this era.

Two further illustrative examples help to cast light on this distinction. The world's first human heart transplant was conducted by Dr Christiaan Barnard of the University of Cape Town at Groote Schuur Hospital in 1967. This stunning achievement brought unprecedented international attention and was used by the government to showcase the country's capacities. The circumstances in which the operation took place were conditioned by the climate of apartheid, not least the existence of first-world medicine in a relatively isolated country, where scientific daring was less constrained by the ethical constraints experienced by rivals in the United States and United Kingdom. But the heart transplant highlighted some basic contradictions of apartheid, not least in respect of cross-racial organ donors and recipients.[144]

While Chris Barnard might have been a poster boy for apartheid, he did not quite fit. He was an Afrikaner from a modest rural background in the Karoo, whose father, Adam, a dominee in the Dutch Reformed Church, preached to the local Coloured community and was paid far less than Dominee Rabie of the white *Boerekerk*.[145] The member of parliament for Beaufort West in Barnard's youth was the apartheid diplomat Eric Louw. For many years Louw and Barnard were honoured in separate wings of the local museum, but the Barnard and the Louw children shared little in common, socially or politically. Although they might be accepted into the Louw household on

[144] S. Horwitz, 'Race, Kidney Transplants, Immunosuppression Research, and White Supremacy under Apartheid, 1960–80', *Canadian Bulletin of Medical History*, 37, 2 (2020), 461–89.

[145] C. Barnard and C. B. Pepper, *One Life* (New York: Bantam 1971), 12–14.

occasion, Barnard felt 'set apart in many ways which could not be hidden'.[146]

The first transplant instantly transformed Barnard into an international hero and soon afterwards he became a celebrity with a playboy reputation. As well as having decidedly unpuritanical tastes, Barnard was suspected of not being a government supporter. The team at Groote Schuur Hospital included his brother, Marius, who later joined the liberal political opposition, as well as English-speaking doctors who were frankly opposed to apartheid. The University of Cape Town, whose medical faculty members worked at Groote Schuur, was generally under suspicion by the government for being anglophile and liberal.

To outward appearances heart transplantation reflected well on the country. However, the racial classification of the donors and recipients raised qualms within race-obsessed South Africa about the implications of bodily borrowings. (This was a time when blood donors and transfusions as well as ambulances were rigidly segregated.) The first two transplant patients (Louis Washkansky and Philip Blaiberg) were Jewish and classified as white. Barnard is said to have enquired whether Denise Darvall, the first donor, was classified Coloured when he was informed that her heart might be suitable for Washkansky. Impatient to go ahead, Barnard would happily have used the heart of a black patient suffering from cardiomyopathy, but Val Schrire, the hospital's chief cardiologist, ruled this out on the grounds that it would look as though blacks were being experimented on.[147] In this instance Darvall's heart was used because she was white, but not because the surgeons were committed to racial categorisation. By the time of the second operation, this issue was bypassed and the heart donor, Clive Haupt, was, according to state classification, Coloured.

The transplants raised all sorts of moral and ethical problems which were by no means expunged by the exhilaration of success. Only years later did it emerge that Barnard injected Darvall's heart with potassium so as to hasten bodily death (she was already pronounced brain dead, but her heart continued to beat strongly). Further questions arose when it emerged that Darvall's kidneys were given to a young

[146] Barnard and Pepper, *One Life*, 14.
[147] D. McRae, *Every Second Counts: The Extraordinary Race to Transplant the First Human Heart* (London: Simon and Schuster, 2006), 173–4.

Coloured boy, Jonathan van Wyk. Soon after the operation Marius Barnard was reported as saying, 'What I am most thrilled about as a South African is that we could use the heart of an Anglican girl to put into the body of a Jewish gentleman and a kidney into a coloured child.'[148] Such hybridity did not sit well with the high-apartheid government and its rigid systems of racial classification.

In contrast to the heart transplantation, Project Coast developed science directly in support of apartheid. This, too, involved a cardiologist, Wouter 'Dr Death' Basson, P. W. Botha's personal physician. Basson was at the centre of a secret military-sponsored chemical and biological warfare programme that became operational from 1981. Among its many projects was the production of nerve agents and tranquillising drugs intended, in combination with tear gas, to aid in crowd control. Clandestine methods of contraception were designed to lower the birth rate of black people; bizarrely, it was hoped to produce a genetically specific bacterial 'bomb' that would only affect them. Lothar Neethling, a chemist and senior policeman with two doctorates who acted as a mentor to Basson, devised toxins known as 'Lothar's potions' which were allegedly used to kill ANC activists.[149]

Although Project Coast was run as Basson's personal fiefdom, it was authorised by senior members of the military and relied on strong links with ARMSCOR as well as the expertise of veterinarians, microbiologists and other scientists working in front research facilities like Roodeplaat Research Laboratories and Delta G Scientific. According to Helen Purkitt and Stephen Burgess, muscle relaxant pills developed by Basson were used in 1982 by the South African military to disable and then kill some 200 SWAPO prisoners.[150] Within South Africa, the security forces administered toxins to some leading members of the anti-apartheid movement in targeted assassination attempts.

Heart transplants continued at Groote Schuur till the 1980s; Chris Barnard withdrew from surgery because of his arthritis. Project Coast

[148] Reported in *The Sunday Gleaner*, 24 December 1967 (online); Horwitz, 'Race, Kidney Transplants'.

[149] M. du Preez, 'SA's own bemedalled Dr Mengele is dead', *IOL News*, 13 July 2005. www.iol.co.za/news/politics/sas-own-bemedalled-dr-mengele-is-dead-248047

[150] Based on H. E. Purkitt and S. F. Burgess, *South Africa's Weapons of Mass Destruction* (Bloomington: Indiana University Press, 2005); evidence to the Truth and Reconciliation Commission.

was only finally wound down in the 1990s, and Basson managed to evade conviction on over sixty counts in a long-running trial that was concluded in 2002. In 2013 Basson was eventually convicted on charges of unprofessional conduct by the Health Professions Council of South Africa, but two years later he was revealed to be teaching University of Stellenbosch students at a Cape Town hospital.

As in the case of the development of nuclear weapons, Project Coast did not have much direct impact on political outcomes. It nevertheless exemplifies the extent to which elements of the late-apartheid military state were prepared to go in order to defend white supremacy. Project Coast was clearly an extreme example of science in the service of apartheid. Yet, as this chapter has shown, there were other instances – such as optical astronomy and environmental conservation – where science and technology were tangential to apartheid or supported it only incidentally by flattering national pride and prowess. Barnard's heart transplant programme is another such example, though it raised awkward moral and ethical problems about apartheid's core assumptions.

7 *Big Science and Indigenous Knowledge: Post-Apartheid South Africa and the African Renaissance*

The dramatic political transition after 1990 marked a major change in South Africa's internal institutional landscape and signalled the country's renewed international acceptability at the United Nations, the Commonwealth, and the Organisation of African Unity. This virtuous circle validated both the new democracy and the long histories of anti-apartheid struggle fought by these supra-national organisations. The passage from pariah state to paragon of human rights and reconciliation also opened routes for new scientific initiatives and imaginings.

In this context the decision to decommission the country's nuclear arsenal was remarkable – and not only because South Africa became the first country in the world to take this step. It also pointed to the patterns of grand gesture and prudential deal-making that now seem more characteristic of the transition than the politics of reconciliation. Quite simply, the governing National Party had no more use for a bomb and did not wish to entrust such weaponry to its successors. For its part, the African National Congress (ANC) government was eager to capitalise on global approbation and so turn nuclear swords into ploughshares. A major exception was the government's decision to divert part of the peace dividend into a highly controversial and deeply corrupt international arms deal, signed in 1999, which was justified at the time in terms of South Africa's role as a continental peacekeeper.[1]

In 1998, Deputy President Thabo Mbeki gave a major speech to launch his idea of an African Renaissance. This flagship policy, which served as the leitmotif of his presidency, projected him as a philosopher-statesman by sounding a strong message of pan-African leadership from the south. State-led developmentalism lay at the heart of Mbeki's confident, progressive message, which saw science and

[1] W. M. Gumede, *Thabo Mbeki and the Battle for the Soul of the ANC* (Cape Town: Zebra Press, 2007), 172; Andrew Feinstein, *After the Party: Corruption, the ANC and South Africa's Uncertain Future* (London: Verso, 2010).

technology acting as an important agent of social transformation. These themes were explored in a major conference on the African Renaissance held the following year, attended by 470 delegates. A key academic proponent was the medical scientist and first black chair of the Medical Research Council, Malegapuru William Makgoba, who edited the conference proceedings.[2]

Mbeki gave a rousing opening keynote address to the conference. The opening chapter of the resulting African Renaissance volume, contributed by Dialo Diop, cited palaeoanthropologist Phillip Tobias to the effect that Africa was the home of the first hominids, and it was where spoken language, stone tools and use of fire originated.[3] Diop affirmed that as well as being responsible for the Stone Age revolution, Africa was the home of many other scientific and technical firsts, including hieroglyphics, the domestication of plants and animals, and 'the first scientists in mathematics, astronomy, medicine, architecture and so on'. William Gumede, biographer of Mbeki, saw his commitment to the African Renaissance as a 'rallying cry to unite South Africans – black and white – behind a vision of the future'.[4] There are compelling parallels, and differences, between Mbeki's mobilisation of science and philosophy in pursuit of patriotism and developmental progress and the white South Africanism expounded by Jan Smuts more than half a century before.[5]

This chapter seeks to understand post-apartheid science in terms of Mbeki's dual vision of global leadership, on the one hand, and Africanisation, on the other. It explores the tensions between his modernising vision and his appeal to indigenous knowledge systems as part of a novel plan for human and societal renewal. We focus first on the commitment to science in post-apartheid South Africa, forcefully expressed in 1996 when the Mandela government published a White Paper on science and technology, 'Preparing for the 21st Century'. In emphasising the necessity for innovation in a globalising world, the document advocated scientific knowledge as a means of achieving democratic social and economic transformation. This would include fundamental science as well as more widespread scientific

[2] M. W. Makgoba (ed.), *African Renaissance: The New Struggle* (Sandton and Cape Town: Mafube and Tafelberg, 1999).

[3] D. Diop, 'Africa: Mankind's Past and Future', in Makgoba, *African Renaissance*, 3–4.

[4] Gumede, *Thabo Mbeki and the Battle for the Soul of the ANC*, 256.

[5] These are explored in Dubow, 'Global Science, National Horizons'.

education in which black South Africans would be fully engaged. Applied scientific research, it was hoped, could play a major role in providing solutions to systemic inequality in the country.

Secondly, we discuss the re-emergence of interest in, and research on, indigenous and local knowledge by scientists and academics responding to the changing political climate. In fields such as botany, for example, much of this was framed by an incorporative approach – accepting, even celebrating, hybridity and pluralism in the construction of knowledge. We also illustrate the practice of local knowledge in the veterinary sphere. Yet in debates about understanding and treating HIV/AIDs, 'Western' scientific and indigenous 'African' ideas were mobilised as antagonistic approaches. Our analysis focuses on the polarisation of public discussion, not least because Mbeki espoused a critique of biomedical science. Finally, we explore some major scientific projects in post-apartheid South Africa, from palaeontology to astronomy. While Africanist rhetoric was part of Mbeki's message, we suggest that state-led developmentalism lay close to the heart of his, and the ANC's, initially confident, progressive project.

New Institutional Foundations

Scientists were quick to seize the possibilities offered by the country's new international acceptability and they proved adept in using at least some of the humanist elements of the African Renaissance in doing so. In 1994 George Ellis, a University of Cape Town cosmologist, wrote a substantial report on behalf of the Royal Society which surveyed the country's scientific landscape as a totality. The document sought to balance the need to sustain areas of scientific advantage with research that was 'to the ultimate benefit and lasting advantage' of society as a whole and not merely those who were already privileged. Looking forward to the millennium, Ellis noted that the South Africa of the twenty-first century 'will be part of the global community', affected by developments in Africa and the rest of the world. It was 'unlikely to become a "great power" ... because of limited human and natural resources' but should instead seek to 'find an appropriate niche where its resources can be maximised through cooperation'.[6] As the country

[6] G. Ellis, *Science Research Policy in South Africa. A Discussion Document for the Royal Society of South Africa, May 28, 1994* (Cape Town: Royal Society of South Africa, 1994), 13–14.

was relatively advantaged compared with the rest of Africa in respect of skills and infrastructure, it should seek to take advantage of its favoured geographic position in the southern hemisphere.

In 1994 Mandela's new government created a new Ministry of Arts, Culture, Science, and Technology, initially headed by Ben Ngubane, a medical doctor and former health minister in KwaZulu-Natal. Mandela wrote the foreword to a Canadian-sponsored investigation into the future of science and technology in a democratic South Africa. Here he recalls the advice of Dr Thomas Odhiambo, head of the African Academy of Sciences, about the importance of 'S&T'. Odhiambo said that many African countries were only now realising 'that they would have been further along the path of economic development had they recognized, following independence, the importance of investments in this area.'[7] Continued support for the CSIR, notwithstanding its involvement in apartheid military research, was one outcome of Mandela's cautious approach to rebuilding.

In 1996, the Mandela government published its White Paper on science and technology, 'Preparing for the 21st Century'. This reiterated themes in Ellis's discussion document, including the importance of sustaining world-class 'curiosity-driven research'. Science, it insisted, should not be reduced to utilitarian outcomes. Rather than pursuing a 'formula-driven approach', basic research was deemed to be fundamental. There were 'cultural reasons' for supporting 'flagship' sciences like physics and astronomy alongside applied research. Neglecting them 'would be to take a negative view of our future – the view that we are a second-class nation, chained forever to the treadmill of feeding and clothing ourselves'.[8]

This aspirational statement was very much a product of South Africa's optimistic new democracy. In the spirit of inclusivity, the humanities and social sciences were also regarded as essential to the innovative culture that the government sought to encourage by creating an 'enabling environment'.[9] A key role was assigned to the

[7] N. Mandela, 'Foreword', in M. van Ameringen (ed.), *Building a New South Africa. Vol. 3 Science and Technology Policy* (IDRC, Ottawa [1995]).

[8] Government White Paper on Science and Technology: Preparing for the 21st Century, Dept. Arts, Culture, Science and Technology, 4 September 1996, www.gov.za/sites, chapter 2.7; South Africa, Green Paper on Science and Technology: Preparing for the 21st Century (Pretoria, 1996)

[9] White Paper on Science and Technology, chapter 4.1.

proposed National Research Foundation for the disbursement of grants and funding. This would augment the grant-giving functions of specialised bodies like the Medical Research Council, the Human Sciences Research Council, and the Foundation for Research Development.

The White Paper, following the logic of a 1996 Green Paper on science and technology which pushed for a 'national system of innovation', recommended the creation of a special Innovation Fund to subsidise research and development. The skewed priorities characteristic of the late-apartheid era were to be redressed by new fiscal and tax incentives to invest in research and development, and the private sector was enjoined to do more to sustain innovation. Crucially, the concentration of research resources in the defence industries would be reversed. A new partnership with the profitable arms industry was to be secured by leveraging military technology for civilian purposes. Through such means the military would 'shake off negative perceptions associated with its role in the South Africa of the past'. The Atomic Energy Corporation should likewise refocus on fundamental research as a consequence of the abandonment of the country's nuclear weapons programme.[10] No reference was made in the White Paper to the role or importance of indigenous knowledge.

Reinvigorating South African science and reorienting it away from the secretive military-industrial complex that underpinned the late-apartheid state, commanded significant support within universities as well as government. In a climate of widespread revulsion at the excesses and perversions perpetrated by President P. W. Botha's securocrats, there was an overwhelming willingness to transcend lines of colour and political affiliation and, by so doing, work for the common good.

The period of high apartheid in the 1960s and early 1970s had seen substantial advance in scientific capacity. Money flowed freely towards white English- and Afrikaans-speaking tertiary institutions as well as the CSIR and technologically oriented parastatals. In respect of the migration of highly skilled scientific personnel, the country experienced an estimated net gain of 20 per cent between 1965 and 1985 –

[10] White Paper on Science and Technology, chapter 8.4–8.6; South Africa's Green Paper, 51, 69.

followed by a sharp reversal leading to an outflow of skills.[11] E. C. Reynhardt, a physicist based at the University of South Africa, calculated that in the period 1967–75 there was a real increase in research and development spending of 8 per cent per annum. His analysis of South African publication rates in the natural sciences (gleaned from the Science Citation Index) claimed an increase four times greater than the average for the rest of the world. From 1977 to 1980, however, there was a discernible downward trend in publication rates. This was attributed to the decline in expenditure and to intellectual isolation. The numbers of graduates in the natural sciences also declined over this period.[12]

This trajectory, if accurate, closely mirrored the performance of the South African economy. After sustained growth in the 1960s, the 1973 oil crisis proved a profound economic shock, while the revival of political activism in the streets and on the factory floor began to erode the basis of the apartheid system. A detailed metrical study suggested that South African international research productivity, including the social sciences, reached a peak (around 3,500 per year) in 1987, followed by a lengthy plateau until 2000. Government spending on research and development peaked at 1.04 per cent in 1987, whereupon it declined to 0.4 per cent in 1994, the year of South Africa's first democratic election.

The distorting effects of the late-apartheid military-industrial complex are well illustrated in the case of chemistry. The evisceration of the National Chemical Research Laboratory from the mid-1980s, one of the key research institutions of the CSIR, was a blow to fundamental science and to new academic careers as well. To some extent this was offset, as intended by the state, by growing employment possibilities for applied chemists in industry and organisations like the Council for Mineral Technology (MINTEK) and SASOL.[13] But the overall erosion in research capacity was registered by the CSIR's downgrading of its Physics Research Laboratory at the same time. Some physicists were able to find employment in the expanding nuclear sector (civil and

11 South Africa's Green Paper on Science and Technology: Preparing for the 21st Century (Pretoria, 1996), 22.

12 E. C. Reynhardt, 'Universiteitsnavorsing in die Fisiese Wetenskappe in Perspektief', *SAJS*, 78 (1982), 393–9.

13 R. Haines, 'Chemistry', in R. Diab and W. Gevers (eds.), *The State of Science in South Africa* (Pretoria: Academy of Science of South Africa, 2009), 106.

military) until the 1990s, but many became redundant as a result of the changing strategic and political environment. One indication of this decline – though it may also signal a new focus on international publishing priorities – was the collapse in December 1993 of the *South African Journal of Physics*, founded in 1978.[14]

By contrast, the immediate post-apartheid era proved beneficial to sciences such as astronomy, biodiversity and medicine. Expenditure on research and development more than doubled to 0.95 per cent in 2006–7. The number of research publications soared to an annual total of 7,500 in 2010. The introduction of a new state funding formula in 1994, and the inclusion of South African journals in International Scientific Indexing (ISI) in 2009, had a markedly positive effect.[15] Almost 90 per cent of this increase was in civil research.[16] A new government department of Science and Technology was created, which soon outlined ambitious plans for the restructuring of research and innovation. The Royal Society and the Akademie vir Wetenskap en Kuns continued in existence, largely catering to their existing academic constituencies, but now more peripheral.

George Ellis's Report had called for the formation of an Academy of Science – as opposed to umbrella organisations representing sectoral scientific constituencies – in order to represent the interests of science as a whole and to help formulate national policies. In 1996 a new national organisation was established, the Academy of Science of South Africa (ASSAf), which was given statutory recognition in 2001–2. The advent of ASSAf involved repealing the 1959 legislation governing the Akademie vir Wetenskap en Kuns, which was so closely associated with the Afrikaner establishment.[17]

Notwithstanding its official status, ASSAf was independent of state control. It bore a mandate to promote broad-spectrum scientific activity, including the social sciences and humanities. In 2009, ASSAf took over control of the *South African Journal of Science*, now seen as a

[14] H. Moraal, 'Physics and Astronomy', in Diab and Gevers (eds.), *The State of Science in South Africa*, 84–6.

[15] A. Pouris, 'Science in South Africa: The Dawn of a Renaissance?', *SAJS*, 108 (2012), 1–6.

[16] J. Mouton and W. Gevers, 'Introduction' in Diab and Gevers (eds.), *The State of Science in South Africa*, 45, 50–1; South Africa's Green Paper on Science and Technology. Preparing for the 21st Century (Pretoria, 1996), 9.

[17] Academy of Science of South Africa Act, 2001, *Government Gazette*, 440, 23151 (2002), www.gov.za/sites; Ellis, *Science Research Policy*, 255.

'national asset', from the much-diminished South African Association for the Advancement of Science, historical home of anglophone science. Unlike the older scientific academies, ASSAf sought to take a more activist and inclusive approach to the needs of post-apartheid South Africa. The statutory National Research Foundation (established in 1999) was tasked with distributing funds and served as a conduit for research institutions, 'centres of excellence', and individual researchers. Following its absorption of some of the functions abandoned by the CSIR in the 1980s, the rise of the NRF can be seen as the fulfilment of various attempts since the 1930s to promote national research and infrastructure in an integrated and targeted manner.[18]

One important initiative undertaken by ASSAf was a detailed metrical analysis of the pattern of South African scientific publications and their international impact. The report, produced in 2006, sought to support national and local publications that were indexed in recognised international bibliometric databases. It was calculated that there were some 16,000 active South African scientific researchers who, collectively, produced 3,500 ISI-listed papers per annum (a declining but near-equivalent number of publications were produced in non-ISI journals). ASSAf proposed a research funding formula to support accredited local publications.[19]

The funding environment was conducive to such ideas. State support of research and development increased in real terms by 52 per cent between 2001 and 2012.[20] Expenditure on social sciences and humanities research also benefited, rising from 12 per cent of gross expenditure on research and development in 2005 to around 20 per cent in 2014 – quite high by international standards.[21] A new National

[18] National Research Foundation Act, *Government Gazette* 397, 19017 (1998), as amended by the Science and Technology Laws Amendment Act, 2014, 586, 37594 (2014), D. Walwyn and L. Cloete, 'Draft White Paper on Science, Technology and Innovation Neglects to Prioritise Issue of Performance and Human Capability', *SAJS*, 114, 11/12 (2018), DOI:10.17159/sajs.2018/5679

[19] Academy of Science of South Africa, *Report on a Strategic Approach to Research Publishing in South Africa* (Pretoria, 2006), xvii, 1.

[20] D. Walwyn and L. Cloete, 'Universities are Becoming Major Players in the National System of Innovation', *SAJS*, 112, 7/8 (2016), DOI:10.17159/sajs.2016/20150358.

[21] N. Molotja and G. Ralphs, 'A Critical Review of Social Sciences and Humanities R&D Expenditure in South Africa, 2005–2014', *SAJS*, 114, 7/8 (2018), www.sajs.co.za.

Institute for the Humanities and Social Sciences, linked to the Department of Higher Education and Training, was created in 2013. With an annual budget of around R150 million and a broad mandate, the Institute was designed to promote a research culture and contribute to the achievement of social development and equity.[22]

The ANC government invested ambitiously in scientific research. But deep and persisting patterns of inequity, competing political imperatives, and over-ambitious policy frameworks resting on fragile institutional foundations, are sources of instability and weakness. Tensions centre on the need to satisfy constituencies that seek at once to satisfy basic developmental needs while remaining 'globally competitive'. Here, the politics of race and representation are never far below the surface. In the realm of scientific policy, this has often expressed itself in a commitment to advancing universal knowledge while at the same time according due recognition to indigenous African knowledge. Whereas 'indigenous knowledge' was not mentioned in the 1996 White Paper on Science and Technology, the need to promote and transfer 'indigenous technology' did feature prominently in the 1998 National Research Foundation Act, including in the title of the legislation.[23] The imprint of Thabo Mbeki's thinking about the African Renaissance was clearly evident by this time. Deep, historical fault lines of race – and class – underlie this ambiguity.

The African Renaissance and Indigenous Knowledge

Mbeki's call for research and the mobilisation of indigenous knowledge was backed up with funding and some institutional framework. An indigenous knowledge programme was directed by Mogege Mosimege, who specialised in mathematics education, at the CSIR and the Department of Science and Technology from the late 1990s. Mosimege adumbrated the value of incorporating 'ethnomathematics' in scientific education, for example, by using African board games such

[22] *National Institute of the Humanities and Social Sciences Annual 1017/18 Annual Report*, www.nihss.ac.za.

[23] Act 23 of 1998: 'To provide for the promotion of research, both basic and applied, and the extension and transfer of knowledge in the various fields of science and technology and indigenous technology; and for this purpose to provide for the establishment of a National Research Foundation; and to provide for incidental matters.'

as *morabaraba*.[24] In 2002 the journal *Indilinga: African Journal of Indigenous Knowledge Systems* was launched. Herbert Vilakazi, formerly professor of sociology at the universities of Transkei and Zululand, wrote the introductory article for the journal on 'African Indigenous Knowledge and Development Policy', arguing that African development strategies should not be based on Western influences and ideas.[25] Elsewhere he declared that the intellectuals responsible for 'building the new Africa' should first educate 'themselves about the principles and patterns of African civilisation, whose knowledge they have largely lost'.[26] In 2006 – in the context of continued debates about the treatment of HIV/AIDS – Vilakazi was appointed by Mbeki to head a presidential task team on traditional medicines.

Mbeki called for the appreciation and preservation of African knowledge and cultural practices, and these included medical traditions. Although, as we shall see, Mbeki's view on HIV/AIDS collided with orthodox medical ideas, it would be misleading to suggest that the proponents of indigenous knowledge envisioned, and practised, a clear binary division between 'Western' and 'African' cosmologies and epistemologies. A conspicuous feature of the new South Africa was in fact cultural borrowing and exchange. For example, Archbishop Tutu's promotion of the 'rainbow nation' encapsulated a conscious hybridity. Whereas the ideology of scientific 'South Africanism' in the first decades of the twentieth century embraced a common white identity at the expense of Africans, the South Africanism of the 1990s sought determinedly, though sometimes scratchily, to find commonality across lines of colour and culture.

Ethnobotany provides one good example, not least because it encompasses dynamic understandings of indigenous knowledge that link to anthropology, environmental activism and local medical practice. There has been a long tradition in which botanists, observing or compiling lists of South African plants, recorded African uses, particularly as medicines (Chapter 1). Ludwig Pappe, the first official colonial

24 N. Nkopodi and M. Mosimege, 'Incorporating the Indigenous Game of Morabaraba in the Learning of Mathematics', *South African Journal of Education*, 29, 3 (2009).

25 H. W. Vilakazi, 'African Indigenous Knowledge and Development Policy', *Indilinga*, 1, 1 (2002), 1–5.

26 H. W. Vilakazi, 'The Problem of African Universities', in Makgoba (ed.), *African Renaissance*, 204.

botanist in the Cape Colony, mentioned such uses by Khoekhoe and Dutch settlers at the Cape in the mid-nineteenth century (Chapter 2).[27] Nineteenth-century medical training included botanical input; physicians from James Barry to William Atherstone were interested in the medicinal potential of plants. Buchu (*Agathosma* species), widely used by the Khoekhoen, achieved global recognition for its healing properties.[28]

In a country where abortion was for a long time illegal, Khoekhoe and African experts were consulted by white as well as black women for herbal abortifacients.[29] Andrew Smith, a teacher at Lovedale School, published on eastern Cape medicinal species in 1895.[30] The missionary Henri Junod worked closely with African informants in Mozambique and the Transvaal in his research and collection of plants. Plant uses were also recorded in dictionaries of African languages, produced largely by missionaries, in the late-nineteenth and early-twentieth centuries.

In 1932, at a time when scientists in other disciplines were largely incurious about indigenous knowledge, J. M. Watt and Maria G. Breyer-Brandwijk published the first edition of their encyclopaedic volume on medicinal plants with comments on African use. Watt, who had qualified in Edinburgh, was professor of pharmacology at Wits and his co-author a researcher in phytochemistry. They saw their book as 'dedicated to the progress of science in South Africa' and believed that efficacious popular remedies and knowledge should be recorded before they disappeared.[31] They were funded and assisted both by the university and the government Division of Plant Industry in the Department of Agriculture. The second edition of 1962 was greatly extended to include up-to-date scientific, especially phytochemical,

27 Pappe, *Florae Capensis Medicae Prodromus*.

28 C. H. Low, 'Different Histories of Buchu: Euro-American Appropriation of San and Khoekhoe Knowledge of Buchu Plants', *Environment and History*, 13, 3 (2007), 333–61.

29 H. Bradford, 'Herbs, Knives and Plastic: 150 Years of Abortion in South Africa', in T. Meade and M. Walker (eds.), *Science, Medicine and Cultural Imperialism* (New York: Macmillan, 1991), 120–47.

30 A. Smith, *A Contribution to South African Materia Medica, Chiefly from Plants in Use among the Natives* (Lovedale: Lovedale Institution Press, 1888).

31 J. M. Watt and M. G. Breyer-Brandwijk, *The Medicinal and Poisonous Plants of Southern Africa* (Edinburgh: E & S Livingstone, 1932).

references from around the world, juxtaposed with recorded African local knowledge.[32]

This text was a repository of information for later popular books – for example, that by Ben-Erik van Wyk and others on *Medicinal Plants of South Africa* (1997) and van Wyk and N. Gericke's *People's Plants* (2000).[33] The authors wrote self-consciously in a 'new era' and noted that productive collaborations were being formed between scientists and African communities around plant knowledge. They were interested in the scientific documentation of indigenous knowledge and also in validating and popularising such knowledge. *People's Plants* covered a very wide range of uses, including food, crafts and building, and recorded African names. In their section on medicine the authors explicitly stated that 'indigenous African medicine coexists with Western allopathic medicine' and that 'not all medicinal plant use can be explained from the reductionist perspective of Western science'.[34] Van Wyk and Gericke's text is a celebratory approach in keeping with the transition to a post-apartheid South Africa. It represented an acceptance of more relativist approaches to knowledge and aimed to salvage information 'before it is irretrievably lost to future generations', and to 'raise awareness of the role plants play in people's daily lives'.

The most detailed regional discussion of African plant knowledge appeared in articles written by Tony Dold and Michelle Cocks at Rhodes University. They, too, were explicitly celebratory in their *Voices from the Forest* (2012), arguing both for the centrality of plants in Xhosa culture and for the importance of African knowledge in preserving biodiversity in the long term.[35] Dold also participated in an illustrated, popular book aimed at schoolchildren.[36]

32 J. M. Watt and M. G. Breyer-Brandwijk, *The Medicinal and Poisonous Plants of Southern and Eastern Africa: Being an Account of their Medicinal and Other Uses, Chemical Composition, Pharmacological Effects and Toxicology in Man and Animal* (Edinburgh: E & S Livingstone, 1962).

33 B.-E. van Wyk, B. van Oudtshoorn and N. Gericke, *Medicinal Plants of South Africa* (Pretoria: Briza Publications, 1997); B.-E. van Wyk and N. Gericke, *People's Plants: A Guide to Useful Plants of Southern Africa* (Pretoria: Briza Publications, 2000).

34 Van Wyk and Gericke, *People's Plants*, 119.

35 T. Dold and M. Cocks *Voices from the Forest: Celebrating Nature and Culture in Xhosaland* (Johannesburg: Jacana, 2012).

36 S. Zukulu, et al., *Medicinal and Charm Plants of Pondoland* (Pretoria: South African National Biodiversity Institute, 2012).

Scientific discussion of the medicinal properties of plants shades into a connected literature on the commercial exploitation of indigenous plants, and especially of threatened species. Busy traditional medicine (*muthi*) markets thrive in Durban, around Warwick Road, and in Johannesburg at Mai Mai and Faraday Street; there are many smaller venues and street sales across the country.[37] The gathering of *muthi*, sometimes by poor African rural women, for whom this enterprise provides some supplementary income, requires knowledge of plant species.[38] In some cases the market for such plants has been international. During the 1950s, *Dioscorea sylvatica*, a wild yam called Elephant's Foot, or *ingwevu* in Zulu, was sought by pharmaceutical companies such as Boots in Britain.[39] It was identified by US scientists as a valuable base material for diosgenin, an ingredient for hormone medicines such as cortisone and the contraceptive pill. While Mexico was the main source of supply, yam-hunters scoured parts of South Africa and around 6,000 tons of Elephant's Foot (perhaps 200,000–300,000 plants) were extracted from the kloofs of Mpumalanga and KwaZulu-Natal, processed in a Johannesburg factory and exported. In 1960 the Natal Parks Board, concerned about the longer-term survival of the species, clamped down on harvesting. By 1963, the need for the plant material was bypassed by synthetic compounds.

Recent examples of international use include *Hoodia gordonii*, thought to be effective for slimming, and *Pelargonium sidoides*, sought by German pharmaceutical companies in the early twenty-first century for medical treatment of respiratory diseases.[40] The discovery and

[37] V. L. Williams, K. Balkwill and E. T.F. Witkowski, 'Unraveling the Commercial Market for Medicinal Plants and Plant Parts on the Witwatersrand, South Africa', *Economic Botany*, 54, 3 (2000), 310–27.

[38] T. Kepe, 'Medicinal Plants and Rural Livelihoods in Pondoland, South Africa: Towards an Understanding of Resource Value', *International Journal of Biodiversity Science & Management*, 3, 3 (2007), 170–83.

[39] W. Beinart and R. Beinart, '"From Elephant's Foot . . . to Cortisone": Boots Pure Drugs Company and *Dioscorea sylvatica* in South Africa, c. 1950–1963', *South African Historical Journal*, 71, 4 (2019), 644–75; V. L. Williams et al., 'Forest Elephant's Foot: Dioscorea sylvatica Eckl.', *SANBI Red List of South African Plants*, www.redlist.sanbi.org/species.

[40] L. A. Foster, *Reinventing Hoodia: Peoples, Plants, and Patents in South Africa* (Seattle: University of Washington Press, 2017); Z. Nokwanda Msomi and S. Matthews, 'Protecting Indigenous Knowledge Using Intellectual Property Rights Law: The Masakhane *Pelargonium* Case', *Africanus: Journal of Development Studies*, 45, 1 (2015), 62–77.

exploitation of *Hoodia*, a bitter succulent used by generations of San (Bushmen) people to stave off hunger and thirst, serves as an excellent example of how indigenous knowledge and rights intersected with business enterprise in the new South Africa. In 1996 an appetite suppressant molecule in *Hoodia*, known as P57, was patented by the CSIR, which had conducted research into indigenous plants since the 1960s. The pharmaceutical company Phytopharm, in collaboration with Pfizer, gained the licence to produce a 'natural' anti-obesity drug. However, the San were initially overlooked as financial beneficiaries because Pfizer claimed that they were 'extinct'. Such attitudes displayed a distinct whiff of exploitative bioprospecting – the practice of extracting biochemical resources and indigenous knowledge to produce commercially valuable compounds of advantage to pharmaceutical companies.

An energetic campaign ensued, guided by Roger Chennells, a human rights lawyer and land claims specialist. He foregrounded the irony that an anti-obesity drug produced for Westerners utilises the indigenous knowledge of some of the world's most deprived people. An apology from the CSIR's bioprospecting programme for ignoring the San's interests was made at the 2002 World Summit on Sustainable Development in Johannesburg. In 2003, following tough negotiations, San elders and representatives of the CSIR signed an agreement whereby substantial royalties from the drug would be paid into a Hoodia Benefit Trust in recognition of indigenous intellectual property rights.[41] This outcome, by which the interests of indigenous people were recognised in a deal brokered between a government agency and a leading international corporation, is a signal example of how the commercialisation of indigenous or traditional knowledge by scientific companies can potentially work to mutual advantage. *Hoodia* extract pills are now widely available online and some products in their advertising make a virtue of the plant's use for centuries by South Africa's indigenous peoples.

[41] A. Barnett, 'In Africa the Hoodia Cactus Keeps Men Alive', *The Observer*, 17 June 2001; 'South Africa: Marginalised San Win Royalties From Diet Drug *hoodia gordonii*, www.hoodiadietpills.org; 'Native Healers Want Drug Company Cooperation' by D. Craft, www.unl.edu/rhames/courses/current/hoodia.htm; Minutes of 'Workshop on Benefit Sharing with Developing Countries: The case of human genetic resources', Cape Town, May 2004, www.wits.ac.za/bioethics.

Despite this realignment, South Africa has found it difficult to legislate to protect indigenous knowledge and avoid bioprospecting.[42] In respect of plants, the National Environmental Management: Biodiversity Act of 2004 established some framework. Not only did it aim to conserve biological diversity and impose sustainable controls over biological resources, it also provided for the 'fair and equitable sharing of benefits derived from bio-prospecting related to indigenous biological resources'. This was to be implemented through permits and benefit-sharing agreements. In the same year, the government adopted an indigenous knowledge policy for the 'recognition, promotion, development, protection and affirmation of IKS'. The Intellectual Property Laws Amendment Act of 2013 aimed to give wider recognition, but it proved difficult to define the concept of 'indigenous' and to specify the relevant stakeholders and communities within the terms of a broader intellectual property law that was intended primarily to regulate formally registered patents, designs and trademarks.

Specific legislation was also attempted in two versions of a Protection, Promotion, Development and Management of Indigenous Knowledge Systems Bill (2015–16). The Bills envisaged registration of a wide range of knowledge, but it exists as diverse and changing practice with little written down. Claims to ownership by individuals, and by various layers of community, are often contested, especially because such knowledge, by virtue of being dynamic and hybrid, can bridge national boundaries. For example, any indigenous ownership of a plant such as rooibos, widely developed as a specialist tea and grown commercially over a long period, is difficult to establish. It also needs to be acknowledged that both smallholders and commercial enterprises in South Africa have been beneficiaries of American plants over centuries; in a sense their debt to native and settler Americans, who domesticated and developed those plants, is incalculable. This would include maize, tomatoes, prickly pear, avocado pear, some species of bean and pumpkin.

Discussions about ethnobotany in South Africa signalled the country's entry into wider global debates about the rights and knowledge systems of 'first nation' peoples. In the 1980s Darrell Posey, an

[42] L.-A. Tong, 'Aligning the South African Intellectual Property System with Traditional Knowledge Protection', *Journal of Intellectual Property Law and Practice*, 12, 3 (2017), 179–90; Foster, *Reinventing Hoodia*.

American ethnoentomologist, headed a project documenting traditional biological knowledge in Brazil and worked with local academics and museums to hold a congress of ethnobiology.[43] This was translated onto a global stage at the Rio Earth Summit of 1992 in a parallel 'earth parliament' with indigenous representatives. Posey argued that indigenous knowledge held the key to the preservation of biodiversity; it required recording not only as a traditional system but as a dynamic resource. He aimed to mobilise indigenous people, champion their intellectual property rights, and provide platforms that would transform the politics of biodiversity. A parallel set of arguments evolved in Indian debates, spearheaded by Vandana Shiva.[44] Her approach was more strongly linked to an ecofeminist and anti-capitalist agenda focusing on communities that were more fully inserted into capitalist economies. Such arguments were also to be found in African and Australian contexts.

Michelle Cocks developed Posey's ideas, based on detailed fieldwork among Xhosa-speaking communities in the Eastern Cape: there too, she argued, indigenous knowledge contributed to biodiversity conservation. This work gives anthropological depth to a sharp inversion of long-held ideas that African systems of agriculture, and particularly their tendency to overstock on commonage, cause environmental degradation.[45] Cocks and others argued to the contrary that communal systems of landholding, together with deeply set local methods of management, would be more likely to promote ecological sustainability. She applied such ideas about biocultural diversity not only to more self-sustaining societies, but also to communities 'whose lifestyles have been affected and transformed by modernity'.[46] In this respect, she takes her place among scholars who propose an environmentalism of the poor. As they argue, communities who depend deeply on local

[43] D. Posey, *Indigenous Knowledge and Ethics: A Darrell Posey Reader* (New York: Routledge, 2004).

[44] V. Shiva, *Staying Alive: Women, Ecology and Survival in India* (London: Zed Press, 1988).

[45] C. M. Shackleton, 'Are the Communal Rangelands in Need of Saving?, *Development Southern Africa*, 10, 1 (1993), 65–78; Beinart, 'African History and Environmental History'.

[46] M. Cocks, *Wild Resources and Cultural Practices in Rural and Urban Households in South Africa: Implications for Bio-cultural Diversity Conservation* (Grahamstown: Institute of Social and Economic Research, Rhodes University, 2006), 8.

natural resources for medicinal plants, fuelwood, housing material, pastures, food supplements and craftwork will use their knowledge to protect their environments even in the context of rapid social change and commodification. Cocks also suggested that ideas about biodiversity were interconnected with strong nature-based religious beliefs.

Such arguments need to be explored in the context of a broader analysis of environmental change. Neither environmental degradation in African reserves – probably at their low point from the 1920s to the 1960s (see Chapter 4) – nor subsequent recovery was the result largely of local knowledge.[47] Causes of this environmental stabilisation are more likely to have been widespread withdrawal from smallholder arable agriculture; decline in livestock numbers; declining dependence by most rural families on rural resources; and climate change conducive to bush encroachment. Such factors, together with conservationist controls, are likely to have resulted in a denser bush and grass cover in some African-occupied communal areas.[48]

In spheres like these, there is clearly a significant interpenetration of local or indigenous and scientific knowledge. Such interactions speak to debates about the equivalence of knowledge. Should a hierarchical distinction be made between indigenous, local (or vernacular) knowledge and scientific knowledge?[49] In a recent discussion, Wilson Akpan argues against such a distinction.[50] He also questions the tendency to relegate local knowledge to the sphere of 'ways of doing things'. Such knowledge is apt to be killed by empathy, a 'target of caring thought and quiet disdain'. Like the anthropologist Paul Richards and the historian Toyin Falola, he invokes the dynamism and power of hybrid indigenous knowledge.[51] For Akpan, all knowledge is equivalent though it can operate on the basis of different kinds of rationality.

47 M. T. Hoffman, 'Changing Patterns of Rural Land Use and Land Cover in South Africa and Their Implications for Land Reform'. *JSAS*, 40, 4 (2014), 707–25.

48 B. J. Wigley, W. J. Bond and M. Timm Hoffman, 'Thicket Expansion in a South African Savanna under Divergent Land Use: Local vs. Global Drivers?', *Global Change Biology*, 16, 3 (2010), 964–76.

49 For 'vernacular', see Beinart, *Rise of Conservation*; Jacobs, *Birders of Africa*.

50 W. Akpan, '"Local" Knowledge, "Global Knowledge", "Development Knowledge": Finding a New Balance in the Knowledge Power Play', *South African Review of Sociology*, 42, 3 (2011), 116.

51 P. Richards, *Indigenous Agricultural Revolution* (London: Harper Collins, 1985); T. Falola, *The Power of African Cultures* (Rochester: Rochester University Press, 2003).

'Advanced societies', he argues, need local knowledge and traditional systems need knowledge from outside, 'unlocking the knowledge-power that lies underutilized in all human communities'.[52] While we agree that it is important to study and understand these interactions, and that all knowledge systems are porous and dynamic, we think it is nevertheless also important to understand the distinctive basis of scientific knowledge.

Local Knowledge in Practice

The rediscovery and validation of indigenous knowledge can be part of an attempt to contest Western scientific hegemony. Even when research is not linked to such advocacy, it explores everyday practice and the role of hybrid local knowledges in ordering environmental, agricultural and medical domains for millions of rural people. The practice of local knowledge is still widespread in South Africa in respect of human and animal healing. Both popular and academic literature record traditional medical practice and its imbrication with scientific ideas. African practitioners, sometimes using hybrid approaches, such as Khotso Sethuntsa, achieved high profile and some wealth during the 1960s and 1970s.[53] Credo Mutwa promoted a broader, popular statement of African ways of knowing in *Indaba, My Children* (1964) and established a cultural garden in Soweto with sculptures and symbols said to be rich in African mythology.[54] This was a notably successful instance of self-reinvention by someone who had incurred the wrath of student activists in 1976 because he was seen as legitimising apartheid conceptions of separate development. Mutwa has now re-established a cultural village known as Kwakhaya Lendaba (the Place of Stories) as a popular destination on the Soweto tourist map.

In the field of academic anthropology, Harriet Ngubane wrote the most systematic analysis of African medical ideas in South Africa, *Body and Mind in Zulu Medicine* (1977), exploring the multifaceted character of Zulu thinking about disease and healing.[55] Brought up as

52 Akpan, '"Local" Knowledge', 125.

53 F. Wood and M. Lewis, *The Extraordinary Khotso: Millionaire Medicine Man from Lusikisiki* (Johannesburg: Jacana, 2007).

54 C. Mutwa, *Indaba: My Children* (Johannesburg: Blue Crane Books, 1964); *My People: The Writings of a Zulu Witch-Doctor* (London: Penguin, 1971).

55 H. Ngubane, *Body and Mind in Zulu Medicine: An Ethnography of Health and Disease in Nyuswa-Zulu Thought and Practice* (London: Academic Press, 1977).

a Roman Catholic and educated at St Francis College, Mariannhill, Ngubane immersed herself in Zulu thought, which she explicated in a book discussing matters like sorcery, ancestors and pollution. Her treatment of healing strategies addressed ideas of causation, as well as plant medicines believed to cure specific conditions, 'administered independently of any ritual context', and those with specific symbolic and protective functions.[56] She researched as a Zulu-speaking insider, but also an academic outsider, 'belonging and not belonging'.[57]

Ngubane's thesis was researched in the comparative context of Cambridge anthropology and her book written during a post-doctoral fellowship in Oxford. She was innovative in the South African context for analysing African healing as an overarching and coherent system of thinking. Reflecting back on her research in 1981, Ngubane argued that many African users of indigenous medicine were prepared to adopt biomedical methods as well, whereas practitioners of scientific medicine would routinely berate patients if they consulted traditional healers. Flint noted a similar porosity in the nineteenth century (Chapter 2), but saw it operating in both directions. Ngubane also took care to distinguish between legitimate diviners, who possess religious authority and remain deeply embedded within indigenous rural networks, and 'self-appointed healers' in urban areas, whose numbers include 'not a few charlatans'.[58]

African indigenous knowledge has found some support in the sphere of veterinary medicine, where state funding has been available for scientific investigations into African plant usage. Onderstepoort Veterinary Institute, long at the heart of government-sponsored scientific research, is now home to the Jotello Soga Ethnoveterinary Garden of African medicinal plants, named in honour of South Africa's first – and for many years only – black veterinary surgeon (see Chapter 3). At the turn of the twenty-first century, scientists at the University of Pretoria researched a number of African plant remedies in search of effective chemical compounds.[59] They aimed to discover the details of

[56] Ngubane, *Body and Mind*, 109. [57] Ngubane, *Body and Mind*, 4.

[58] H. Ngubane, 'Aspects of Clinical Practice and Traditional Organization of Indigenous Healers in South Africa', *Social Science and Medicine*, 15B (1981), 365.

[59] J. N. Eloff, 'Ethnoveterinary Use of Southern African Plants and Scientific Evaluation of Their Medicinal Properties', *Journal of Ethnopharmacology*, 119 (2008), 559–74.

local usages in order to investigate their potential for pharmaceuticals. In this sense their work carried on the tradition of Watt and Breyer-Brandwijk. Despite promising tests, the costs of systematic development into drugs was prohibitive, and in most instances there were already remedies available from pharmaceutical companies for the diseases that new plant compounds could potentially treat.

Patrick Masika and his team at the University of Fort Hare have focused on local practices and medicinal plants, both to record usage and discuss the efficacy of therapies.[60] Academic proponents of traditional medicines argue that their cheapness and availability mean they could be ideal solutions for treating some common, endemic diseases, as long as ingredients are not overexploited. They are beginning to develop a scientific and experimental, rather than just a local and practical, dimension to ethnoveterinary research. Shirley Motaung, at Tshwane University of Technology, and colleagues are investigating medicinal plants for their potential in developing treatments for human bone and cartilage regeneration.[61] She has launched a biotech company to manufacture plant-based medicinal products.

While traditional medical practices followed people from the countryside into the cities and have become a nationwide phenomenon, African local veterinary knowledge is more restricted to the 18 million or so people in the rural and dense-settlement districts of the former homelands and areas newly transferred in land reform schemes.[62] Although there is an increasing volume of academic research on local veterinary knowledge, it has barely been incorporated into state practice. After 1994, the ANC government devolved some veterinary provision to the provinces. Provincial departments of agriculture employed animal health technicians to assist with disease control and dipping. Yet many livestock owners, black and white, were largely left to their own devices when deciding how to prevent or treat diseases.

60 P. J. Masika, A. Sonandi and W. van Averbeke, 'Tick Control by Small-Scale Farmers in the Central Eastern Cape Province, South Africa', *Journal of the South African Veterinary Association*, 68, 2 (1997), 45–8; P. J. Masika, W. van Averbeke and A. Sonandi, 'Use of Herbal Remedies by Small-scale Farmers to Treat Livestock Diseases in Central Eastern Cape Province, South Africa', *Journal of the South African Veterinary Association*, 71, 2 (2000), 87–91.

61 F. N. Alaribe, M. J. Maepa, N. Mkhumbeni, S. C. K. M. Motaung, 'Possible Roles of *Eucomis autumnalis* in Bone and Cartilage Regeneration: A Review', *Tropical Journal of Pharmaceutical Research* 17 (4), 741–9.

62 Beinart and Brown, *African Local Knowledge.*

Land reform has gradually increased the areas available to African livestock owners, and most African beneficiaries use newly acquired land to run livestock. Perhaps counter-intuitively, animal health is of growing importance and interest to a greater number of people, notwithstanding the country's rapid urbanisation over the last century.

Throughout this period, African farmers were exposed to the practice, if not always the theories, of biomedicine, including dipping, vaccinations and quarantine restrictions.[63] In interviews with African livestock owners between 2008 and 2012, mostly older men, Beinart and Brown found that their informants had appropriated some of these scientifically based strategies. The terms 'local' or 'vernacular' rather than 'indigenous' seemed the best description of the hybrid knowledge that resulted – treatments of disease are also often locally specific. Therapeutic pragmatism is evident in respect of animal as well as human disease.[64] Interviews reveal that the great majority of livestock owners used plant medicines. Local treatments shared certain characteristics, for example widespread use of plant laxatives, especially aloe solutions, well known for this property. Livestock owners resorted to pharmaceutical products, such as vaccines, antibiotics and dipping chemicals, when they could afford them. Animals were also treated with household standbys such as Epsom salts, cleaning fluids and Coca-Cola (which is believed to dissolve plastic bags in the digestive systems of cattle).[65]

Even when they regularly used biomedicines, few of the livestock owners interviewed fully accepted biomedical explanations. For example, many doubted that ticks transmitted specific diseases. They saw ticks as troublesome and they had African-language names for the most common tick-borne diseases such as gall sickness (anaplasmosis) and heartwater. They also made post-mortem observations of the organs of animals following slaughter for consumption. But they understood the named diseases largely as a consequence of seasonal environmental changes or other environmental causes such as the state

[63] W. Beinart and C. Bundy, *Hidden Struggles in Rural South Africa* (Johannesburg and London, 1987), 191–221.

[64] Feierman and Janzen (eds.), *The Social Basis of Health and Healing*; Flint, *Healing Traditions*; L. Schumaker, D. Jeater and T. Luedke, 'Introduction. Histories of Healing: Past and Present Medical Practices in Africa and the Diaspora', *JSAS*, 33, 4 (2007), 707–14.

[65] Beinart and Brown, *African Local Knowledge*.

of the veld, poisonous plants, drought, or intense rainfall and muddy kraals. Some also professed beliefs that widows and menstruating women could precipitate disease, undermine the fecundity of their animals, or bring misfortune to the cattle kraal.

By and large, local veterinary knowledge is not very effectively circulated. It is recorded in only a limited number of inaccessible academic theses and publications. Old men say that the youth are not interested, and the youth respond that old men are reluctant to share their knowledge. Most women are excluded from the transmission of knowledge around cattle diseases, though gender is less of a constraint in dealing with small stock and chickens. Even among its exponents, knowledge of cures and plant medicines tends to be fragmented and uncertain. Individual livestock owners in the Eastern Cape commonly know a limited number of core remedies and resort to specialists for difficult conditions such as problems in animal reproduction. Within the same villages, individuals interpret and treat animal diseases in different ways. Efficacy of treatment may be judged by observation – for example, the rapidity and scale of excretion, the decline in symptoms or a successful birth.

Livestock days, organised by government animal health technicians and sometimes addressed by representatives of the pharmaceutical companies, are an important source of spreading knowledge about biomedicines. Many livestock owners acknowledge that biomedicines are quicker and stronger: antibiotics such as terramycin are widely seen as a cure-all. Insecticides, both chemical dips and paint-ons such as Deadline, were valued for combating ticks because plant medicines seldom seemed effective. Some in North West Province believed that a solution of aloes killed ticks, but scientific testing in laboratories has not produced clear results. There has been little testing of plant medicines in the field.

Certain maladies, especially those seeming to cause general weakness and lethargy, were attributed to what might be termed 'ambient supernatural forces' with specific names such as *umkhondo* in isiXhosa and *mohato* in Setswana.[66] There is usually no malevolent individual involved, as in witchcraft; rather, these illnesses are seen to arise from agents such as dew on the grass or supernatural creatures, or from pollution by women. When a cause cannot be found, *mohato* is

[66] Beinart and Brown, *African Local Knowledge*, chapters 7 and 8.

sometimes invoked. Biomedicines are unlikely to be seen as adequate for such cases.

Animal diseases and treatments have remained an important area of local or indigenous knowledge until recent years because so many rural African families still have livestock. Treating animals is a major area of local practice that draws together observation of symptoms, diagnosis, environmental understanding and plant knowledge. This is to some extent a dynamic field, incorporating new ideas and treatments, but repositories of such knowledge are also fragmented and there is uncertainty about diagnosis and cures. The balance within hybrid and pluralistic approaches appears to be shifting unevenly towards biomedical ideas and treatments, constrained to some degree by cost for poor livestock owners. Discriminatory, gendered taboos are probably weakening and women in some families have been handling livestock, including cattle, for decades.

Scientific research into local veterinary knowledge, such as that by Patrick Masika, is beginning to seep into the training of extension workers in the Eastern Cape. Yet despite some receptivity and interest, government vets and animal health technicians are generally reluctant to incorporate traditional knowledge, preferring to rely on scientific diagnosis and biomedicines. It is one thing to understand local practices, and recognise their prevalence, another to advocate them. Government officers point to the variability of plant medicines, unpredictable outcomes, lack of testing and efficacy, the limits of local knowledge especially in relation to new diseases, and problems that might arise from its deployment in state-sponsored medical strategies. On the whole, popular demand from the state is for the reintroduction of chemical dipping or similar tick prophylaxis, as well as inoculation against anthrax. At the same time some veterinary officials are increasingly interested in reintroducing traditional Nguni cattle breeds, which are smaller but less susceptible to a number of diseases.

HIV/AIDS

In the case of HIV/AIDS, scientific diagnosis and treatment came into direct and dramatic conflict with alternative and traditional approaches – and at high human cost.[67] AIDS was first diagnosed in

[67] In a huge literature, N. Nattrass, 'Aids and the Scientific Governance of Medicine in Post-Apartheid South Africa', *African Affairs*, 107, 427 (2008),

South Africa in 1982, shortly after it was recognised as a new disease. Distinctive features of the disease in South Africa, and Africa more generally, were its rapid spread through the heterosexual population and in particular to poorer African communities. While knowledge about the disease, its causes and its origins, was limited in the early 1980s, it soon became clear that AIDS was potentially life-threatening and there was little chance of a cure. As it weakened the immune system, sufferers frequently succumbed to other diseases, such as tuberculosis, which remained stubbornly entrenched in South Africa. Such characteristics made the causes and sequence of the disease opaque to many South Africans, especially those with limited scientific education.

By 1984 the disease could be diagnosed by testing; this technique became widely available in South Africa over the next few years. Modes of diffusion were also clarified – particularly the exchange of bodily fluids through sexual intercourse, mother-to-child transmission, and the use of infected needles. Although by 1985 government and NGO campaigns for prevention had been launched especially in the UK and Europe, public education was by no means so widespread or effective in African countries. Blame and stigma spread rapidly, and the means of prevention and treatment were contested. HIV/AIDS represented a challenge for scientists; moreover, even when greater scientific certainty was established, this knowledge was not easily transmitted and accepted.

AIDS encouraged racialised suspicions and conspiracies in South Africa, just as in the United States, where the idea that the disease was devised in a laboratory as a 'bioweapon' gained currency. Medical specialists and media made careless statements about African sexual practices; this in turn led to a defensive reaction, and the disease was sometimes portrayed as a Western attempt to undermine African society. Some black South Africans saw AIDS as an apartheid plot linked to the biological warfare schemes of Wouter Basson or as part of an effort to control black fertility. In 1988, Mzala, writing in the South African Communist Party journal *Sechaba*, characterised the theory that AIDS was a zoonosis that had moved from animals to people in

157–76; and N. Nattrass, *Mortal Combat: AIDS Denialism and the Struggle for Antiretrovirals in South Africa* (Pietermaritzburg: University of KwaZulu-Natal Press, 2007) have been a good guide. See also N. Nattrass, *The AIDS Conspiracy: Science Fights Back* (Johannesburg: Wits University Press, 2012).

Africa as 'racist prejudice'. The Wits Sociology of Work Programme sponsored a study in 1989 that highlighted the vulnerability of mine workers to AIDS if it were to get a hold. Cyril Ramaphosa of the National Union of Mineworkers asked them not to publish, saying: 'It's offensive, it focuses on the private life of the members, indeed it's racist.'[68] In 1991 *Drum* magazine carried an article from a US journal which posed the rhetorical question: 'Is AIDS a conspiracy against Blacks?'[69]

In South Africa, medical education around the disease in the 1980s was slow and 'doom-laden'.[70] Legislation in 1987 allowed the Minister of Health to isolate and quarantine people who tested positive. Measures were also taken to exclude foreign migrant mineworkers diagnosed with the disease. The rising number of cases in hospitals such as Baragwanath, serving Soweto, caused concern. There was discussion as to whether the families and sexual contacts of those diagnosed should be informed. Because of the uncertainty about modes of transmission, some of those identified as carriers of AIDS were shunned.

By 1990 more sustained efforts at medically based action were developing through such organisations as the National Medical and Dental Association (NAMDA), founded in 1982 by a network of anti-apartheid medical professionals.[71] A series of articles on AIDS appeared in the *South African Medical Journal* and Quarraisha Abdool Karim, recently qualified and returned from a fellowship in the United States, was appointed to head an AIDS research unit under the Medical Research Council. She later spearheaded an AIDS programme in the Mandela government. In 1990, the ANC made a film on the dangers of AIDS that was circulated both to MK camps in exile and within the country. That same year the ANC held a conference in Maputo aimed at developing equitable access to health, which brought

[68] M. Mbali, '"The New Struggle": A History of AIDS Activism in South Africa, 1982–2003', DPhil thesis, University of Oxford (2009), 122. See also M. Mbali, *South African AIDS Activism and Global Health Politics* (London: Palgrave Macmillan, 2013).

[69] H. Phillips, 'HIV/AIDS in the Context of South Africa's Epidemic History' in K. D. Kauffmann and D. L. Lindauer (eds.), *AIDS and South Africa: The Social Expression of a Pandemic* (Basingstoke: Palgrave Macmillan, 2004), 34.

[70] R. Hodes, *Broadcasting the Pandemic: A History of HIV on South African Television* (Cape Town: HSRC Press, 2014), 22.

[71] Mbali, *South African AIDS Activism.*

together exiled and internal health activists. On this occasion ANC exiles, aware of the rapid spread of AIDS in other African countries, gave the disease more priority. Chris Hani, chief of staff of MK, warned of the 'untold damage' and 'suffering' that it could cause.[72] The conference committed itself to a community-based approach that could lay the basis for more equitable access to health and a national health service; preventive education and testing for AIDS would be expanded through such structures.

In 1992, Nelson Mandela opened the Durban conference of the National AIDS Convention of South Africa, jointly organised by the government and the ANC. Karim and others presented detailed epidemiological analyses of the incidence of the disease. South African prevalence, so far as it was recorded, was still substantially lower than in central and eastern Africa, where it had reached over 20 per cent in a few countries such as Uganda, Zimbabwe, Zambia and Malawi. In South Africa it was less than 5 per cent among adults and about 2 per cent among pregnant women, who were by then being regularly tested at public antenatal clinics. Medical scientists estimated that 95 per cent of infection was through heterosexual activity and that women and youths were particularly susceptible. At the time, the disease seemed manageable, albeit threatening; two key ANC health activists, Manto Tshabalala-Msimang and Nkosazana Dlamini-Zuma, who became Minister of Health in Mandela's government (1994–9), played prominent roles in developing an AIDS strategy for the ANC. In the early and mid-1990s, it seemed that a scientifically based and socially sensitive campaign of prevention could emerge, and in 1994 the national budget for AIDs, as part of the primary health care programme, was doubled.

From the mid-1990s a number of factors coincided to create divisions and uncertainties in the South African AIDS prevention and treatment programmes. Firstly, HIV prevalence increased exponentially, as did the number of deaths from AIDS. Department of Health statistics showed rates of 10 per cent maternal prevalence by 1995 and 24 per cent by 2000.[73] Despite the commitment of many NGOs and medical networks, South Africa was clearly unable to control the

[72] Mbali, 'The New Struggle', 132.
[73] J. D. Lewis, 'Assessing the Demographic and Economic Impact of HIV/AIDS' in Kauffmann and Lindauer (eds.), *AIDS and South Africa*, 100.

epidemic. HIV/AIDS was beginning to cast a shadow on the democratic transition.

In 1995 Mbongeni Ngema, creative director of the popular musical *Sarafina!*, was commissioned to produce *Sarafina 2* as an AIDS education vehicle. It proved to be dubiously funded and, in the opinion of critics, artistically lacking and ineffective in its message. The ANC was defensive in the face of widespread criticism. Far more contentious was the issue of treatment. Pharmaceutical companies were testing antiretroviral drugs for AIDS in the 1990s. In 1996 scientists at the University of Pretoria tried a chemical compound, based on an industrial solvent, which they called Virodene, and claimed that it had promising properties as a cure for AIDS.[74] These tests bypassed the Medicines Control Council (MCC), which should have sanctioned them. The MCC was sceptical about the claims, found that the drug contained a toxic substance, and opposed further testing.

Minister of health Nkosazana Dlamini-Zuma, nevertheless promoted the drug and sponsored a presentation to cabinet in 1997 at which extravagant claims were made, including testimony from two AIDS sufferers. Dlamini-Zuma attempted to persuade the MCC to change its mind and allow further testing. Deputy President Mbeki supported the initiative and went public with a newspaper article in March 1998, accusing the MCC of obstruction. Dlamini- Zuma said that she thought that patients, dying from AIDS, with little alternative, had the right to take Virodene and that the government had the responsibility to see that it was more fully tested. The MCC, suspected of being under the control of cautious whites from the old guard, was accused of refusing patients what might have been a 'mercy treatment to which they are morally entitled'.[75] Dlamini-Zuma argued that 'the cruel games of those who do not care should not be allowed to set the national agenda'. Peter Folb of the MCC responded that the Council had solicited more than seven independent reports both locally and internationally, which unanimously backed their decision not to allow human testing of Virodene.

Folb was replaced in March 1998 and tests went ahead; Dlamini-Zuma was keen that the government should come up with an effective

[74] J. Myburgh, 'The Virodene Affair (I)', *Politicsweb*, 17 September 2007: 'The secret history of the ANC's response to the HIV/AIDS epidemic'.

[75] Myburgh, 'The Virodene Affair'.

and cheaper drug, bypassing the pharmaceutical companies. The press alleged that the ANC had a financial interest in the drug. Karim, who had by this time left government, uneasy about policy directions, later said: 'There was this sense that this drug would be the thing that offset the perception . . . of Africans as substandard and less than capable. All eyes were upon [the ANC] and the expectations were very high and they were really trying to find their feet but they didn't want to exercise caution. This was driven by this need to show the world: "Yes, Africans can do this. We can do this." Virodene became our redemption.'[76]

Analogous claims were made for Kemron, an ineffective AIDS drug produced in Kenya which was touted as a miracle cure in America by the Nation of Islam's head, Louis Farrakhan.[77] In a parallel development, South Africa passed the Medicines and Related Substances Control Amendment Act of 1997. This enabled the government to source generic drugs from countries such as India and Brazil that could undercut the prices charged by major pharmaceutical companies. A number of these companies lobbied strenuously against the Act and then took the government to court. The case was dropped only in 2001. Thus, at the same time as the government was promoting Virodene, it was locked in a long battle with major pharmaceutical companies over drug pricing.

Opinion was not divided simply by race. William Makgoba, a Durban- and Oxford-trained immunologist, deputy vice-chancellor at Wits, and from 1999 head of the Medical Research Council, as well as leading advocate of the African Renaissance, called for the curtailment of tests of Virodene on humans. Coming from one of the key intellectuals associated with new Africanist approaches, his was a significant intervention. Hopes for the drug did not fade immediately. While a trial planned in Botswana did not go ahead, Virodene was tested in Tanzania with funding from South Africa. In March 2002, the drug was reported to be of no value in delaying the onset of AIDS.

By 1998, the antiretroviral drug AZT, long in development, became more affordable, and Nevirapine was shown to be effective against mother-to-child transmission. Uganda instituted the use of

[76] Myburgh, 'The Virodene Affair'; V. van der Vliet, 'South Africa Divided against AIDS: A Crisis of Leadership' in Kauffman and Lindauer (eds.), *AIDS and South Africa*, 86.

[77] Nattrass, *The AIDS Conspiracy*, 58ff.

antiretrovirals, but Dlamini-Zuma rejected these on the grounds that they were not cost-effective. In the midst of this increasingly polarised and politicised medical debate, the Treatment Action Campaign (TAC) was launched in December 1998. Led by Zackie Achmat, an HIV-positive anti-apartheid campaigner, this civil society action group grew rapidly into a social movement committed not only to securing state-funded treatment for sufferers but also to combating the stigma associated with AIDS. The TAC espoused global scientific understandings of the disease and attempted to insert these ideas in the popular imagination of all South Africans.

In this context Mbeki gradually moved to a position that was sceptical of conventional scientific understandings of HIV/AIDS. He became a convinced AIDs dissident after he acceded to the presidency in 1999. Explanations for Mbeki's motives are debated.[78] He was influenced by a dissident lobby in the United States and elsewhere that contested the link between HIV and AIDS and disputed the effectiveness of antiretrovirals. He established a council on AIDS in which those who supported such views, such as leading international AIDS denialists Peter Duesberg and David Rasnick, were strongly represented. Mbeki produced a long and angry paper entitled 'Castro Hlongwane', which not only criticised the scientific work on AIDS and the motives of allegedly greedy pharmaceutical companies, but also condemned racist mischaracterisations of African masculinity and sexuality. Mbeki argued that there was a 'studied and sustained attempt to hide the truth about diseases of poverty'; he saw black people as 'victims', who 'fully understand the legacy of centuries-old and current racism on our society and ourselves'.[79]

Mbeki's views distilled an approach to science that could be found in much of the critical, anti-colonial academic literature touching on themes in medical and environmental history. His hostility to Western or Eurocentric medical ideas was thus influenced by postcolonial and postmodern critiques of Western science, coupled with mistrust of capitalist pharmaceutical corporations.[80] Some

[78] D. Fassin, *When Bodies Remember: Experiences and Politics of AIDS in South Africa*, 46; Nattrass, *The AIDS Conspiracy*, 86–7.

[79] van der Vliet, 'South Africa Divided', 88.

[80] S. Robins, '"Long Live Zackie, Long Live": AIDS Activism, Science and Citizenship after Apartheid', *JSAS*, 30, 3 (2004), 651–72; R. Kowalenko, 'Thabo Mbeki, Postmodernism, and the Consequences', *South African Journal*

historians of medicine in Africa had long reacted against the idea of benevolent Western treatments bringing unquestioned benefit to Africa. They viewed science as coercive, insensitive to local knowledge and practice, and frequently unsuccessful or counterproductive – for example, in controlling trypanosmiasis or combating malaria. Charles Geshekter, an unapologetic AIDS dissident and critic of 'conventional assumptions' who served on Mbeki's AIDS Advisory Panel, drew on his expertise as an African historian. He framed disease in terms of poverty and underdevelopment, leading him to dismiss AIDS discourse as 'one of the greatest medical fallacies of our times' and as a 'sex panic'.[81]

'Plagues throughout history have spawned conspiracy theories', Virginia van der Vliet has noted, 'and HIV/AIDs ... produced a bumper crop'.[82] Medical sciences were seen as racist and imperialistic in their discourse and practice and, at least in Mbeki's eyes, inadequate and dishonest in explaining the scourge that assailed largely black South Africans. In turn, activists and scientists arguing for treatment saw the consequences of Mbeki's position as akin to 'genocide by omission'.[83] Within a few years, Mbeki was being accused of causing not only the infection of babies but also the death of 250,000–300,000 people who might have been saved by the early roll-out of antiretrovirals. For the title of her book on AIDS prevention, Catherine Campbell used the mordant words of satirist Pieter-Dirk Uys: 'in the old South Africa we killed people. Now we are just letting them die.'[84]

Manto Tshabalala-Msimang took over as Mbeki's minister of health in 1999 and compounded the errors of her department. She supported the position taken by her predecessor and focused on preventive messages, sexual education, distribution of free condoms, and nutrition. She also rejected findings that antiretrovirals were effective, describing

of Philosophy, 34, 4 (2015), 441–61; J. Youde, 'The Development of a Counter-Epistemic Community: AIDS, South Africa, and International Regimes', *International Relations*, 19, 4 (2005), 421–39.

81 C. L. Geshekter, 'The Globalization of AIDS: On Using History to Criticise Public Health Hysteria', paper presented to 5th Annual Conference of the Historical Society, University of North Carolina, Chapel Hill (2006), www.immunity.org.uk.

82 van der Vliet, 'South Africa Divided', 85.

83 van der Vliet, 'South Africa Divided', 86.

84 C. Campbell, *'Letting Them Die': Why HIV/AIDS Prevention Programmes Fail* (Oxford: James Currey, 2003).

them as 'poison'. Although professional organisations, provincial health departments, NGOs and activists, some with access to foreign funds, began planning and even implementing the distribution of anti-retrovirals by the beginning of the twenty-first century, the lack of government commitment retarded any national provision.[85]

A legal victory by the TAC in 2002 and a cabinet revolt against Mbeki and Tshabalala-Msimang gradually facilitated the universal roll-out of antiretrovirals from 2003. Even then, however, when the ministry became more involved, much depended on non-governmental initiatives – Medecins Sans Frontières in Cape Town and the Eastern Cape, or the 'Summertown project' in a mining town near Johannesburg. At this time, it was clear that disease prevalence in South Africa was outstripping that of neighbouring countries: millions of deaths were predicted and, without treatment, life expectancy was projected to fall from sixty-five to fifty-five years. Some of the worst predictions may have been alarmist, as they were not the result of national surveys but were based on the testing of women at antenatal clinics. Younger women were a particularly vulnerable cohort, among whom national prevalence, measured at such clinics, had risen to over 25 per cent.

Although Jonny Steinberg detected a 'besieged and aggressive nativism' in the reaction to antiretrovirals, Mbeki's stance in relation to drugs was more an attack on foreign pharmaceutical companies than an espousal of African remedies.[86] Yet many traditional healers, as well as foreign vendors of untested medicines, claimed that they could make some contribution to treating AIDS and a few made very large claims. Steinberg records a herbalist in Lusikisiki, not a qualified *sangoma*, who offered plant mixtures as a cure for AIDS. As in the case of ARVs, a trifecta of components was included: one said to be for TB, one for diarrhoea and one for headaches – 'when you mix them together it cures AIDS'.[87]

There was clearly significant demand for such medicines. As noted above, even when biomedicines were used, African patients often practised medical pluralism. A plant popularly called in English the African potato (*Hypoxis hemerocallidea*), long favoured for a number

[85] J. Steinberg, *Three Letter Plague: A Young Man's Journey through a Great Epidemic* (Johannesburg: Jonathan Ball, 2008); Campbell, *'Letting Them Die'*.

[86] Steinberg, *Three Letter Plague*, 92.

[87] Steinberg, *Three Letter Plague*, 222.

of ailments by traditional healers, was recommended by the Department of Health for AIDS treatment, although not tested.[88] A Dutch nurse concocted a mixture called Africa's Solution which was supposedly made of the African potato, olive leaf and grapefruit seed to be taken with garlic. Tshabalala-Msimang promoted a range of nutritional solutions, including vitamin therapies developed by Matthias Rath, an entrepreneur with a dubious record.[89] The Health Department was reluctant to intervene or test traditional, complementary or alternative medicines for AIDS although the MCC did act against *Ubimbi*, sold by Siphiwe Hadebe, as a cure.[90] Traditional healers publicly declared their support of Tshabalala-Msimang

In KwaZulu-Natal a former lorry driver, Zeblon Gwala, marketed a herbal cure called *uBhejane* (rhino) from 2006. The treatment was apparently revealed to Gwala in a dream by his late grandfather who had also been a traditional healer.[91] In this case, Gwala made specific claims that the solution 'increases your CD4 count and reduces the viral load until it disappears' and advised patients to stop taking antiretrovirals. Tshabalala-Msimang supported this cure, as did Herbert Vilakazi, head of the presidential task force on traditional medicine. *New Yorker* journalist Michael Specter interviewed Vilakazi at the time, reporting that he shared the view that antiretrovirals were 'damaging' and possibly 'toxic'. He thought that a cure for AIDS was more likely to be found in African medicines.[92] Vilakazi echoed Mbeki's ideas in general terms. 'Orthodox medicine has reached a dead end,' he claimed. 'There is walking evidence and evidence that comes from a lab. There are plenty of people who rely on walking evidence. A person has terrible arthritic pains. An old man gives them herbs and they get relief. We can't say that the only thing

88 E. Mills et al., 'African Herbal Medicines in the Treatment of HIV: *Hypoxis* and *Sutherlandia*. An Overview of Evidence and Pharmacology', *Nutritional Journal*, 4, 19 (2005).
89 Nattrass, *The AIDS Conspiracy*, 78.
90 Nattrass, 'Aids and the Scientific Governance of Medicine', 167.
91 Hodes, *Broadcasting the Pandemic*, 134–36; Adrian Flint, 'Traditional Healing, Biomedicine and the Treatment of HIV/AIDS: Contrasting South African and Native American Experiences', *International Journal of Environmental Research and Public Health*, 12, 4 (2015), 4321–39.
92 M. Specter, 'The Denialists: The Dangerous Attacks on the Consensus about H.I.V. and AIDS', *New Yorker*, 12 March 2007.

that matters comes in reports from Western labs.'[93] *UBhejane* was tested and found to have no positive effect on the treatment of HIV/AIDS. Though not harmful, it was expensive and created confusion.

These controversies left a deep legacy. However, since the turn of the century, scientific analyses have secured much wider support. The TAC made a major impact and focused its message on 'living positively', securing treatment, and the scientific validity of antiretrovirals.[94] It avoided debates about race and the origins of the disease. The TAC developed a regular, successful TV programme called *Beat It!* where those with AIDs could articulate their position to large audiences, confront stigmatisation, and make the scientific language of testing and treatment familiar. Early audiences reached 500,000 and by 2009 episodes had been seen by an estimated 14 million viewers.[95] Radio programmes, telephone helplines and websites expanded the coverage. Such sustained campaigning, together with the expansion of school and university education, the growing national testing programme, and the roll-out of antiretrovirals helped to create and sustain a new 'health citizenship', especially among younger people. They increasingly appealed to a public scientific imagination about disease, which contributed to a fundamental change in South Africa's popular consciousness. Howard Phillips talks of the 'biomedicalization of South African society' such that ARVs often became the first resort for treatment.[96]

Under Aaron Motsoaledi, the strongly committed and effective minister of health from 2009, South Africa developed by far the biggest antiretroviral treatment programme in the world. By 2016 about 3.5 million were receiving treatment out of an estimated 6 million people living with HIV. India was next with less than a million in treatment. In 2017 life expectancy at birth was estimated at 61.2 years for males and 66.7 years for females, only a little lower than in the 1980s before HIV/AIDS struck as an epidemic disease.[97]

[93] Specter, 'The Denialists'.

[94] Hodes, *Broadcasting the Pandemic*; Robins, 'Long Live Zackie, Long Live'; J. Steinberg, 'Re-examining the Early Years of Anti-retroviral Treatment in South Africa: A Taste for Medicine', *African Affairs*, 116, 462 (2017), 60–79.

[95] Hodes, *Broadcasting the Pandemic*, 6.

[96] Phillips, 'HIV/AIDS', 44.

[97] Stats SA, Mid-year population estimates, 2017, on web.

Quite why Thabo Mbeki became so personally invested in AIDS denialism remains difficult to explain. Mark Gevisser offers a helpful discussion.[98] In many other contexts, Mbeki was an enthusiastic moderniser who supported scientific initiatives: he was sometimes criticised as a remote technocrat. Yet he was affronted at racial stereotypes around black sexual behaviour. A residual Marxism found expression in his suspicion of 'Big Pharma' companies and his insistence that AIDS was a disease of the poor. Mbeki's scepticism was shaped by a particular reading of anticolonial and postcolonial critiques of Western knowledge. His self-perception as an intellectual as well as a politician also undoubtedly played a role and his personal commitment to the African Renaissance could in such contexts open the way for alternative ideas about science and policy. According to this vision, Africa should appropriate scientific frameworks on its own terms – and, in this case, on his terms. His vision proved attractive to some, but it had steep costs.

Figure 7.1 Thabo Mbeki, 'AIDS is not someone else's problem. It is my problem. It is your problem'.
Credit: Ulrich Baumgarten/Getty Images

[98] M. Gevisser, *Thabo Mbeki: The Dream Deferred* (Johannesburg: Jonathan Ball, 2007), chapter 41.

'I am an African': Mbeki's Modernising Vision in Relation to Astronomy and Palaeontology

In a celebrated speech, delivered on the occasion of the confirmation of the new South African Constitution in 1996, Deputy President Mbeki affirmed that he owed his being to the Khoekhoen, 'the blood of the Malay slaves who came from the East' and also to 'the migrants who left Europe to find a new home on our native land'. 'Being part of all these people, and in the knowledge that none dare contest that assertion,' he declared: 'I am an African.' The speech was widely interpreted at the time as a profession of racial inclusivity. Yet closer inspection of the lyrical text reveals nagging doubts. 'At times, and in fear,' he confessed, 'I have wondered whether I should concede equal citizenship of our country to the leopard and the lion, the elephant and the springbok, the hyena, the black mamba and the pestilential mosquito.' These doubts came to the fore in the HIV/AIDS crisis.

At his most confident, Mbeki was drawn to being a scientific visionary and proved fully equal to the task of reconciling an outward-looking Africanism with ambitious scientific universalism. This was evident in his administration's enthusiastic sponsorship of astronomy and palaeoanthropology, two disciplines with long pedigrees in South Africa that flourished in the post-apartheid democratic era.

In 2002 the Department of Science and Technology identified four fields of scientific research in which South Africa was seen to have a strongly competitive geographic advantage: astronomy, human palaeontology, biodiversity, and Antarctic research. It also added two further areas in which the country had an 'obvious knowledge advantage', namely indigenous knowledge and deep-mining technology.[99] In line with the aspirational 1996 government White Paper on science and technology, 'Preparing for the 21st Century', the development of southern skies astronomy became a national priority. Momentum was provided through skilled lobbying by astronomers with strong 'struggle' credentials, such as Bernie Fanaroff and Rob Adam, who joined the new government in senior positions. Leading South African astronomer Patricia Whitelock has rightly observed that astronomy was a particular beneficiary from new national funding arrangements,

[99] See Government of South Africa, 'South Africa's National Research and Development Strategy' (Pretoria, Department of Science and Technology, August 2002), eqdb.nrf.ac.za.

overseen by the CSIR's Foundation for Research Development and its successor body, the National Research Foundation. With the advent of the post-1994 'rainbow nation', astronomers moved to position their ambitious research projects as showpieces of national science. Whitelock shows that the South African astronomical community was highly effective in taking advantage of the post-apartheid political climate in order to pursue 'visionary projects'.[100]

The South African Astronomical Observatory (SAAO) used its optical telescopes in Sutherland to provide southern hemisphere data to the Whole Earth Telescope project, an international collaborative scheme involving multiple observatories focusing on a single object at a point in time. Sutherland's telescopes were ideally placed to record the impact in 1994 of Comet Shoemaker-Levy 9 on Jupiter, which was televised live, drawing 'an extraordinarily large South African audience'.[101] Surveys of ultraviolet rich stars resulted in cooperation between the SAAO, the Australian Astronomical Observatory, and the COSMOS photographic facility based in Edinburgh. Research into variable stars, focusing on the physical processes of their evolution, was pursued with notable success by South African astronomers. In the field of radio astronomy HartRAO's 26-metre apparatus, originally devoted to the tracking of American satellites, became an important centre for geodesy and for imaging distant cosmic radio sources. It was also the progenitor of new, expanding, ambitions in radio astronomy.

In 1998 the government agreed to fund half of a large 10-metre optical telescope, comprising a lattice of hexagonal mirrors, as part of an international and regional consortium. The scientific case for funding was first made by the director of the SAAO, Robert Stobie, together with Ian Glass and David Buckley.[102] The National Research Foundation stated in 1999 that its 'two primary missions' were to provide a state-of-the-art facility for local and international astronomers and astrophysicists, and to overcome and redress past government policies that 'dislocated the majority of South Africans from science, engineering and

[100] P. A. Whitelock, 'Optical Astronomy in Post-Apartheid South Africa: 1994 to 2004' in A. Heck (ed.), *Organizations and Strategies in Astronomy*, vol. 5 (Dordrecht: Springer, 2004).

[101] Whitelock, 'Optical Astronomy', 46, 47; B. Warner, 'Astrophysics in South Africa', *Transactions of the Royal Society of South Africa*, 63, 1 (2008), 61–7.

[102] R. S. Stobie, I. S. Glass and D. A. H. Buckley, *Southern African Large Telescope: A Proposal for Funding* (SAAO, 1993), SAAO archives SALT box.

technology education'.[103] Based on the Hobby-Eberly telescope in Texas and known as SALT (the Southern African Large Telescope), this facility has provided South African astronomers with outstanding capacity in the fields of optical photometry and high-speed spectroscopy. By calling it 'Southern' its protagonists signalled a willingness to encourage regional African cooperation in the subcontinent and so to share – or capitalise on – the dividend resulting from the country's transition to democracy. The SALT project was strongly supported by the NRF under the directorship of microbiologist Khotso Mokhele, who was instrumental in gaining government support for astronomy and facilitating South Africa's re-entry into international science.[104]

Figure 7.2 South African Large Telescope, SAAO Sutherland, showing lattice mirror and locally engineered piping.
Credit: Saul Dubow

[103] NRF, 'Salt Collateral Benefits Plan' by Tom Sebring, Bob Stobie and Khotso Mokhele, 19 April 1999 'final version', SAAO archives, SAAO Executive Meetings 1977–99 box.

[104] Warner, 'Astrophysics in South Africa', 62–4; de Kock and Moraal (eds.), *Physics in South Africa*; www.wits.ac.za/alumni/distinguished-graduates/honorary-degree-citations/khotso-mokhele.

Mbeki attended the inauguration of SALT in 2005 and proclaimed that 'this giant eye in the Karoo' (or 'gigantic African eye') will 'tell us as yet unknown and exciting things about ourselves'; it would 'provide all humanity with the largest and most modern single optical-infrared telescope in the Southern Hemisphere, while enabling our country and continent to remain among the front ranks of those involved in astronomy'.[105] He drew a direct connection between the cosmic potential of SALT and the imminent inauguration of the Cradle of Humankind site at Sterkfontein: for South Africans, as human beings, Mbeki argued, it was appropriate 'to continue the search for the origins of the infinite beginnings of the universe ... in the very geographic space that gave birth to homo sapiens'.[106]

This association had in fact been made before, albeit in a very different political context. Announcing its intention to consolidate South African astronomical activity at Sutherland in 1973, the CSIR spoke then of the small Karoo town, mostly known to scientists as a region rich in fossil deposits, as 'a new name in the world of astronomy'. Here 'the science of the stars will meet the science of the earth'.[107] The Sutherland region of the Karoo basin had long been associated with 'mammal-like reptiles' or therapsids, first discovered and identified by Andrew Bain in the 1840s (Chapter 2). Their fossil traces are now seen as evidence of global faunal and floral interchange, and hence as further evidence of continental drift. It is likely that these mammal-like reptiles were the fossil deposits to which the CSIR was referring, rather than accumulating evidence of our human progenitors, for in the age of high apartheid, suggestions about ancient human habitation long preceding the arrival of white colonists remained a source of political as well as of religious unease. Evolution was not taught in South African schools, and when the subject did surface Christian-nationalist precepts could filter knowledge to support perverse racist distortions – like the notion that whites were created by God, whereas blacks were descended from apes.[108]

105 'Speech of the President of South Africa, Thabo Mbeki, at the official inauguration ceremony of the Southern Africa Large Telescope (SALT): Sutherland, Northern Cape', 10 November 2005, www.thepresidency.gov.za; D. Buckley, 'SALT: "Gigantic African Eye"', *Quest*, 2, 2 (2005), 6–9.

106 Buckley, 'SALT: "Gigantic African Eye"'.

107 *Scientiae*, 14, 2 (1973), 6.

108 Durand, 'Paleontologie', 335–6.

The direct connection that Mbeki drew between the cosmic potential of SALT and human origins fitted perfectly with his integrative conception of the African Renaissance. Mbeki proudly drew attention to the new visitor centre at Maropeng – Setswana for 'the place where we once lived' – which was dedicated to publicising and popularising the UNESCO-branded Cradle of Humankind World Heritage Site. Since the mid-1930s this vast palaeontological storehouse has been yielding hugely important australopithecine fossils dating back 3.5 million years. Phillip Tobias, the grand old man of South African palaeontology and, towards the end of his life, praise-singer of the African Renaissance, was justifiably proud of the World Heritage Site, having done a great deal to bring it about. In a keynote address to a conference on the 'African Renais-Science' in 2002, Tobias captured the spirit of the moment by claiming that South Africa's hominid australopithecines were the progenitor of all living humans, that 'Africa gave the world its first culture', and that southern African hominids were thus 'not a southern African aberration, but a pan-African revelation'. In a similar vein, Tobias, a skilled wordsmith as well as a polymath, addressed Mbeki at a Cabinet meeting in 1998: 'Africa is the birthplace of humanity. Instead of African Renaissance, you should be talking about African Naissance.'[109]

For many years Tobias had presided over the palaeontological collection assembled at Wits by Raymond Dart – and also over Dart's complex legacy. Tobias was a figure of world renown in the field of palaeoanthropology and most likely one of South Africa's best-known scientists. An engaging synthesiser and communicator with enormous breadth of knowledge, including in the new field of population genetics, he was also a committed liberal opponent of apartheid. A lecture he delivered at Wits in honour of the politically outspoken medical doctor Raymond Hoffenberg was published in pamphlet form as *The Meaning of Race* (1961) and expanded in 1970.[110]

In line with the post-war UNESCO critique of biological race, Tobias was concerned here to challenge attempts to correlate brain size and brain matter with race and intelligence. In 1968 his doctoral

[109] Kuljian, *Darwin's Hunch*, 236.

[110] P. V. Tobias, *The Meaning of Race* (Johannesburg: South African Institute of Race Relations, 1961), expanded in P. V. Tobias, 'Brain-size, Grey Matter and Race: Fact or Fiction?', *American Journal of Physical Anthropology*, 32, 1 (1970), 3–25.

student Hertha de Villiers published the results of her groundbreaking statistical study, based on ranges of variation in crania rather than working from an idealised definition of particular 'types', and showing no differences in brain sizes.[111] Such work fed into a broader interdisciplinary commitment by historians and social scientists in South Africa to avoid offensive racial categorisations and language associated with entrenched racial prejudices.

On a visit to Paris in 1951, Tobias was shown the bodily remains of Sara Baartman, including her brain and genitalia, which had been deposited at the National Museum of Natural History following her death in Paris in 1815 and dissection by the leading French anatomist and naturalist, Georges Cuvier.[112] The story of how Sara Baartman was exhibited in Britain and France in order to demonstrate Khoesan racial and sexual difference was prompted in part by revelations in Stephen Jay Gould's critique of race, *The Mismeasure of Man* (1981). These caused public outrage in South Africa and by the 1990s there were frequent calls for Baartman's remains to be repatriated to South Africa. Tobias now became a key interlocutor between the South African and French governments. Mandela's personal intervention added to the moral and political pressure, and in 2002 Sara Baartman was reburied near Hankey in the Eastern Cape in an emotional ceremony that incorporated Khoekhoe ritual. Mbeki was present at Baartman's burial. He delivered an oration in which he castigated European scientists for her monstrous and barbaric treatment: 'She was ferried to Europe as an example of the sexual depravity and the incapacity to think of the African woman in the first instance and the African in general.'[113] It is worth noting that this angry denunciation of racism was delivered at precisely the moment that Mbeki was so publicly engaged in the HIV/AIDS controversy.

The South African geneticist Himla Soodyall, who argues that some of the oldest mitochondrial DNA lineages found in modern human

[111] Kuljian, *Darwin's Hunch*, 118–19; Dubow, *Scientific Racism*, 60–1.

[112] On Tobias's role, see Kuljian, *Darwin's Hunch*; more generally S. J. Gould, *The Mismeasure of Man* (London: W. W. Norton & Co., 1st edn, 1981) and 'The Hottentot Venus' in *The Flamingo's Smile: Reflections in Natural History* (New York: W. W. Norton & Co., 1985); C. Crais and P. Scully, *Sara Baartman and the Hottentot Venus: A Ghost Story and a Biography* (Princeton, NJ: Princeton University Press, 2008).

[113] T. Mbeki, 'Speech at the Funeral of Sarah Bartmann, 9 August 2002', www.sahistory.org.za.

populations are retained in Khoekhoe populations, was given the task of opening the old jars containing Baartman's remains for burial (in the process she was nearly overcome by the preservative fumes). Soodyall was by then a prominent public figure, her work adding credence to 'Out of Africa' theories of human evolution, which were the focus of the 'Cradle of Humankind'.[114] In 2005 she was honoured with the state award of the bronze Order of Mapungubwe for 'outstanding contributions in the field of science'.

This inclusive spirit was celebrated by the opening of a new Origins Centre by Mbeki at the University of the Witwatersrand in 2006. An impressive museum devoted to hominin evolution over two million years, it has special emphasis on stone tools and artefacts as well as rock art and more general visual and material culture of the Khoesan. In 2019 the University of Cape Town renamed the Jameson Hall, its main assembly space, after Baartman. Popular enthusiasm for rock art and the deep history of humanity is reflected in widely available artefacts and souvenirs, testimony to the reorientation of science and heritage in South Africa's public sphere.

The resurgence of public interest in human origins as a showpiece for the African Renaissance was a remarkable development. In 1993 Phillip Tobias's retirement had raised serious questions about the future of palaeoanthropological research at the University of the Witwatersrand and its unique fossil collections. There was little or no direct state support for either, but the advent of the new South Africa proved to be an epiphany, and fresh corporate funding became available. In 1994, Ron Clarke from Wits began to review old boxes from Sterkfontein containing fossil material. In one of these, he found bones of a foot which he correlated with other remains from the same individual. Tobias gave it the charming name 'Little Foot'. The structure of the foot suggested an ability to walk upright as well as to live in trees. Three years later Clarke found further related material in a box of monkey fossils housed in the Wits Anatomy Department. This prompted him to re-excavate a cave at Sterkfontein.[115]

[114] H. Soodyall and T. Jenkins, 'Unravelling the History of Modern Humans in Southern Africa: The Contribution of Genetic Studies' in P. Bonner, A. Esterhuysen and T. Jenkins (eds.), *A Search for Origins: Science, History and South Africa's 'Cradle of Humankind'* (Johannesburg: Wits University Press, 2007), 78–90.

[115] Kuljian, *Darwin's Hunch*, chapters 19 and 21.

Through a mixture of fortune and skill Stephen Motsumi and Nkwane Molefe, who worked with Clarke, recovered the remaining part of the foot just two days after commencing their search. By 1998 the full skeleton was excavated. Little Foot was presented to the world in December 1998 in the presence of Thabo Mbeki. It attracted enormous media attention as the most complete example of *Australopithecus* ever recovered. If Clarke's estimate of its age is right – more than 3.5 million years – its significance also rests on its being a much older species than Raymond Dart's type specimen, *A. africanus*, and older too than the celebrated 'Lucy' (*A. afarensis*) found in Ethiopia in 1974.[116] The disputed dating of the Little Foot – other estimates suggest it is more like 2–2.5 million years old – is important in positioning it as a potential human ancestor.[117]

Even more spectacular discoveries were made by Tobias's student Lee Berger, who eventually succeeded him as the supremo of palaeoanthropology at Wits after a protracted internecine struggle with Clarke. By contrast with the painstaking Clarke, Kansas-born Berger is a media-savvy extrovert who styles himself an explorer as much as a scientist. Berger sees little point in publishing in traditional scientific journals and prefers to get his results out directly to the public by way of e-publications. He views palaeoanthropology as a 'competitive sport' and actively embraces controversy and publicity. Having tried to fire Clarke for taking too much time to reveal Little Foot (whose existence Clarke hid from Berger in an act of prudential skulduggery), Berger re-emerged centre stage in 2008 with the discovery of bones in the Malapa Cave, near Sterkfontein, that appear to be intermediate between *Australopithecines* and *Homo habilis*. Berger has named the 2-million-year-old fossils *Australopithecus sediba* (from the Sotho word for 'spring' or 'well').

In a series of papers written since 2010, Berger has claimed that *sediba* is a more direct ancestor of modern humans than *africanus*. With sponsorship from *National Geographic* (where he is 'explorer-in-residence') Berger went on to excavate an underground cave system, known by recreational cavers as Rising Star. He engaged six slightly built women cavers – dubbed 'underground astronauts' – to navigate

[116] R. Clarke, 'Latest Information on Sterkfontein's *Australopithecus* Skeleton and a New Look at *Australopithecus*', *SAJS*, 104 (2008), 443–9.

[117] Clarke, 'Latest Information'; M. Balter, '"Little Foot" Fossil Could Be Human Ancestor', *Science*, 14 March 2014, www.sciencemag.org/news/2014/03.

the narrow system whose twists and turns have been given adventurous, playful, names. In 2013 a trove of 1,550 fossil parts, amounting to some fifteen skeletons, the largest assembly ever found in Africa, were retrieved from a remote chamber dubbed Dinaledi (stars). *Homo naledi*, Berger argues, is a new species within the genus *Homo*, who lived around 300,000 years ago and so overlapped with modern *Homo sapiens*. Their presence, in a chamber that is very difficult to access, has led him to postulate that the bodies of *H. naledi* were purposely buried 'down the chute'. If so, and the claim remains disputed, the inference is that these small-brained bipedal hominins behaved communally and evinced complex, ritualised behaviour.[118]

At the 2015 ceremony at Wits where Berger 'gave' *Naledi* to the world, an effusive Cyril Ramaphosa, then deputy president of South Africa, predicted that *Naledi* would 'catch the imagination' of humankind and inspire a retelling of the 'story of our own common ancestry'. Wits vice-chancellor Adam Habib has also stressed the discovery of *Naledi* as a contribution to 'a common humanity'.[119]

Biodiversity

Of the four areas of distinct scientific advantage identified by the government for support, astronomy and palaeontology have perhaps made the most spectacular scientific advances in post-apartheid South Africa. Both speak directly to global audiences and both work in different dimensions of 'deep time'. There have been equally important gains in biodiversity science, which, in the form of palaeobotany, also extends deep into the remote past. As we have argued, South Africa was site of a long history of botanical and ecological research bridging disciplines.

Brian van Wilgen, an influential environmental researcher based formerly at the CSIR, argues (with colleagues) that a key foundation

118 L. R. Berger et al., '*Homo naledi*, a new species of the genus *Homo* from the Dinaledi Chamber, South Africa', *eLife*, 10 September 2015; R. L. Holloway et al., 'Endocast Morphology of *Homo naledi* from the Dinaledi Chamber, South Africa', *PNAS*, 115, 22 (2018), 5738–43; P. Williams, 'Digging for Glory', *New Yorker*, 27 June 2016; M. Greshko, 'Cave Explorers Find New Fossils of Mysterious Human Relative', *National Geographic*, 13 September 2017; J. Shreeve, 'This Face Changes the Human Story. But How?', *National Geographic*, 10 September 2015.

119 www.wits.ac.za/homonaledi; www.wits.ac.za/homonaledi.

stone in such scientific endeavour was laid by an interdisciplinary committee report chaired by C. L. Wicht in 1945.[120] Concerns about the future of fynbos after widespread fires in the Cape Floristic Region resulted in the South African Royal Society appointing experts from different fields to collate knowledge. This gave impetus to a series of long-term research experiments, based initially at Jonkershoek near Stellenbosch, that were amongst the earliest to study the interrelationship between vegetation, water and fire, in river catchments. These investigations nurtured increasingly complex research into ecosystems that in many respects anticipated and underpinned the concept of biodiversity, popularised in the 1980s and 1990s. Following Wicht's report, innovative ideas were developed about the effects of fire on fynbos, as its role in the reproduction of some species was elucidated.[121] Systematic research began on the impact of invasive species, particularly Australian acacia or wattle. The ground was laid for new strategies of managing fire and also of alien species.

The concept of bioinvasion has been questioned by Africanists because it seemed to focus too much on the primacy of indigenous species in a world where vegetation had become irredeemably hybrid in most inhabited zones.[122] In South Africa, all farmers, and poor rural people in particular, used alien plants from maize and prickly pear to black wattle and pine. But the idea of defending indigenous biodiversity gained impetus at the time of the political transition, as part of a global project. This was a new language through which to promote environmental conservation, associated in the past with the white minority. In the 1990s, biodiversity and conservation research accelerated with new institutional frameworks and innovative policy directions.

Biodiversity conservation expanded in a number of broad fields. A newly constituted SANParks took up the legacy of protected areas. The board was transformed to include far more diverse representation and the first black director of Kruger, David Mabunda, was appointed

120 van Wilgen et al., 'Ecological Research and Conservation Management'.
121 Pooley, *Burning Table Mountain.*
122 J. Guyer and P. Richards, 'The Invention of Biodiversity: Social Perspectives on the Management of Biological Variety in Africa', *Africa*, 66, 1 (1996), 1–13; W. Beinart, *Plant Transfers, Bio-invasions and Biocultural Diversity: Perspectives from Africa* (Nehru Memorial Museum and Library Occasional Paper, New Series 42, New Delhi, 2014).

in 1998. Although such parks had also been associated in the past with the removal and exclusion of African people, the ANC committed itself to expanding their area, developing marine reserves, and pursuing more inclusive strategies. National parks increased in number and area. Provinces also have powers to reserve land: Cape Nature in the Western Cape energetically expanded protected areas and tourism from about 320,000 ha in the early 1990s to 900,000 ha in 2017.[123] The process of amalgamating the old Botanical Research Institute, formerly under the Department of Agriculture, together with the national botanical gardens and herbariums, was completed. The South African National Biodiversity Institute (SANBI) was founded and given robust powers in the National Environmental Management: Biodiversity Act of 2004 – absorbing earlier legislation and providing for comprehensive national biodiversity management.

Government and universities expanded long-standing expertise in researching and managing biological invasions. The biological eradication campaign against opuntia species (Chapter 5) was initially aimed to protect livestock farmers who found that these plants were damaging their animals, but by the 1980s, broader ecological arguments were canvassed: invasive species destroyed indigenous biological diversity. Such concerns materialised in the wide-ranging Conservation of Agricultural Resources Act (CARA) of 1983, which consolidated the lists of 'weeds and invader plants', categorised them according to their danger, and made stronger legal provision for their eradication. Opuntia was classed among the most undesirable and it became illegal to handle, sell the fruit, or spread them. Given that black women in poor rural communities picked and sold the fruit, this Act threatened significant social repercussions, but it was not much enforced because the plant was still widely used.

When an Agricultural Research Council was established in 1990 as a parastatal, various sub-sections of the Department of Agriculture were transferred to it, including the Plant Protection Research Institute, itself an amalgam of the former entomology, plant pathology and weeds sections. This body was responsible for a wide range of insect introductions in order to control invasives. A successful example was the American weevil, *Stenopelmus rufinasus*, released in 1997 to counter the American red water fern, *Azolla filiculoides*, which clogged dams,

[123] Cape Nature, *Annual Report*, 2016–17.

reservoirs and water systems.[124] The five species of weevils that eat invasive black wattle seeds, introduced to South Africa by 2011, made less impact. Hopes focussed instead on an Australian midge or gall fly that attacks flowers and stops the formation of seeds.[125] These are the ideal solution because they stop reproduction without affecting the trees of smallholders and plantation owners. South Africa became one of the leading countries in the use of biocontrols with perhaps 106 agents, mainly insects, introduced over the century up to 2014.[126] Exotic insects have, paradoxically, been used to increase indigenous biodiversity. South African insects were also exported, notably dung beetles to help control livestock manure in Australia.

Nevertheless, bioinvasions were seen to become more rather than less of a threat. Van Wilgen and Richardson, two leading South African scientists in this field, argued that: 'human communities and natural ecosystems worldwide are under siege from a growing number of destructive invasive alien species. . . . These species erode natural capital, compromise ecosystem stability, and threaten economic productivity.'[127] Despite growing expertise and a sense of urgency, the approach to exotics was uneven because many were also useful plants. For example, Australian saltbush (*Atriplex* species), classified as a weed, was planted for most of the twentieth century as a fodder for sheep in the semi-arid districts of South Africa. It does not generally invade rapidly, and some farmers consider that it enhances pastures. Similarly, American mesquite (*Prosopis* species) was deliberately

124 M. P. Hill, 'Biological Control of Red Water Fern, *Azolla filiculoides* Lamarck (Pteridophyta: Azollaceae), in South Africa' in T. Olckers and M. P. Hill (eds.), *Biological Control of Weeds in South Africa (1990–1998). African Entomology Memoir*, 1 (1999), 119–24.

125 B. W. van Wilgen, C. Dyer, J. H. Hoffmann, 'National-scale Strategic Approaches for Managing Introduced Plants: Insights from Australian Acacias in South Africa', *Diversity and Distributions*, 17, 5 (2011), 1060–75; J. Hoffman, 'Tiny Gall Fly Has a Big Impact on Invasive Black Wattle', www.invasives.org.za (2014).

126 J. Hoffman, 'Celebrating a Century of Weed Biological Control in South Africa', paper delivered to XIV International Symposium on the Biological Control of Weeds, Kruger National Park, South Africa (2014).

127 D. M. Richardson and B. W. van Wilgen, 'Invasive Alien Plants in South Africa: How Well do We Understand the Ecological Impacts?, *SAJS*, 100 (2004), 45–52; B. W. van Wilgen, D. M. Richardson et al., 'The Economic Consequences of Alien Plant Invasions: Examples of Impacts and Approaches to Sustainable Management in South Africa', *Environment, Development and Sustainability*, 3 (2001), 145–68.

introduced in the late nineteenth century as a fodder and shade plant in the driest pastoral districts of South Africa and was still being praised in the mid-twentieth century as 'probably amongst the most important fodder trees introduced into South Africa'.[128] While the pods have high nutritional value, the seeds, like those of prickly pear, survive ingestion so that livestock are active agents in spreading the plant. Following rampant invasion in some areas, scientists favoured eradication, but this has not been pursued systematically.[129] American lantana, ubiquitous in India and Australia as well as in South Africa, became highly invasive on forest fringes and roadsides.[130] Its unripe berries are toxic to livestock, but birds eat them when ripe and spread the plant that sustains them.

The CARA, amended in certain respects, remained basically intact following the political transition in 1994. However, the emphasis around some aspects of biodiversity conservation shifted because of the changing political context and because some scientists espoused social priorities. The new political moment opened up rethinking of a wide range of environmental responsibilities with unpredictable results.[131] For example, state forests were increasingly privatised. By contrast, water legislation, which for a long time favoured the rights of rural landowners, moved in the opposite direction towards greater state control. Brutsch and Zimmermann, authorities on prickly pear and its eradication, were sensitive to the value of the plant for poor Eastern Cape communities and became increasingly interested in its utilisation, particularly the cultivated spineless variety for the commercial production of fodder and fruit. They also argued that *O. ficus-indica*, at least, was unlikely to erupt again because it was held in check

[128] E. E. M. Loock. 'Three Useful Leguminous Fodder Trees', *Farming in South Africa*, 2, 250 (1947), 7–12.

[129] M. T. Hoffman, S. Todd, Z. Ntshona and S. Turner, *Land Degradation in South Africa* (Pretoria: Department of Environmental Affairs and Tourism, 1999).

[130] S. A. Bhagwat, et al., 'A Battle Lost? Report on Two Centuries of Invasion and Management of *Lantana camara* L. in Australia, India and South Africa', *PLoS ONE*, 7, 3: e32407 (2012).

[131] F. Khan, 'From Conservation to Environmental Justice: Trends in the Relationship between the State and Environmental Civil Society Organisations in South Africa' in S. Ravi Rajan and L. Sedrez (eds.), *The Great Convergence: Environmental Histories of Brics* (Delhi: Oxford University Press, 2018); Carl Death, 'Environmental Movements, Climate Change, and Consumption in South Africa', *JSAS*, 40, 6 (2014), 1215–34.

by imported insects and proposed removing it from the most dangerous category of weed in CARA and decriminalising its use.[132]

By contrast, scientists involved in an innovative new programme called Working for Water were ardent eradicationists – although they were also concerned to accommodate political change. Guy Preston at the University of Cape Town moved into the Department of Water Affairs in Mandela's government under the charismatic ANC lawyer, Kader Asmal. Together they gained sufficient political support to put Working for Water into practice. Scientific writing on invasive plants not only linked them to agricultural and ecological costs, but also condemned their thirst for water in a drought prone country. In this context, Australian acacia (wattle), hakea and eucalyptus trees were identified as the major culprits. Acacias were initially imported into the nineteenth-century Cape to stabilise sandy soil and into Natal for the leather- tanning and timber industry. Some of these Australian plants, such as the majestic blue gum, were largely planted for shade and aesthetic reasons. Many of them became self-spreading invaders.

The campaign against invasive acacias gained momentum from the 1970s as scientists and public opinion increasingly came to value the western Cape's indigenous fynbos vegetation. Exotic Australian trees not only crowded this out, they were blamed for more frequent and more intense fires that killed off some fynbos species.[133] Working for Water drew on ideas such as the protection of indigenous biodiversity, the enhancement of water supplies and the control of fire, to justify state expenditure on the eradication of exotic plants.[134] This programme, it was argued, would improve water supplies throughout the country.

132 M. O. Brutsch and H. G. Zimmermann, 'The Prickly Pear (Opuntia Ficus-Indica [Cactaceae]) in South Africa: Utilization of the Naturalized Weed, and of the Cultivated Plant', *Economic Botany*, 47, 2 (1993).

133 Richardson and van Wilgen, 'Invasive Alien Plants'; CSIR Environmentek and Institute for Plant Conservation, University of Cape Town (multiple authors), *An Assessment of Invasion Potential of Invasive Alien Species in South Africa, Final Report* (Stellenbosch: CSIR, 2004); A. H. M. Görgens and B. W. van Wilgen, 'Invasive Alien Plants and Water Resources in South Africa: Current Understanding, Predictive Ability and Research Challenges', *SAJS*, 100 (2004), 27–33; J. and J. L. Comaroff, 'Naturing the Nation: Aliens, Apocalypse and the Postcolonial State', *JSAS*, 27, 3 (2001), 627–51.

134 M. P. de Wit, D. J. Crookes and B. W. van Wilgen, 'Conflicts of Interest in Environmental Management: Estimating the Costs and Benefits of a Tree Invasion', *Biological Invasions*, 3 (2001), 167–78.

Working for Water, launched in 1995, included a public works and job creation campaign, which provided an added incentive for the ANC government seeking to expand employment for poor people in rural districts.[135] Within a few years the programme employed over 30,000 people and brought much-needed income in areas of high unemployment; by 2014 an estimated 180,000 people had been employed. Working for Water in many respects provided environmental and social gains at relatively little cost to government and attracted international interest. It took limited account of local use of exotics such as black wattle (*Acacia mearnsii*), a major target in the eastern provinces of Mpumalanga, KwaZulu-Natal and the Eastern Cape.[136] Historically, the tree had been planted around African homesteads as a quick-growing source of building timber, fencing, firewood and medicine. But with the advent of concrete and mud bricks, as well as electricity, diminished usage was giving new impetus to invasion. And, as in the case of popular AIDS education, Working for Water became a vehicle for spreading scientific ideas – in this case about biodiversity – at the grass roots.

The 'Big Science' of Astronomy

Optical and radio astronomy mark the pinnacle of South Africa's commitment to pure science and 'big science'. In the former field, the launch of the Southern African Large Telescope (SALT) in 2005 greatly expanded capabilities at Sutherland. Radio astronomy lagged some way behind. Development of an array of sixty-four radio antennae telescopes, known as MeerKAT – a double pun signalling an extension of the prototype seven-dish Karoo Array Telescope (KAT) – began in the Carnarvon region of the Karoo. Innovative local engineering allowed MeerKAT, in the words of one of its key protagonists,

[135] P. Woodworth, 'Working for Water in South Africa: Saving the World on a Single Budget?', *World Policy Journal* (Summer 2006), 31–43; C. Marais, B. W. van Wilgen and D. Stevens, 'The Clearing of Invasive Alien Plants in South Africa: A Preliminary Assessment of Costs and Progress', *SAJS*, 100 (2004), 97–103.

[136] Beinart, *Plant Transfers, Bio-invasions*; M. Aitken, H. Rangan and C. A. Kull., 'Living with Alien Invasives: The Political Ecology of Wattle in the Eastern Highveld, Mpumalanga, South Africa', *Etudes Ocean Indien*, 42, 3 (2009), 115–40.

Bernie Fanaroff, 'to deliver more with less'.[137] In 2010 the first of these antennae was linked, via very-long-baseline interferometry, to Hartebeesthoek. Michael Cherry, who wrote frequently on South African science for the journal *Nature*, calculated that the allocation of R1.9bn for the construction of MeerKAT, for the period 2009–12, amounted to 14 per cent of the Department of Science and Technology budget and three times the annual contribution to the NRF.[138]

MeerKAT was the precursor to an even more ambitious global project, known as the Square Kilometre Array (SKA) so named because of the total size of its 1 million m^2 collecting area. Headquartered at Jodrell Bank outside Manchester and collecting enormous quantities of data from dishes in South Africa and Australia, the SKA, when completed in the 2020s, promises to be the largest radio telescope in the world. In peering back into the dark age of the cosmos to the era after the Big Bang, it seeks to answer fundamental cosmological questions: 'What is Dark Energy? Was Einstein right about gravity? What is the nature of dark matter? Can we detect gravitational waves? When and how did the first stars and galaxies form? What was the origin of cosmic magnetism? How do Earth-like planets form? Is there life, intelligent or otherwise, elsewhere in the Universe?'[139]

Also unknown are the precise funding arrangements of the international mega-project, which have recently been estimated at €1.8 billion.[140] South Africa's share of the expenditure is difficult to calculate, as are the implications of so much state investment in one mega-project for other areas of scientific research. The country's success in securing its key place in the SKA consortium involved intense competition with Australia, a rivalry that goes back 200 years and that was eventually resolved by co-hosting the multinational project between the two southern hemisphere countries. Within South Africa, the SKA was heavily promoted by reference to its developmental potential. It is promised that Africa will be 'integrated into European and global science; a new generation of African scientists and technicians will receive advanced training; the infrastructure of very fast computing

[137] Interview with Bernie Fanaroff by Saul Dubow, Cape Town, 5 September 2019.
[138] M. Cherry, 'South African Science: Black, White and Grey', *Nature*, 463 (2010), 727.
[139] prepska.skatelescope.org.
[140] A. Ahuja, 'World's Biggest Telescope Will Chart the Cosmic Dawn', *Financial Times*, 18 March 2019.

and big data will benefit the country; local communities will be able to take advantage of educational and employment opportunities'.[141]

Whether the benefits of the SKA will in fact be seen by local communities remains uncertain. Already, there is evidence of local discontent. The area designated as an 'Astronomy Advantage Area' covers a vast area of the Northern Cape. Far from being a pristine desert, the Karoo is a populated area, and has been for millennia. Enabling legislation allows for regulatory restrictions over matters like electric fencing, radio masts, farm machinery, and the use of mobile phones. The cosmic signals being measured by the SKA can be fifteen orders of magnitude (1,000,000,000,000,000) weaker than those of a cell phone, so any interference is a major problem.[142] In Hawaii, in a striking illustration of colliding global and indigenous cosmologies, first nation peoples have objected strongly to the erection of a giant optical telescope on the sacred site of Mauna Kea. By contrast, in South Africa local and global tensions currently focus on modernity's distribution of material benefits.

Much of the promotional literature concerning the SKA evokes the idea that the enormous commitment to the science of cosmology represents the fulfilment of deep human and universal curiosity about the heavens. From 2014, the SKA sponsored a 'Shared Sky' exhibition which featured a collaborative painting from Wajarri aboriginal artists in Western Australia and a collaborative quilt from South African indigenous artists. It was said to have been stimulated by 'essentially identical views of the night sky to the peoples that have lived there for tens of thousands of years', coupled with the observation that the 'night sky is an increasingly scarce natural resource that belongs to and is shared by all humanity.'[143] The use of the singular term 'sky' in preference to 'skies' underlines the presumed identity between Australian and South African first nations.

While the link made between Australian and South African indigenous peoples embodies the spirit of scientific cooperation between these two countries, such claims obscure more than they reveal. A recent archaeological and ethnological study points out that the SKA's

[141] Dubow, '200 Years of Astronomy in South Africa'.
[142] 'Astronomy Geographic Advantage Act', www.ska.ac.za; *Government Gazette*, 621, 40706, 21 March 2017; I. Vegter, 'SKA: Be Grateful Karoo Residents Didn't Object', *Daily Maverick* (online), 29 May 2012.
[143] 'The Shared Sky Concept', www.skatelescope.org/shared-sky

120,000 km^2 'astronomy reserve' in the Karoo 'coincides almost exactly with a region described by |Xam prisoners of the mid-nineteenth century as |Xam-ka !au, their homeland'.[144] The authors question whether the South African San Council, which represents Bushman groups from the Kalahari region, is well placed to speak for the |Xam and their descendants who still populate the area around Carnarvon in the Karoo.

More fundamentally, they question the underlying conceit that the cosmological universe of the |Xam bears much relationship to that of modern science. The extraordinary oral archive assembled by Wilhelm Bleek and Lucy Lloyd, based on long interviews with San prisoner-informants working on the Cape Town breakwater, offers rich insights into the ways in which San people like ||Kabbo conceived of the Karoo universe. As the authors of the study observe: 'Whereas modern scientific cosmology admits a universe that is enormous, far-flung, and remote – in which human life, and especially that of the individual, is minuscule and on the whole utterly inconsequential – the understanding emerging from the stories of the |Xam is of a sky that is nearby, and filled with "persons" of consequence to the living. The sky is indeed a part of the landscape, enmeshed "with" earth and its inhabitants, animate and inanimate, rather than being separated distantly "above".'[145]

This insight illustrates a key theme we have explored in this chapter, namely that indigenous or vernacular knowledge systems articulate unevenly with universal scientific knowledge. They do not exist in a state of binary opposition. At certain moments, as in research related to health and ethnobotany, they can be complementary; in the case of HIV/AIDS they became explicitly oppositional; in the cosmology of stargazing they exist in parallel universes, other than for reasons of feel-good inclusiveness. Context matters a great deal.

So too, does political motivation. The Mandela and Mbeki governments gave a significant boost to science, especially those forms of inquiry that arose out of South Africa's geographic position and history, defined in terms of scale and deep time: astronomy, palaeoanthropology, and long-run studies of biodiversity and climate change.

144 J. Parkington, D. Morris, J. M de Prada-Samper, 'Elusive identities: Karoo |Xam descendants and the SKA' *JSAS* 45, 4 (2019), 729, 730.

145 Parkington et al., 'Elusive identities', 733.

Remarkably, there is at least a superficial concordance between the scientific ambitions of South Africa's leading philosopher-statesmen, Jan Smuts and Thabo Mbeki. Both promoted distinctive forms of South Africanism: the former ethnically inclusive as regards whites, and focused on membership of the Commonwealth; the latter inclined to Africanism yet also inflected with a strong sense of global belonging. For Smuts, the South Africanisation of science entailed an emphasis on discovering evolutionary divergence and the emergence of different kinds of humans; by contrast, in the era of the African Renaissance, palaeoanthropology sought to demonstrate the emergence of a common humankind centred on the continent of Africa.

Smuts lost office in 1948 to the National Party in part because he was portrayed as being too slippery ('*slim*') and as having lost touch with ordinary Afrikaners in pursuit of his ambitions as a world statesman and leader of the British Commonwealth. Mbeki lost out to his successor, Jacob Zuma, in part because he was seen as a modernising technocrat who was unable to connect with the ANC's increasingly impatient mass base. Mbeki's attempt to promote the African Renaissance, while also developing South Africa's presence in international science, was a bold and distinctive endeavour to bring together sharply divergent tendencies in the rainbow nation. His presidency ended in bathos and humiliation. But Mbeki's dual vision continues to reverberate more than a generation after the formal end of apartheid.

The example of the iThemba pharmaceutical company – isiZulu for 'hope' – is illustrative of this duality. Set up in 2009 as a drug discovery venture by a group of ambitious South African, American and British scientists and entrepreneurs, with financial support from the South African government, iThemba sought to develop innovative drugs to tackle endemic diseases such as malaria, HIV/AIDS and TB. It was motivated by the desire to find 'African solutions for African problems' and to do so by utilising the best of African and overseas talent as well as first-rate equipment.

Anne Pollock's ethnography of the fledgling company, which folded in 2015, does not dwell on its failure – this is not unusual for start-up biomedical ventures. Rather, she focuses on the contradictions that constrained its new South African 'socioscientific imaginary' and its ambition to challenge North–South inequities. She draws a number of broader conclusions. One of these is that under apartheid there was 'significant innovative science' directed to benefiting the white minority, notwithstanding the country's relative isolation. In the post-apartheid era, by contrast, South African science has become 'at once more inclusive nationally and regionally but more marginal globally. The transition from the colonial and apartheid models to one in which science would serve (the majority of) the people has been a rocky one – and it remains incomplete.'[146] This seems a fair judgement.

[146] A. Pollock, *Synthesising Hope: Matter, Knowledge, and Place in South African Drug Discovery* (Chicago, IL: University of Chicago Press, 2019), 15.

Afterword

We have aimed to write a book that speaks to multiple audiences, starting from the proposition that a synoptic view of the scientific imagination in South Africa over 300 years has much to offer specialist as well as general readerships. Firstly, for those interested primarily in South Africa, our argument is that scientific ideas and practices have irreversibly shaped society, and that they should be understood as an intrinsic element of political power, economic growth, and social change. In spheres from human and veterinary medicine to mining and agriculture, from the technologies of transport to those of water and energy, science has distinctively moulded South Africa's modernity. It has done so directly through the exercise of techno- and bio-power – and also indirectly through theories and rhetorics of exclusion and inclusion, progress and entitlement, regeneration and degeneration.

Because science is no one thing and holds out the promise of universal improvement, it has also been deployed to persuade conquered groups, white as well as black, that rational progress was the route to broad social advancement. A belief in the power of science was crucial to the ways in which colonists and settlers in southern Africa held power and also the manner in which they came to think of themselves as belonging to Africa: the scientific imagination became key to acquiring a sense of indigeneity and to asserting privileged claims to citizenship.

For all the systematic efforts to exclude them, black South Africans by no means stood outside or beyond the rapid changes brought by scientific innovation. Science may often have objectified black people as subjects of study, but it also shaped their everyday lives as peasants, workers and participants in a colonised world. These all-encompassing effects were unavoidable. Almost everyone had to inoculate against smallpox (and later their cattle against anthrax) and dip their livestock. Innumerable African men were immersed in the advanced technologies of the deep level mines. Many black South Africans were incorporated as patients, nurses, medical auxiliaries and (to a lesser extent) as doctors into the practices of biomedicine. Especially after the Second World War, black people were increasingly absorbed into a modern consumer society. Key African nationalists invoked a progressive, modernist, political programme in which scientific modes of thinking held out the possibility to improve the future well-being of black South Africans. These themes remain muted in recent historiography: while there are valuable historical texts illustrating the influence of science – for example on firearms, medicine, nursing, wildlife, the 'biometric state' and fire – our book suggests further routes for thinking about, and expounding, the role of science in writing a general history.[1]

By coincidence, international interest in South Africa peaked towards the end of each of the last three centuries. These moments were accompanied by intense scientific activity and engagement. By the late eighteenth century, the Cape and its hinterland was recognised as a major source of information about species new to European science and featured in the great integrative task of mapping and classifying the natural world. Its rich botanical and zoological inheritance was a particular source of fascination. By the close of the nineteenth century the country had become a locus of applied geology, chemistry, and engineering, as well as infrastructural investment to develop the gold mines and to fight Britain's largest war since Crimea. The mineral revolution depended on rapid technological advance in order to extract resources deep within earth's crust; mapping outcrops and underlying geological structures in turn opened up fundamental questions about the history of the earth's continents as well as animal and human evolution.

The twentieth century was bookended by two deeply contested political struggles, leading to major realignments: the 'new' South Africa which brought about the Union of South Africa (1910), and the transition to a democratic, non-racial new South Africa in 1994. Ethnic contestation between English- and Afrikaans-speakers dominated party politics in the first half of the twentieth century, and science was both a marker and a maker of white political imaginaries at this time. In the second half of the twentieth century apartheid came to define domestic and international debates about South Africa's future. As popular resistance opened up the possibility of transforming the country into a democratic society and reintegrating a pariah state into the world community of nations, so science, technology and development was called upon in new ways.

[1] Storey, *Guns, Race and Power*; Digby, *Diversity and Division in Medicine*; Carruthers, *National Park Science*; Marks, *Divided Sisterhood*; Breckenridge, *Biometric State*; Pooley, *Burning Table Mountain.*

This period also engendered intense scientific imagining. The first two presidents of the new South Africa, Nelson Mandela and Thabo Mbeki, articulated a vision of South Africa as a modern, scientific nation able to hold its own in the world. With different emphases, both identified science, and the scientific imagination more generally, as central to the future of a modern, postcolonial, African society.

Our second envisaged audience comprises those with interests in the history of science and the formative disciplines and theories that underpinned it. Although we are not ourselves trained in this field, our hope is that specialists will find interest in an integrated history of a deeply fractured country, whose geographic and political isolation from Europe has meant that it has been rather neglected as a zone of scientific activity. Consider, for instance, the country's treatment in the state-of-the-art *The Cambridge History of Science*. This project, designed to cover disciplinary developments in the field as a whole, only touches on South Africa. The Cape barely features in the extended discussions of European science in volume 4 on the eighteenth century, at a time when we see intense European engagement with its indigenous human and natural world.[2] Astronomy is mentioned briefly in 'Sciences and Voyages of Discovery'. Chapters address 'Non-Western Traditions' in India, China, Japan, the Islamic world and Spanish America, but not in Africa. South Africa is largely absent even in volume 6, which covers some of the themes and disciplines, such as geology, zoology, botany, palaeontology and environmental sciences, that we consider to have been national strengths over the long term.[3] In the chapter on 'Discovery and Exploration', Roy Macleod, an Australian author strongly aware of the significance of the global periphery, is largely silent on the Cape – save for a brief reference to its Observatory.[4]

No doubt, there are other countries and societies which also merit greater attention. We cannot assert with confidence that authors writing very largely about scientific innovation and texts in the core Western countries, should bring Southern Africa in from the outer peripheries of the history of

[2] R. Porter (ed.) *The Eighteenth Century*, volume 4 of *The Cambridge History of Science*, general editors D. C. Lindberg and R. L. Numbers (Cambridge: Cambridge University Press, 2008).

[3] P. J. Bowler and J. V. Pickstone (eds.), *The Modern Biological and Earth Sciences*, volume 6 of Lindberg and Numbers (eds.) *The Cambridge History of Science* (2009). In T. M. Porter and D. Ross, *The Modern Social Sciences* volume 7 (2008), O. Sichone has a chapter on 'The Social Sciences in Africa', 466–81.

[4] R. Macleod, 'Discovery and Exploration' in Bowler and Pickstone (eds.), *The Modern Biological and Earth Sciences*, 34–59.

science and technology. Our evidence does not prove that the country, more than others, ought to be seen as an important contributor to scientific innovation. But we do argue that the Cape and South Africa were significant reference points and contributors in fields such as zoology, botany and geology. Africa's role in zoology requires fuller mention than a stuffed giraffe. There is potential to include what may seem to be the scientific margins through concepts such as peri-centricity and poly-centricity.

South Africa was and is a middle-income country. At its peak around 1960 – when the great post Second World War commodity boom reached its height – its GDP ranked globally at 18–20th; in 2019 it is around 33–35th. Yet it was not an average, middling country. Its position on the shipping route to the east was key to the Portuguese and then the Dutch maritime empires. By the nineteenth century, Cape and Natal ports linked expansive shipping networks bridging the Atlantic and Indian oceans. They also served as important access points to the African hinterland, bringing whites into conflict with indigenous societies and British imperialism into contestation with Afrikaners as well as rival German and Portuguese empires. The development of science and technology was an important, and thus far an underestimated, dimension of such processes.

In the process of familiarising ourselves with the range and depth of scientific activity in South Africa, we have been surprised by the richness of the field. This has become apparent through writing a nationally based history over the long term – an unusual project in contemporary work on the history of science. We provide examples where scientists contributed to significant findings in zoology and botany, palaeontology and anthropology, astronomy and earth sciences. We are aware of the dangers of overstating the country's significance in the history of science more generally. Yet our overview reveals many examples of adaption, innovation and impact.

Take medicine and veterinary science. South Africa was probably the locus of the first successful Caesarean section operation in the Anglophone world, and it was performed by the first British woman to qualify as a physician. Christiaan Barnard performed the first heart-transplant operation at Groote Schuur hospital at a time when only a very few specialists conceived of this as being in the realms of the possible. Veterinary scientists developed serums and vaccines against rinderpest, anthrax and horsesickness. Physicians connected with the mining industry expanded treatment of lung diseases, including an early 'iron lung' (1918). Outbreaks of polio in the 1940s and 1950s prompted concerted research that placed South Africa at the forefront of work on immunisation at the point where Jonas Salk developed his famous vaccine. The fact that wartime outbreaks of polio disproportionately affected whites, the availability of an ample supply of vervet monkeys used in the production of a vaccine, as well as large-scale public support and the resources of public laboratories, were important factors in this response.[5]

Adaptation rather than invention *de novo* was a frequent process and economic necessity provided the prompt for urgent transfers of ideas and technology. The development of the MacArthur Forrest method of gold extraction on the Witwatersrand, and the production of dynamite on an unprecedented scale, enabled the world's largest gold mining industry. During the Second World War, scientists working with Basil Schonland devised from scratch, at pace, and in secret, a system of radar to defend the country's long coastline. The mineral-energy complex underpinned important developments in techno-politics during the height of the apartheid era: lack of oil reserves, cheap coal, national security and the drive for self-sufficiency hastened the expansion of proprietary oil-from-coal technology, as well as the enrichment of domestic uranium sources to harness nuclear power and produce nuclear weapons.

Thirdly, our text is for scientists, who are far larger in number than historians. Some have turned their hands to the history of fields such as astronomy, botany and medicine. We have drawn on carefully researched texts by Ian Glass and Brian Warner on astronomy, J. C. de Villiers on surgeons in the South African war, and on the many contributors to A. C. Brown's *History of Scientific Endeavour in South Africa*.[6] Our perspective on specific sciences is, necessarily, illustrative rather than systematic because we seek to make more general arguments; this may require tolerance from specialists looking for a more sequential analysis of disciplines and developments in which they are absorbed. However, by following individual careers, exploring wider institutional networks, and moving between specialisms, we attempt to explore broader connections and to show unexpected relationships and homologies emerging out of diverse networks and histories. We treat scientists as social beings as well as individual practitioners in order to demonstrate the ways in which they and their institutions were influenced by political contexts and affinities.

Perhaps our most critical audience, fourthly, will be those who do not need persuading that science is irredeemably embedded in colonial, racial and patriarchal power relations. This constituency (which

[5] Gear, 'Backround to the Poliomyelitis Vaccination Campaign', 501–2.
[6] Glass, *De la Caille*; de Villiers, *Healers, Helpers and Hospitals*; Brown (ed.), *History of Scientific Endeavour.*

happens to be the one in which we trained) may wish to see a more overarching study of knowledge, one that highlights African indigenous and local knowledge over the long term. It was suggested to us, for example, that we start with Khoekhoe society and the medieval African kingdom of Mapungubwe. We decided against this approach early in our planning; it is a huge challenge, in part addressed in fields like archaeology, which has recorded in detail the material remains of indigenous endeavour but less so ideas and culture.[7]

We have incorporated discussion of indigenous and local knowledge, especially when it became significant in scientific debates and political conflict. In broad terms, we argue that science-based progressivism has had many different political shades, encompassing co-optation, incorporation, and coercion. Our analysis illustrates that scientific ideas were by no means universally accepted, especially when they were perceived to be part of coercive state policies. Yet we reject the assumption that resistance to scientific progress can automatically be mapped onto race. For example, we remind readers that a major late nineteenth-century political crisis was precipitated amongst Afrikaners by enforcement of compulsory dipping to combat scab in sheep. Similar conflicts over biomedical approaches created tension in the South African war concentration camps. In both cases, disagreements about the cause and treatment of diseases were exacerbated by broader conflicts over British power.

In the middle years of the twentieth century, prescriptions for the Betterment of African rural areas, on the basis of scientifically conceived conservationist interventions, helped to trigger rural rebellions. However, rural African livestock owners, largely older men, gradually came to accept dipping as a means of tick control. ANC president A. B. Xuma was a practising doctor and a strong advocate of progressive social medicine in urban as well as rural areas. He was followed by James Moroka, also a medical doctor. The next ANC president, Nobel laureate Albert Luthuli, was both a chief and the chair of the Natal and Zululand Bantu Cane Growers' Association, in which capacity he arranged for the purchase of a tractor to help the organisation improve their members' productivity.[8]

During the transition from apartheid in the 1990s, debates around the relationship of Western and indigenous knowledge were sharpened, but not always neatly along lines of race. While the promise of the 'rainbow

[7] Mitchell, *Archaeology of Southern Africa*.
[8] R. Vinson, *Albert Luthuli* (Athens: Ohio University Press, 2018), 29.

nation' – an explicitly hybrid conception – encouraged new interest in fields like ethnobotany, the overwhelming historical 'whiteness' of science was undeniable. The intensity of debates about the causes and treatments of HIV/AIDS at this time left few in doubt about the continuing potential for racialised arguments about science. Yet in this case also, many African people accepted biomedical approaches, which were widely advocated by NGOs such as the Treatment Action Campaign.[9]

Thabo Mbeki's 'African Renaissance' was ambivalent in its approach, sometimes emphasising counter-narratives to 'Western' science. Nevertheless, ANC governments since 1994 have largely recognised the legacy and value of South African scientific work. Three white South African scientists, as well as Hamilton Naki who worked with Barnard, were amongst the six inaugural recipients of the Mapungubwe award, South Africa's highest honour, in 2002. They included Schonland, early architect of the CSIR. The ANC gave science a significant boost, yet sought to reorient it in new directions by retracting from the nuclear ambitions and military priorities of the Afrikaner nationalist government. Political transition provided a peace dividend and encouraged a focus on social, technological, and medical research with the potential to address poverty and disease. Enhanced government funding also supported areas of past strength, in fields that allowed scope for imaginative thinking such as astronomy, palaeoanthropology and biodiversity.

While the emigration and retirement of scientific professionals, particularly doctors and engineers, has depleted South Africa's skills base, post-apartheid science has become more connected and multi-national. Notably, recent debates about the 'decolonisation' of science have been more apparent in the humanities and social sciences than in university faculties of science. Moreover, the background of those doing science in South Africa is changing for the better. Many more black South Africans are being trained and are finding opportunities in government, state enterprises and universities. A steady stream of scientists has come from African countries because South African institutions have advantages with respect to working conditions and global connectivity. In this context, science can be an important vehicle for universalist modes of thinking and practice.

Leading African scientists continue to train in European and North American universities. For example, archaeologist Shadreck Chirikure,

[9] Nattrass, *Mortal Combat*; Hodes, *Broadcasting the Pandemic*; Steinberg, 'Re-examining the Early Years of Anti-retroviral Treatment'.

who studied at the University of Zimbabwe and University College, London, heads a path-breaking Archaeology Materials Laboratory at the University of Cape Town, focussing on metallurgy.[10] Chemist Kelly Chibale, director of a major research centre on Drug Discovery and Development (H3D) at the University Cape Town, trained in Zambia and Cambridge. His own work explores new anti-malarial drugs.[11] Tebello Nyokong of Rhodes University, also a chemist and a specialist in photodynamic therapy for cancer treatment, was one of the first two women scientists to receive the Mapungubwe award. Trained in Lesotho and in Canada, Nyokong has become a spokesperson for African women in science: 'My main driving force is my passion for Africa. We are viewed as a continent of war and famine, and not as one that is a major player in Science and Technology. In my small way, I wish to change this perception'.[12] In this way South African-based science remains strongly connected to expanded global networks and serves to anchor ambitions for African practitioners.

As we completed this book the Covid-19 pandemic arrived in South Africa and as the book enters the final stages of production there is troubling news of a new viral variant, B.1.351, which has mutated in South Africa with the potential to spread rapidly. The disease and the government's response underline global interconnectedness.

Initially there was widespread concern that the virus could devastate a country with a weak public health system and a population disproportionately affected by HIV/AIDs, tuberculosis, and other infectious diseases. Much of the population lives in crowded conditions with little possibility of effective social distancing. Such fears are exacerbated by visions of societal fragmentation and contested biopolitics.

After the first case of Covid was officially confirmed on 5th March, 2020, likely transmitted by a traveller returning from Italy, infection rates doubled daily. Despite levels that remained well below those in the northern hemisphere, President Ramaphosa closed the borders and implemented a lockdown on 27 March – including a ban on alcohol sales – before any deaths were recorded. This proved relatively successful. Impressive levels of compliance may have drawn on the

10 S. Chirikure, *Indigenous Mining and Metallurgy in Africa* (Cambridge: Cambridge University Press, 2010).

11 J. Okombo and K. Chibale, 'Recent Updates in the Discovery and Development of Novel Antimalarial Drug Candidates', *MedChemComm (RSC Medicinal Chemistry)*, 9, 3 (2018), 437–53.

12 https://nsesafoundation.org/tebello-nyokong/

expanding popular scientific knowledge acquired during the AIDS epidemic. Within a few weeks, however, reports circulated that the promised government food supplies for poor communities, both urban and rural, were inadequate, that corruption was diverting resources, and that people without work could not afford to purchase food.

It is notable that Ramaphosa was strongly influenced by scientific advice, including that by Salim Abdool Karim, internationally known epidemiologist and chair of the health ministry's scientific advisory committee. He and his wife, Quarraisha, had been key figures in policy formation in the early years of AIDS (chapter 7) only to be sidelined by Mbeki and his supporters. Early evidence from the lockdown, despite the difficulties of enforcement, suggested that transmission had been sharply curtailed and deaths from the disease did not spread as quickly as expected. Mitigating factors may have included the youth of the population and the possibility of beneficial effects from TB treatment or other acquired immunities.

Abdool Karim was quick to question the idea of South African exceptionalism, partly because there was no clear scientific basis for this. There was a danger that rumour, as in the case of AIDS, might quickly outstrip scientifically authorised government prescriptions. The Covid pandemic underlined medical and scientific interdependence as well as the relationship between disease outcomes, social restraints, and faultlines of race and wealth. Notably, the government demonstrated a firm commitment to biomedical approaches and it was able to draw upon significant centres of knowledge and expertise including collaboration between the state and private sectors. In this crisis, the government's response strategies clearly endorse the universality of scientific knowledge and testify to the fact that South Africa's post-apartheid scientists are equipped to participate effectively in the global transfer of knowledge. More broadly, it emphasises the need for the consolidation and building of African scientific capacity.

Writing history is bounded by what has happened. Our text explores the complexity of the colonial era and its indelible legacies. However much they may wish to, it is impossible for South Africans, whether black or white, to imagine themselves outside of this context. Science is also globally embedded in this long history, and we have argued that it both facilitated colonialism (in its broadest sense) and to some degree stood outside such processes. The legacies of specific sciences, from astronomy to zoology, provide dynamic and creative systems of thought and method. We have sought to illustrate the interest and originality of earlier ideas and to connect these to some of the new routes along which scientific thinking is travelling. Scientific work in multiple sites with diverse backing is important in order for South Africa to address its social and environmental problems. The scientific imagination needs to be nurtured in the country, along with the capacity to think creatively about history and society.

Select Bibliography

Abbreviations:
JSAS – Journal of Southern African Studies
SAJS – South African Journal of Science

African Regional Scientific Conference Johannesburg 1949, 2 vols. (Pretoria: Government Printer, 1949).

Akpan, W. '"Local" Knowledge, "Global Knowledge", "Development Knowledge": Finding a New Balance in the Knowledge Power Play', *South African Review of Sociology*, 42, 3 (2011), 116–27.

Allibone, T. E. 'Basil Ferdinand Jamieson Schonland 1896–1972', *Biographical Memoirs of Fellows of the Royal Society*, 19 (1973), 629–53.

Atherstone, W. G. 'From Graham's Town to the Gouph', *Cape Monthly Magazine*, April 1871, 304–10 and July, 42–51.

Atmore, A. and P. Sanders. 'Sotho Arms and Ammunition in the Nineteenth Century', *Journal of African History*, 12, 4 (1971), 535–44.

Austin, B. *Schonland: Scientist and Soldier* (Bristol: Institute of Physics, 2001).

Bain, A. G. 'On the Geology of South Africa', *Transactions of the Geological Society of London*, series 2, 7 (1856), 175–92.

Bank, A. *Bushmen in a Victorian World: The Remarkable Story of the Bleek–Lloyd Collection of Bushman Folklore* (Cape Town: Double Storey, 2008).

Barnard, C. and C. B. Pepper, *One Life* (New York: Bantam, 1971).

Barrow, J. *An Account of Travels into the Interior of Southern Africa in the Years 1797 and 1798* (London: T. Cadell and W. Davies, 1801).

Basalla, G. 'The Spread of Western Science', *Science*, 156, 3775 (1967), 611–22.

Basson, N., *Passage to Progress: The CSIR's Journey of Change 1945–1995* (Johannesburg: Jonathan Ball, 1996).

Bayly, C. A., *Imperial Meridian: The British Empire and the World 1780–1830* (London: Longman, 1989).

Beinart, W. 'African History and Environmental History', *African Affairs*, 99, 395 (2000), 269–302.

Plant Transfers, Bio-invasions and Biocultural Diversity: Perspectives from Africa (Nehru Memorial Museum and Library Occasional Paper, New Series 42, New Delhi, 2014).

The Rise of Conservation in South Africa: Settlers, Livestock and the Environment, 1770–1950 (Oxford: Oxford University Press, 2003).

'Soil Erosion, Conservationism and Ideas about Development: A Southern African Exploration, 1900–1960', *JSAS*, 11, 1 (1984), 52–83.

Twentieth-Century South Africa (Oxford: Oxford University Press, 2001).

Beinart, W. and R. Beinart. '"From Elephant's Foot . . . to Cortisone": Boots Pure Drugs Company and *Dioscorea sylvatica* in South Africa, c. 1950–1963', *South African Historical Journal*, 71, 4 (2019), 644–75.

Beinart, W. and K. Brown. *African Local Knowledge and Livestock Health: Diseases and Treatments in South Africa* (Oxford: James Currey, 2013).

Beinart, W., K. Brown and D. Gilfoyle. 'Experts and Expertise in Colonial Africa Reconsidered: Science and the Interpenetration of Knowledge', *African Affairs* 108, 432 (2009), 413–33

Beinart, W. and L. Wotshela. *Prickly Pear: The Social History of a Plant in the Eastern Cape* (Johannesburg: Wits University Press, 2011).

Bennett, B. M., 'Margaret Levyns and the Decline of Ecological Liberalism in the Southwest Cape, 1890–1975', *South African Historical Journal*, 67, 1 (2015), 64–84.

Berger, L. R., et al. '*Homo naledi*, a new species of the genus *Homo* from the Dinaledi Chamber, South Africa', *eLife*, 10 September 2015.

Boucher, M. *Spes in Arduis: A History of the University of South Africa* (Pretoria: University of South Africa, 1973).

Bowler P. J. and J. V. Pickstone (eds.), *The Modern Biological and Earth Sciences*, volume 6 of *The Cambridge History of Science* (Cambridge: Cambridge University Press, 2009).

Breckenridge, K. *Biometric State: The Global Politics of Identification and Surveillance in South Africa, 1850 to the Present* (Cambridge: Cambridge University Press, 2014).

Brown, A. C. (ed.), *A History of Scientific Endeavour in South Africa* (Cape Town: Royal Society, 1977).

Brown, K. 'Political Entomology: The Insectile Challenge to Agricultural Development in the Cape Colony, 1895–1910', *JSAS*, 29, 2 (2003), 529–49.

'Tropical Medicine and Animal Diseases: Onderstepoort and the Development of Veterinary Science in South Africa 1908–1950', *JSAS*, 31, 3 (2005), 513–29.

Brutsch, M. O. and H. G. Zimmermann. 'The Prickly Pear (Opuntia Ficus-Indica [Cactaceae]) in South Africa: Utilization of the Naturalized

Weed, and of the Cultivated Plant', *Economic Botany*, 47, 2 (1993), 154–62.

Buckley, D. 'SALT: "Gigantic African Eye"', *Quest*, 2, 2 (2005), 6–9.

Burchell, W. J., *Travels in the Interior of Southern Africa 2 vols.* (London: Batchworth Press, 1953). Reprint of 1822–24 edition.

Burke, P. *A Social History of Knowledge* (Cambridge: Cambridge University Press, 2000).

Burrows, E. H., *A History of Medicine in South Africa* (Cape Town: A. A. Balkema, 1958).

Campbell, C. *'Letting Them Die': Why HIV/AIDS Prevention Programmes Fail* (Oxford: James Currey, 2003).

Cannon, W. F., 'John Herschel and the Idea of Science', *Journal of the History of Science*, 22, 2 (1961), 215–39.

Carnegie Corporation archives, Columbia University, New York.

Carruthers, J. 'Conservation and Wildlife Management in South African National Parks 1930s–1960s', *Journal of the History of Biology*, 41, 2 (2008), 203–36.

The Kruger National Park: A Social and Political History (Pietermaritzburg: University of Natal Press, 1995).

National Park Science: A Century of Research in South Africa (Cambridge: Cambridge University Press, 2017).

'Trouble in the Garden: South African Botanical Politics, *ca.* 1850–1950', *South African Journal of Botany*, 77, 2 (2011), 258–67.

Cartwright, A. P. *The Dynamite Company* (Cape Town: Purnell, 1964).

The Industrial Development Corporation of South Africa Limited (Johannesburg, 1971).

Chambers, D. and R. Gillespie. 'Locality in the History of Science: Colonial Science, Technoscience, and Indigenous Knowledge', *Osiris*, 15, 1 (2000), 221–40.

Chetty, S. 'David Draper: The Making of a South African Geologist', *Historia*, 63, 2 (2018), 17–39.

Chirikure, S. *Indigenous Mining and Metallurgy in Africa* (Cambridge: Cambridge University Press, 2010).

Christie, R. *Electricity, Industry and Class in South Africa* (Basingstoke: Palgrave Macmillan, 1984).

Clarke, R. 'Latest Information on Sterkfontein's *Australopithecus* Skeleton and a New Look at *Australopithecus*', *SAJS*, 104 (2008), 443–9.

Cocks, M. L. *Wild Resources and Cultural Practices in Rural and Urban Households in South Africa: Implications for Bio-cultural Diversity Conservation* (Grahamstown: Institute of Social and Economic Research, Rhodes University, 2006).

Coetzee, J. M. *White Writing: On the Culture of Letters in South Africa* (New Haven, CT: Yale University Press, 1988).

Comaroff, J. and J. Comaroff, 'Theory from the South: Or, how Euro-America is Evolving toward Africa', *Anthropological Forum*, 22, 2 (2012), 113–31.

Cook, H. *Matters of Exchange: Commerce, Medicine and Science in the Dutch Golden Age* (New Haven, CT: Yale University Press, 2007).

Cowling, R. M., et al. 'Levyns' Law: Explaining the Evolution of a Remarkable Longitudinal Gradient in Cape Plant Diversity', *Transactions of the Royal Society of South Africa*, 72, 2 (2017) 184–201.

Critchell, J. T. and J. Raymond. *A History of the Frozen Meat Trade: An Account of the Development and Present Day Methods of Preparation, Transportation, and Marketing of Frozen and Chilled Meats* (London: Constable, 1912).

Cullinan, P. *Robert Jacob Gordon 1743–1795: The Man and His Travels at the Cape* (Cape Town: Struik Winchester, 1992).

Davie, G. *Poverty Knowledge in South Africa: A Social History of Human Science, 1855–2005* (Cambridge: Cambridge University Press, 2015).

Day, J. H. 'Marine Biology in South Africa', in Brown (ed.), *Scientific Endeavour in South Africa*, 86–108.

De Kock, P. R. and H. D. Moraal (eds.). 'Physics in South Africa', ms., SA Institute of Physics, Pretoria, 2011.

De Villiers, J. C. *Healers, Helpers and Hospitals* (Pretoria: Protea Boekhuis, 2008).

De Wit, M. P., D. J. Crookes and B. W. van Wilgen. 'Conflicts of Interest in Environmental Management: Estimating the Costs and Benefits of a Tree Invasion', *Biological Invasions*, 3 (2001), 167–78.

Deacon, H., H. Phillips and E. van Heyningen (eds.). *The Cape Doctor in the Nineteenth Century: A Social History* (Leiden: Brill, 2004).

Delius, P. and S. Schirmer. 'Soil Conservation in a Racially Ordered Society: South Africa 1930–1970', *JSAS*, 26, 4 (2000), 719–42.

Delta. 'Stone Implements in South Africa', *Cape Monthly Magazine*, (New Series, October and December 1870), 236–9, 365–6.

Diab, R. and W. Gevers (eds.). *The State of Science in South Africa* (Pretoria: Academy of Science of South Africa, 2009).

Digby, A. *Diversity and Division in Medicine: Health Care in South Africa from the 1800s* (Oxford: Peter Lang, 2006).

Dlamini, J. *Safari Nation: A Traveling History of the Kruger National Park* (Athens: Ohio University Press, 2020).

Douglass, A. *Ostrich Farming in South Africa* (London: Cassell, Petter, Galpin, 1881).

Draper, M. 'Zen and the Art of Garden Province Maintenance: The Soft Intimacy of Hard Men in the Wilderness of KwaZulu-Natal, South Africa, 1952–1997', *JSAS*, 24, 4 (1998), 801–28.

Du Preez, M. and J. Dronfield. *Dr James Barry: A Woman ahead of Her Time* (London: One World, 2016).

Du Toit, A. L. Papers, BC722, University of Cape Town.

The Geology of South Africa (Edinburgh: Oliver and Boyd, 1926).

Our Wandering Continents: An Hypothesis of Continental Drifting (London: Oliver and Boyd, 1937).

Du Toit, J. T., K. H. Rogers and H. C. Biggs (eds.). *The Kruger Experience* (Washington: Island Press, 2003).

Dubow, S. '200 Years of Astronomy in South Africa: From the Royal Observatory to the "Big Bang" of the Square Kilometre Array', *JSAS* 45, 4 (2019), 663–87.

A Commonwealth of Knowledge: Science, Sensibility, and White South Africa 1820–2000 (Oxford: Oxford University Press, 2006).

'Global Science, National Horizons: South Africa in Deep Time and Space', *The Historical Journal* 63, 5 (2020), 1079–1106.

'Human Origins, Race Typology and the Other Raymond Dart', *African Studies*, 55, 1 (1996), 1–30.

'Introduction' in *The Rise and Fall of Modern Empires*, volume II, S. Dubow (ed.), (Farnham and Burlington, 2013).

'Racial Irredentism, Ethnogenesis, and White Supremacy in High-Apartheid South Africa', *Kronos*, 41, 1(2015), 236–64.

South Africa's Struggle for Human Rights (Jacana: Auckland Park, 2012).

Scientific Racism in Modern South Africa (Cambridge: Cambridge University Press, 1995).

Dubow, S. and A. Jeeves (eds.). *South Africa's 1940s: Worlds of Possibility* (Cape Town: Double Storey, 2000).

Echenberg, M. *Plague Ports: The Global Urban Impact of Bubonic Plague, 1894–1901* (New York: New York University Press, 2007).

Edwards, P. N., and G. Hecht. 'History and the Technopolitics of Identity: The Case of Apartheid South Africa', *JSAS*, 36, 3 (2010), 619–39.

Ellis, G. *Science Research Policy in South Africa. A Discussion Document for the Royal Society of South Africa, May 28, 1994* (Cape Town: Royal Society, 1994).

Etherington, N. *The Great Treks: The Transformation of Southern Africa, 1815–1854* (London: Pearson Education, 2001).

'Were There Large States in the Coastal Regions of Southeast Africa before the Rise of the Zulu Kingdom?', *History in Africa*, 31 (2004),157–83.

Feierman, S. and J. Janzen (eds.). *The Social Basis of Health and Healing in Africa* (Berkeley: University of California Press, 1992)

Flint, K. E. *Healing Traditions: African Medicine, Cultural Exchange, and Competition in South Africa, 1820–1948* (Athens: Ohio University Press, 2008)

Forbes, V. S. (ed.) *Carl Peter Thunberg, Travels at the Cape of Good Hope 1772–1775* (Cape Town: Van Riebeeck Society, 1986), based on 1793–5 English edition.

Foster, L. A. *Reinventing Hoodia: Peoples, Plants, and Patents in South Africa* (Seattle: University of Washington Press, 2017).

Freund, B. *Twentieth-Century South Africa: A Developmental History* (Cambridge: Cambridge University Press, 2018).

Fuchs, V. and E. Hillary. *The Crossing of Antarctica: The Commonwealth Trans-Antarctic Expedition 1955–58* (London: Cassel, 1958).

Gear, J. 'The South African Poliomyelitis Vaccine', *South African Journal of Medicine*, (23 June 1956), 587–94.

Gevisser, M. *Thabo Mbeki: The Dream Deferred* (Johannesburg: Jonathan Ball, 2007).

Gilfoyle, D. 'Anthrax in South Africa: Economics, Experiment and the Mass Vaccination of Animals, c. 1910–1945', *Medical History*, 50, 4 (2006), 465–90.

'Veterinary Research and the African Rinderpest Epizootic: The Cape Colony, 1896–1898', *JSAS*, 29, 1, (2003), 133–54.

Gill, D. 'Address by Sir David Gill to the SAAAS', *Report of the South African Association for the Advancement of Science* (Cape Town, 1903).

A History and Description of the Royal Observatory, Cape of Good Hope (London: HMSO, 1913).

Gish, S. *Alfred B. Xuma: African, American, South African* (Basingstoke: Macmillan, 2000).

Glass, I. S. *Nicolas–Louis de La Caille, Astronomer and Geodesist* (Oxford: Oxford University Press, 2012).

Glenn, I. *François Le Vaillant's Travels in the Interior of Africa via the Cape of Good Hope*, vol 1 (Cape Town: Van Riebeeck Society, 2007).

Glotzer, R. 'The Career of Mabel Carney: The Study of Race and Rural Development in the United States and South Africa', *International Journal of African Historical Studies*, 29, 2 (1996), 309–36.

Good, A. 'The Construction of an Authentic Text: Peter Kolb's Description of the Khoi', *Journal of Early Modern History*, 10, 1 (2006), 61–94.

Grosskopf, J. F. W., et al. *The Poor White Problem in South Africa* (Stellenbosch: South Africa Report of the Carnegie Commission, 1932), 5 volumes.

Gumede, W. M. *Thabo Mbeki and the Battle for the Soul of the ANC* (Cape Town: Zebra Press, 2007).

Gurdon J. B. and N. Hopwood. 'The Introduction of Xenopus laevis into Developmental Biology: Of Empire, Pregnancy Testing and Ribosomal Genes', *International Journal of Developmental Biology*, 44 (2000), 43–50.

Guy, J. 'A Note on Firearms in the Zulu Kingdom with Special Reference to the Anglo-Zulu War, 1879', *Journal of African History*, 12, 4 (1971), 557–70.

Hamilton, C. *Terrific Majesty: The Powers of Shaka Zulu and the Limits of Historical Invention* (Cambridge MA: Harvard University Press, 1998).

Hammel, T. *Shaping Natural History and Settler Society: Mary Elizabeth Barber and the Nineteenth-Century Cape* (Cham: Palgrave Macmillan, 2019).

'Thinking with Birds: Mary Elizabeth Barber's Advocacy for Gender Equality in Ornithology', *Kronos*, 41, 1 (2015), 85–111.

Harries, P. *Butterflies and Barbarians: Swiss Missionaries and Systems of Knowledge in Southeast Africa* (Oxford: James Currey, 2007).

Work, Culture, and Identity: Migrant Laborers in Mozambique and South Africa (London: James Currey, 1994).

Harries, P. and D. Maxwell (eds). *The Spiritual in the Secular: Missionaries and Knowledge about Africa* (Grand Rapids, MI: Eerdmans, 2012).

Harthoorn, A. M. *The Flying Syringe: Ten Years of Immobilising Wild Animals in Africa* (London: Geoffrey Bles, 1970).

Hatch, F. H. and J. A. Chalmers. *Gold Mines of the Rand: Being a Description of the Mining Industry of the Witwatersrand, South African Republic* (London: Macmillan, 1895).

Hecht, G. *Being Nuclear: Africans and the Global Nuclear Trade* (Johannesburg: Wits University Press, 2012).

Hodes, R. *Broadcasting the Pandemic: A History of HIV on South African Television* (Cape Town: HSRC Press, 2014).

Hoffman, M. T. 'Changing Patterns of Rural Land Use and Land Cover in South Africa and Their Implications for Land Reform', *JSAS*, 40, 4 (2014), 707–25.

Hofmeyr, J. H. 'Africa and Science', *SAJS*, 26 (1929), 1–18.

'Summary of Conclusions', *SAJS*, 39 (1942), 371–80.

Holmes, R. *Scanty Particulars: The Strange Life and Astonishing Secret of Victorian Adventurer and Pioneering Surgeon James Barry* (London: Viking, 2002).

Huigen, S. *Knowledge and Colonialism: Eighteenth-Century Travellers in South Africa* (Leiden: Brill, 2009).

Hunter (Wilson), M. *Reaction to Conquest* (London: Oxford University Press, 1964), first published 1936.

Iliffe, R. 'Science Fictions: The Triumph of the Imagination and the Invention of Scientific Creativity', Inaugural Lecture, University of Oxford, 8.11.2018.

Isaacman, A. F. and B. S. Isaacman. *Dams, Displacement, and the Delusion of Development: Cahora Bassa and Its Legacies in Mozambique, 1965–2007* (Athens: Ohio University Press, 2014).

Jacobs, N. *Birders of Africa: History of a Network* (New Haven, CT: Yale University Press, 2016).

Jasanoff, S. 'Future Imperfect: Science, Technology, and the Imaginations of Modernity', in S. Jasanoff and Sang-Hyun Kim (eds.), *Dreamscapes of Modernity: Sociotechnical Imaginaries and the Fabrication of Power* (Chicago: University of Chicago Press, 2015).

Katz, E. N. 'Outcrop and Deep Level Mining in South Africa before the Anglo-Boer War: Re-examining the Blainey Thesis', *Economic History Review*, 48, 2 (1995), 304–28.

Kauffmann, K. D. and D. L. Lindauer (eds.). *AIDS and South Africa: The Social Expression of a Pandemic* (Basingstoke: Palgrave Macmillan, 2004).

Kepe, T. 'Medicinal Plants and Rural Livelihoods in Pondoland, South Africa: Towards an Understanding of Resource Value', *International Journal of Biodiversity Science & Management*, 3, 3 (2007), 170–83.

Kingwill, D. G. *The CSIR: The First Forty Years* (Pretoria: CSIR, 1990).

Kirby, P. R. *Sir Andrew Smith, M.D., K.C.B.: His Life, Letters and Works* (Cape Town: A. A. Balkema, 1965).

Knight, I. *The Anatomy of the Zulu Army from Shaka to Cetshwayo 1818–1879* (Barnsley: Frontline Books, 2015).

Kolb, P. *Naaukeurige en Uitvoerige Beschryving van de Kaap de Goede Hoop*, 2 volumes (Amsterdam: Balthazar Lakeman, 1727).

The Present State of the Cape of Good Hope (London: W. Innys, 1731).

Kuljian, C. *Darwin's Hunch: Science, Race and the Search for Human Origins* (Auckland Park: Jacana, 2016).

Le Gros Clark, W. E. 'Significance of the Australopithecinae', *Nature*, 157, 4000 (1946), 863–4.

Levyns, M. R. 'Migrations and Origins of the Cape Flora', *Transactions of the Royal Society of South Africa* 37 (1964), 85–107.

Lichtenstein, H. *Travels in Southern Africa in the Years 1803, 1804, 1805 and 1806*, 2 volumes (Cape Town: Van Riebeeck Society, 1928, 1930). Reprint of the original English translation 1812 and 1815.

Lilja, F. *The Golden Fleece of the Cape: Capitalist Expansion and Labour Relations in the Periphery of Transnational Wool Production, c.1860–1950* (Uppsala: Uppsala University, Studia Historica Upsaliensia, 247, 2013).

Lister, M. H. (ed.). *Journals of Andrew Geddes Bain* (Cape Town: Van Riebeeck Society, 1949).

Livingstone, D. N. *Putting Science in its Place: Geographies of Scientific Knowledge* (Chicago: University of Chicago Press, 2003).

Low, C. H. 'Different Histories of Buchu: Euro-American Appropriation of San and Khoekhoe Knowledge of Buchu Plants', *Environment and History*, 13, 3 (2007), 333–61.

MacArthur, J. S. Papers, 1856–1920, Balliol College Archives and Manuscripts, Oxford.

Machens, E. W. *Platinum, Gold and Diamonds: The Adventure of Hans Merensky's Discoveries* (Pretoria: Protea Book House, 2009).

Mackenzie, J. M. (ed.). *Imperialism and the Natural World* (Manchester: Manchester University Press, 1990).

Macleod, R. 'On Visiting the "Moving Metropolis": Reflections on the Architecture of Imperial Science', *Historical Records of Australian Science*, 5, 3 (1980), 1–16.

Mager, A. *Gender and the Making of a South African Bantustan* (Oxford: James Currey, 1991).

Maggs, T. '"My Father's Hammer Never Ceased its Song Day and Night": The Zulu Ferrous Metalworking Industry', *Natal Museum Journal of Humanities*, 4 (1992), 65–87.

Makgoba, M. W. (ed.). *African Renaissance: The New Struggle* (Sandton and Cape Town: Mafube and Tafelberg, 1999).

Malherbe, E. G. Papers, Killie Campbell Library, Durban.

Never a Dull Moment (Cape Town: Timmins, 1981).

Mandela, N. 'Foreword', in M. Van Ameringen (ed.), *Building a New South Africa. Vol.3. Science and Technology Policy* (Ottawa: IDRC, [1995]).

Marks, S. *Divided Sisterhood: Race, Class and Gender in the South African Nursing Profession* (Basingstoke: Macmillan, 1994).

Marks, S. and S. Trapido. 'Lord Milner and the South African State', *History Workshop Journal*, 8, 1 (1979), 50–81.

'South Africa's Early Experiment in Social Medicine: Its Pioneers and Politics', *American Journal of Public Health*, 87, 3 (1997), 452–9

Masika, P. J., W. van Averbeke and A. Sonardi. 'Use of Herbal Remedies by Small-scale Farmers to Treat Livestock Diseases in Central Eastern Cape Province, South Africa', *Journal of the South African Veterinary Association*, 71, 2 (2000), 87–91.

Mbali, M. *South African AIDS Activism and Global Health Politics* (London: Palgrave Macmillan, 2013).

Mbeki, T. 'Speech at the Funeral of Sarah Bartmann, 9 August 2002', www.sahistory.org.za/archive/speech-funeral-sarah-bartmann-9-august-2002.

'Speech of the President of South Africa, Thabo Mbeki, at the official inauguration ceremony of the Southern Africa Large Telescope (SALT): Sutherland, Northern Cape', 10 November 2005, www.thepresidency.gov.za

McCulloch, J. *South Africa's Gold Mines and the Politics of Silicosis* (Suffolk: James Currey, 2012).

McDowell, M. *The National Metrology Laboratory of South Africa* (Pretoria: CSIR, 1997).

Millar, A. K. *Plantagenet in South Africa: Lord Charles Somerset* (Cape Town: Oxford University Press, 1965).

Mitchell, P. *The Archaeology of Southern Africa* (Cambridge: Cambridge University Press, 2002).

Molotja, N. and G. Ralphs. 'A Critical Review of Social Sciences and Humanities R&D Expenditure in South Africa, 2005–2014', *SAJS*, 114, 7/8 (2018).

Moore, P. and P. Collins. *The Astronomy of Southern Africa* (London: Robert Hale, 1977).

Mosenthal, J. and J. E. Harting. *Ostriches and Ostrich Farming* (London: Trübner and Co., 1877).

Murray, B. K. *Wits: The Early Years* (Johannesburg: Witwatersrand University Press, 1982).

Myburgh, J. 'The Virodene Affair (I)', *Politicsweb*, 17 September 2007: 'The secret history of the ANC's response to the HIV/AIDS epidemic'.

Nattrass, N. *The AIDS Conspiracy: Science Fights Back* (Johannesburg: Wits University Press, 2012).

Mortal Combat: AIDS Denialism and the Struggle for Antiretrovirals in South Africa (Pietermaritzburg: University of KwaZulu-Natal Press, 2007).

Ngubane, H. *Body and Mind in Zulu Medicine: An Ethnography of Health and Disease in Nyuswa-Zulu Thought and Practice* (London: Academic Press, 1977).

Okombo, J. and K. Chibale. 'Recent Updates in the Discovery and Development of Novel Antimalarial Drug Candidates', *MedChemComm (RSC Medicinal Chemistry)*, 9, 3 (2018), 437–53.

Orpen, J. M. 'A Glimpse into the Mythology of the Maluti Bushmen', *Cape Monthly Magazine*, July 1874, 1–10 with 'Remarks by Dr. Bleek', 10–13.

Padayachee, V. and R. van Niekerk. *Shadow of Liberation: Contestation and Compromise in the Economic and Social Policy of the African National Congress, 1943–1996* (Johannesburg: Wits University Press, 2019).

Pappe, C. W. L. *Florae Capensis Medicae Prodromus; or an Enumeration of South African Plants used as Remedies by the Colonists of the Cape of Good Hope*, 2nd edition (Cape Town: W. Brittain, 1857).

Parkington, J., D. Morris, J. M de Prada-Samper. 'Elusive identities: Karoo |Xam descendants and the SKA', *JSAS*, 45, 4 (2019), 729–47.

Peires, J. B. *The Dead Will Arise: Nongqawuse and the Great Xhosa Cattle-Killing of 1856–7* (Johannesburg: Ravan Press, 1989).

Penn, N. 'Mapping the Cape: John Barrow and the First British Occupation of the Colony, 1795–1803', *Pretexts*, 4, 2 (1993), 20–43.

'Written Culture and the Cape Khoikhoi: From Travel Writing to Kolb's "Full Description"', in A. Delmas and N. Penn (eds.). *Written Culture in a Colonial Context: Africa and the Americas 1500–1900* (Leiden: Brill, 2014), 171–93

Pettey, F. W. 'The Biological Control of Prickly Pears in South Africa', *Union of South Africa, Department of Agriculture and Forestry, Scientific Bulletin 271* (Pretoria: Government Printer, 1948).

Phillips, E. P. 'A Brief Historical Sketch of the Development of Botanical Science in South Africa and the Contribution of South Africa to Botany', *SAJS*, XXVII (1930), 39–80.

Piers, H. W. 'On the Geology of the Cape Peninsula', *Cape Monthly Magazine*, Sept. 1871.

Player, I. *The White Rhino Saga* (London: Collins, 1972).

Zululand Wilderness: Shadow and Soul (Cape Town: David Philip, 1997).

Polakow-Suransky, S. *The Unspoken Alliance: Israel's Secret Relationship with Apartheid South Africa* (Auckland Park: Jacana, 2002).

Pollock, A. *Synthesising Hope: Matter, Knowledge, and Place in South African Drug Discovery* (Chicago: University of Chicago Press, 2019).

Pooley, S. 'Jan van Riebeeck as Pioneering Explorer and Conservator of Natural Resources at the Cape of Good Hope (1652–62)', *Environment and History*, 15, 1 (2009), 3–33.

Burning Table Mountain: An Environmental History of Fire on the Cape Peninsula (Cape Town: University of Cape Town Press, 2015).

Porter, R. (ed.). *The Eighteenth Century*, volume 4 of *The Cambridge History of Science* (Cambridge: Cambridge University Press, 2008).

Poskett, J. *Materials of the Mind: Phrenology, Race, and the Global History of Science 1815–1920* (Chicago: Chicago University Press, 2019).

Prakash, G. *Another Reason: Science and the Imagination of Modern India* (Princeton, NJ: Princeton University Press, 1999).

Pringle, T. *Narrative of a Residence in South Africa*, vol. 1 (Brentwood: Empire Book Association, 1986), first published 1834.
Purkitt, H. E. and S. F. Burgess. *South Africa's Weapons of Mass Destruction* (Bloomington: Indiana University Press, 2005).
Raj, K. *Relocating Modern Science: Circulation and the Construction of Knowledge in South Asia and Europe, 1650–1900* (Basingstoke: Palgrave Macmillan, 2007).
Robins, S. '"Long Live Zackie, Long Live": AIDS Activism, Science and Citizenship after Apartheid', *JSAS*, 30, 3 (2004), 651–72.
Rookmaaker, K. 'The Slow Recognition of the African Rhinoceros from Hondius to Camper', in S. Huigen and J. Kommers (eds.). *Interpretations of Colonial Representations: Reflections on Alterity, Colonial History, and Intercultural Contact* (Saarbrucken: Verlag fiir Entwicklungspolitik, 2004), 35–54.
Roos, N. 'Alcohol Panic, Social Engineering, and Some Reflections on the Management of Whites in Early Apartheid Society', *Historical Journal*, 58, 4 (2015), 1167–89.
Rousseau, L. *The Dark Stream: The Story of Eugene Marais* (Johannesburg: Jonathan Ball, 1982).
Roux, E. R. *The Veld and the Future: Inaugural Lecture* (Johannesburg: Witwatersrand University Press, 1963).
Harvest and Health in Africa (London: Thomas Nelson, 1942).
Ruskin, S. *John Herschel's Cape Voyage: Private Science, Public Imagination and the Ambitions of Empire* (Aldershot: Routledge, 2004).
Sachs, W. *Black Hamlet* (Baltimore, MD: Johns Hopkins, 1996), reissued with introductions by Saul Dubow and Jacqueline Rose; first published 1937.
Schonland Papers, Churchill College, Cambridge, Cambridge University.
Schonland, B. F. J. 'The South African Association for the Advancement of Science, Its Past and Present', *SAJS*, 49 (October–November 1952), 61–8.
'Hendrik Johannes van der Bijl 1887–1948', *Obituary Notices of Fellows of the Royal Society*, 7, 19 (1950), 26–34.
Seekings, J. 'The Carnegie Commission and the Backlash against Welfare State-Building in South Africa, 1931–1937', *JSAS*, 34, 3 (2008), 515–37.
Shapiro, K. A. 'Doctors or Medical Aids: The Debate over the Training of Black Medical Personnel for the Rural Black Population in South Africa in the 1920s and 1930s', *JSAS*, 13, 2 (1987), 234–55.
Simons, P. B. *Ice Cold in Africa: The History of the Imperial Cold Storage & Supply Company Limited* (Cape Town: Fernwood Press, 1999).

Smith, J. B. L. *Old Fourlegs: The Story of the Coelacanth* (London: Longmans Green, 1956).

Smith, T. 'New Bottles for New Wine: A Pericentric Framework for the Study of the Cold War', *Diplomatic History*, 24, 4 (2000), 567–91.

Smuts, J. C. 'South Africa in Science', *SAJS*, 22 (1925), 1–19.

Snedegar, K. *Mission, Science, and Race in South Africa: A.W. Roberts of Lovedale, 1883–1938* (Lanham: Lexington Books, 2015).

Soga, Veterinary Surgeon [J.F.]. 'Disease "Nenta" in Goats', *Agricultural Journal of the Cape of Good Hope* (29 January 1891), 140–2.

Soodyall, H. and T. Jenkins. 'Unravelling the History of Modern Humans in Southern Africa: The Contribution of Genetic Studies' in P. Bonner, A. Esterhuysen and T. Jenkins (eds.), *A Search for Origins: Science, History and South Africa's 'Cradle of Humankind'* (Johannesburg: Wits University Press, 2007), 78–90.

South Africa, Government White Paper on Science and Technology: Preparing for the 21st Century, Dept. Arts, Culture, Science and Technology, 4 September 1996.

Sparks, S. J. 'Apartheid Modern: South Africa's Oil from Coal Project and the History of a Company Town', DPhil thesis, University of Michigan, 2012.

'Between "Artificial Economics" and the "Discipline of the Market": SASOL from Parastatal to Privatisation', *JSAS*, 42, 4 (2016), 711–24.

Sparrman, A. *A Voyage to the Cape of Good Hope towards the Antarctic Polar Circle Round the World and to the Country of the Hottentots and the Caffres from the Year 1772–1776*, 2 volumes (Cape Town: Van Riebeeck Society, 1976, 1977).

Steinberg, J. 'Re-examining the Early Years of Anti-retroviral Treatment in South Africa: A Taste for Medicine', *African Affairs*, 116, 462 (2017), 60–79.

Three Letter Plague: A Young Man's Journey through a Great Epidemic (Johannesburg: Jonathan Ball, 2008).

Steynberg, A. P. 'Chapter 1: Introduction to Fischer-Tropsch Technology', *Studies in Surface Science and Catalysis*, 152 (2004), 1–63.

Storey, W. K. *Guns, Race and Power in Colonial South Africa* (Cambridge: Cambridge University Press, 2008).

Stuart, J. and D. McK. Malcolm. *The Diary of Henry Francis Fynn* (Pietermaritzburg: Shuter and Shooter, 1969).

Stumpf, W. 'South Africa's Nuclear Weapons Program: From Deterrence to Dismantlement', *Arms Control Today* (December 1995/January 1996), 3–8.

Summers, R. *A History of the South African Museum 1825–1975* (Cape Town: A. A. Balkema, 1975).

Swanson, M. W. 'The Sanitation Syndrome: Bubonic Plague and Urban Native Policy in the Cape Colony, 1900–1909', *Journal of African History*, 18, 3 (1977), 387–410.

Swart, S. 'The Ant of the White Soul: Popular Natural History, the Politics of Afrikaner Identity and the Entomological Writings of Eugene Marais', in W. Beinart and J. McGregor (eds.), *Social History and African Environments* (Oxford: James Currey, 2003), 219–39.

'The Construction of Eugène Marais as an Afrikaner Hero', *JSAS*, 30, 4 (2004), 847–67.

Tamarkin, M. *Volk and Flock: Ecology, Identity and Politics among Cape Afrikaners in the Late Nineteenth Century* (Pretoria: University of South Africa Press, 2009).

Thackeray, M. 'Energy Storage for a Cleaner Planet.' Fellow's seminar, Stellenbosch Institute for Advanced Studies, 7 May 2018, stias.ac.za/2018/05

Tilley, H. *Africa as a Living Laboratory: Empire, Development, and the Problem of Scientific Knowledge, 1870–1950* (Chicago: University of Chicago Press, 2011).

Tobias, P. V. 'Brain-size, Grey Matter and Race: Fact or Fiction?', *American Journal of Physical Anthropology*, 32, 1 (1970), 3–25.

Todd, J. *Colonial Technology: Science and the Transfer of Innovation to Australia* (Cambridge: Cambridge University Press, 1995).

Turrell, R. V. *Capital and Labour on the Kimberley Diamond Fields, 1871–1890* (Cambridge: Cambridge University Press, 1987).

Union of South Africa, *Final Report of the Drought Investigation Commission*, U.G. 49-1923 (Government Printer, Pretoria).

Report of the Native Economic Commission 1930–1932, U.G. 22-1932 (Government Printer, Pretoria).

Van der Watt, L-M. 'Return to Gondwanaland: South Africa, Antarctica, Minerals and Apartheid', *Polar Journal*, 3, 1 (2013), 72–93.

'Out in the Cold: Science and the Environment in the Sub-Antarctic and Antarctic in the Twentieth Century', PhD thesis, Stellenbosch University, 2012

Van Eck, H. J. *Some Aspects of the South African Industrial Revolution* (Johannesburg: South African Institute of Race Relations, 1951).

Van Heyningen, E. *The Concentration Camps of the Anglo-Boer War: A Social History* (Auckland Park: Jacana, 2013).

Van Onselen, C. *The Cowboy Capitalist: John Hays Hammond, the American West and the Jameson Raid* (Johannesburg: Jonathan Ball, 2017).

Van Sittert, L. 'Making the Cape Floral Kingdom: The Discovery and Defence of Indigenous Flora at the Cape ca. 1890–1939', *Landscape Research*, 28, 1 (2003), 113–29.

Van Wilgen, B. M., C. Dyer and J. H. Hoffmann. 'National-scale Strategic Approaches for Managing Introduced Plants: Insights from Australian Acacias in South Africa', *Diversity and Distributions*, 17, 5 (2011), 1060–75.

Van Wilgen, B. M., et al. 'The Economic Consequences of Alien Plant Invasions: Examples of Impacts and Approaches to Sustainable Management in South Africa', *Environment, Development and Sustainability*, 3 (2001), 145–68.

Van Wilgen, B. W., et al. 'Ecological Research and Conservation Management in the Cape Floristic Region between 1945 and 2015: History, Current Understanding and Future Challenges', *Transactions of the Royal Society of South Africa*, 71, 3 (2016), 207–303.

Van Wyk, B-E. and N. Gericke. *People's Plants: A Guide to Useful Plants of Southern Africa* (Pretoria: Briza Publications, 2000).

Vilakazi, H. W. 'The Problem of African Universities', in M. W. Makgoba (ed.), *African Renaissance: The New Struggle* (Sandton and Cape Town: Mafube and Tafelberg, 1999).

Von Wielligh, N. and L. von Wielligh-Steyn, *Die Bom: Suid-Afrika se Kernwapenprogram* (Pretoria: Litera, 2014).

Wagener, P. C. 'The Founding of the SAIP' in P. R. de Kock and H. Moraal (eds.), 'Physics in South Africa', ms., SA Institute of Physics, Pretoria, 2011.

Walwyn, D. and L. Cloete. 'Draft White Paper on Science, Technology and Innovation Neglects to Prioritise Issue of Performance and Human Capability', *SAJS*, 114, 11/12 (2018).

Warner, B. *Astronomers at the Royal Observatory, Cape of Good Hope* (Cape Town: A. A. Balkema, 1979).

'Astrophysics in South Africa', *Transactions of the Royal Society of South Africa*, 63, 1 (2008), 61–7.

Cape Landscapes. Sir John Herschel's Sketches 1834–38 (Cape Town: Juta, 2006)

Royal Observatory, Cape of Good Hope, 1820–31: The Founding of a Colonial Observatory (Dordrecht: Springer, 1995)

Weinberg, S. *A Fish Caught in Time: the Search for the Coelacanth* (London: Fourth Estate, 1999).

Williams, V. L., K. Balkwill and E. T. F. Witkowski. 'Unraveling the Commercial Market for Medicinal Plants and Plant Parts on the Witwatersrand, South Africa', *Economic Botany*, 54, 3 (2000), 310–27.

Worger, W. *South Africa's City of Diamonds* (New Haven, CT: Yale University Press, 1987).

Wulf, A. *The Invention of Nature: The Adventures of Alexander von Humboldt, the Lost Hero of Science* (London: John Murray, 2015).

Wylie, D. *Starving on a Full Stomach: Hunger and the Triumph of Cultural Racism in Modern South Africa* (Charlottesville: University of Virginia Press, 2001).

Xuma, A. B. 'ANC Presidential Address, 14 December 1941', South African History Online, https://www.sahistory.org.za/archive/presidential-address-dr-ab-xuma-14-december-1941.

Index